Computational Heat Transfer and Fluid Mechanics

Computational Heat Transfer and Fluid Mechanics

Editors

Pouyan Talebizadeh Sardari
Goodarz Ahmadi
Kiao Inthavong

MDPI • Basel • Beijing • Wuhan • Barcelona • Belgrade • Manchester • Tokyo • Cluj • Tianjin

Editors

Pouyan Talebizadeh Sardari
Brunel University London
UK

Goodarz Ahmadi
Clarkson University
USA

Kiao Inthavong
RMIT University
Australia

Editorial Office
MDPI
St. Alban-Anlage 66
4052 Basel, Switzerland

This is a reprint of articles from the Special Issue published online in the open access journal *Energies* (ISSN 1996-1073) (available at: https://www.mdpi.com/journal/energies/special_issues/Computational_Heat_Transfer_Fluid_Mechanics).

For citation purposes, cite each article independently as indicated on the article page online and as indicated below:

LastName, A.A.; LastName, B.B.; LastName, C.C. Article Title. *Journal Name* **Year**, *Volume Number*, Page Range.

ISBN 978-3-0365-4543-1 (Hbk)
ISBN 978-3-0365-4544-8 (PDF)

Contents

About the Editors

Pouyan Talebizadeh Sardari

Dr Pouyan Talebizadeh Sardari is a research associate at Brunel University, London. His current research focuses on the combined usages of heat pumps and heat storage systems to provide hot water and space heating. His research areas mainly include the applications of thermal energy storage, renewable energy, computational fluid dynamics, heat pumps and nanotechnology. Currently, he is the associate editor of the journal Heat Transfer Engineering and among the topic editorial board of Energies. So far, he has over 120 publications in archival journals.

Goodarz Ahmadi

Goodarz Ahmadi is a Clarkson Distinguished Professor and Robert R. Hill Professor of Mechanical and Aeronautical Engineering at Clarkson University. He is a Fellow of ASME, ASTFE, ISME, ISCE, and IAAM. His research interests include multiphase flows, particle transport and deposition, turbulent flows, flow control, granular flows, respiratory flows, air pollution and flows through porous and fractured media. He received the 2016 ASME Freeman Scholar Award for Fluid Engineering, the Clarkson University Lifetime Research Achievement Award and the University Distinguished Teaching Award. He has 4 patents, has authored 3 books and has over 750 publications in archival journals. In addition, he also has made over 1250 presentations, including 20 plenary and keynotes presentations at national and international technical conferences. He has been serving as the editor and on the editorial and editorial advisory boards of 12 international journals.

Kiao Inthavong

Dr Kiao Inthavong is an Associate Professor of Mechanical Engineering at RMIT University specializing in the computational modelling of respiratory flows with applications in drug delivery and inhalation toxicology. He leads the CFD Research group at RMIT University and is a co-founder of SCONA (Society for Computational Fluid Dynamics of Nose and Airway), which is aimed at fostering collaboration between clinicians and engineers.

energies

MDPI

Article

Numerical Study on the Cavitation Flow and Its Effect on the Structural Integrity of Multi-Stage Orifice

ɔonghee Lee [1,*], Myungjo Jhung [2], Juneho Bae [1] and Soonho Kang [3]

1 Department of Regulatory Assessment, Korea Institute of Nuclear Safety, Daejeon 34142, Korea; bum5520@kins.re.kr
2 Department of Nuclear Safety Research, Korea Institute of Nuclear Safety, Daejeon 34142, Korea; mjj@kins.re.kr
3 Department of Kori NPP Regulation, Korea Institute of Nuclear Safety, Daejeon 34142, Korea; k728ksh@kins.re.kr
* Correspondence: ghlee@kins.re.kr; Tel.: +82-42-868-0683

Abstract: Flow leakage due to cavitation erosion occurred at the socket welding part downstream of the multi-stage orifice installed in the auxiliary feedwater (AFW) pump recirculation line of the domestic nuclear power plant (NPP). To assess the adequacy of the changed operating flow rate proposed by a domestic NPP operator as the corrective measure concerning the flow leakage in the AFW pump recirculation line, the pattern of the cavitation flow in the eight-stage orifice and the connecting pipe depending on the magnitude of the operating flow rate was predicted by using ANSYS CFX R19.1. Additionally, using ANSYS Mechanical, the structural analysis was conducted under the same operating flow rate condition used for the flow analysis, and the structural integrity was evaluated for the allowable stress. Based on the flow analysis results, it was found that the operating flow rate was the main factor to influence the cavitation behavior inside the multi-stage orifice, and cavitation flow still happened even in the vicinity of the corrected operating flow rate, so it should be necessary to fundamentally review the adequacy of the multi-stage orifice design. On the other hand, the geometric dimensions and arrangement of orifice hole position at the individual stage of the multi-stage orifice may have a significant influence on the characteristics of pressure drop and flow patterns (including cavitation). Therefore, these effects were examined by simulating an analysis model in which the hole diameter of the eighth-stage orifice was changed under the design flow rate condition. As a result of flow analysis, it was found that reducing the hole diameter in the eighth stage orifice resulted in increasing the pressure drop. In relation to the structural integrity of the eight-stage orifice and the connecting pipe, it was found that its integrity could be maintained under the design and operating flow rate conditions.

Keywords: auxiliary feedwater system; cavitation; computational fluid dynamics; in-service testing; multiphase flow; multi-stage orifice

tation: Lee, G.; Jhung, M.; Bae, J.; ɪng, S. Numerical Study on the vitation Flow and Its Effect on the uctural Integrity of Multi-Stage fice. *Energies* **2021**, *14*, 1518. ɔs://doi.org/10.3390/en14061518

ɪdemic Editor: ɪyan Talebizadeh Sardari

eived: 29 December 2020
epted: 6 March 2021
lished: 10 March 2021

lisher's Note: MDPI stays neutral ɪ regard to jurisdictional claims in ·ished maps and institutional affil- ns.

1. Introduction

Nuclear power plant (NPP) operators periodically conduct in-service testing (IST) for pumps that perform safety functions and monitor/evaluate the degree of their vulnerability over time. The auxiliary feedwater (AFW) system, one of the representative IST-related systems, plays a role in supplying the coolant to remove heat from the primary system when the main feedwater (MFW) system cannot be used in an emergency, including a small break loss-of-coolant-accident. As shown in Figure 1, this system consists of the main flow line for injecting coolant from the AFW pump to the steam generator and a recirculation line for the safe operation and performance tests of the AFW pump. The multi-stage orifice is installed in the recirculation line to limit the flow rate at the high discharge pressure of the AFW pump and perform the system decompression function.

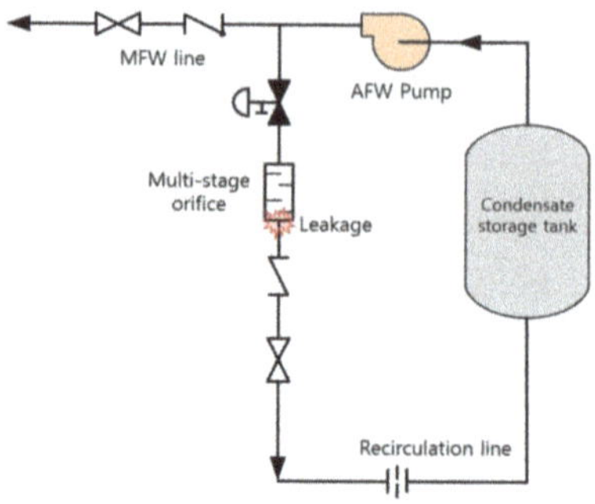

Figure 1. Schematic diagram for the auxiliary feedwater (AFW) pump recirculation line of th domestic nuclear power plant (NPP).

As shown in Figure 2, in the domestic NPP, while the reactor coolant system mair tained the normal operating temperature/pressure and the motor-driven AFW pum supplied the coolant to the steam generator, flow leakage occurred at the socket weldin part downstream of the multi-stage orifice. As a result of analyzing the pipe specimen wit an optical microscope, the domestic NPP operator deduced that the pipe wall thinnin occurred due to erosion, forming a rough surface, and the thinning passed through th pipe and proceeded to the welding part [1]. In conclusion, the cavitation erosion due t the bubbles collapse was suggested as the main cause of the flow leakage at the weldin part [1].

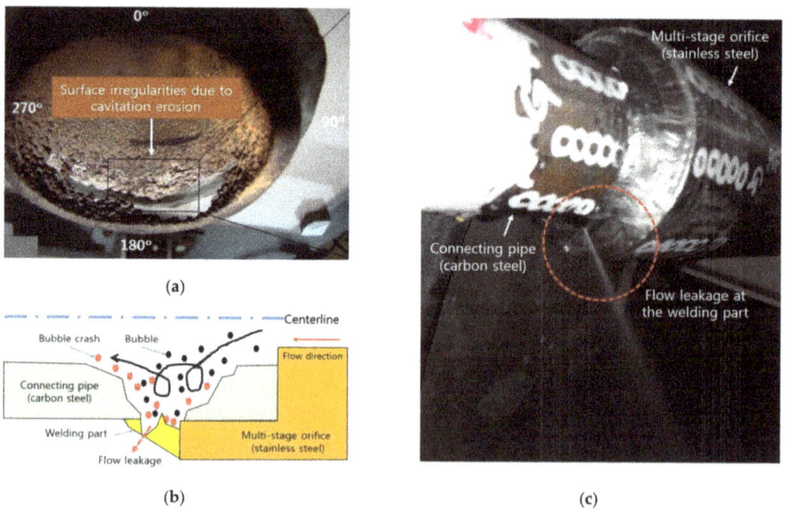

Figure 2. Cavitation erosion as the main cause of the flow leakage at the welding part: (**a**) cro section of the leakage part [1]; Reproduced from [1], Korea Hydro & Nuclear Power: 2018. schematic diagram for the cavitation erosion mechanism [1]; Reproduced from [1], Korea Hydro Nuclear Power: 2018 and (**c**) flow leakage at the welding part [2]. Reproduced from [2], Korea Soci for Fluid Machinery: 2019.

Cavitation may happen inside the orifice or the connecting pipe due to the flow acc eration occurring while passing through the orifice hole and the accompanying pressu drop, and as a result, performance degradation and structural damage of the orifice semblies come about because of high-frequency vibration and material erosion. Cavitati involves complex turbulent multiphase flows, so accurate simulation of these types of flows using available computational fluid dynamics (CFD) software remains a great challenge. In addition, it may be necessary for regulators to appropriately utilize the flow analysis results for the structural analysis in order to make reasonable regulatory decisions when reviewing licensing documents related to the structural integrity evaluation including cavitation effects.

Until recently, there have been very few experimental and numerical studies for the flow characteristics inside the multi-stage orifice. Wang et al. [3] experimentally investigated the inlet and outlet pressure drop characteristics for the multi-stage letdown orifice of the chemical and volume control system (CVCS) and designed the structure of the multi-stage orifice capable of providing a higher pressure drop without cavitation. The pressure drop could be improved by reducing the hole diameter of the orifice inlet or outlet, and the inlet pressure drop control method could produce a higher pressure drop compared to the outlet pressure drop control method. Bai et al. [4] numerically examined the effects of outlet pressure, the circular bead of the orifice inlet, and the change in the shape of the orifice outlet caused by the cavitation erosion on the cavitation characteristics inside the letdown orifice of the CVCS using ANSYS FLUENT. They found that when the inlet pressure was constant, cavitation did not occur in the letdown orifice unless the outlet pressure decreased to a specific value [4]. Once cavitation was formed in the letdown orifice, the degree of cavitation increased as the outlet pressure decreased [4]. The change in the shape of the orifice outlet caused by cavitation erosion significantly enhanced the degree of subsequent cavitation around the outlet of the letdown orifice [4]. Additionally, this effect was reinforced with the increase of shape change [4]. Niyogi et al. [5] confirmed using ANSYS FLUENT whether the performance of an eleven-stage orifice, designed to limit flow rate and prevent cavitation from occurring, was satisfactory. As a result of the simulation, the flow rate passing through the multi-stage orifice was limited to ±8% of the nominal flow rate, and cavitation did not occur. However, even though the analysis model was three-dimensional geometry, two-dimensional axisymmetric flow analysis was performed, and three-dimensional flow analysis was limited to only three stages out of eleven stages. Araoye et al. [6] assessed the effect of the inlet velocity, orifice size, and spacing between orifices on the axial velocity and pressure distribution inside the two-stage orifice using ANSYS FLUENT. They found that the flow characteristics downstream of the two-stage orifice was qualitatively similar to that of a single-stage orifice in terms of the presence of recirculation and reattachment zone, and the shear layer region while some different flow structures, for examples a jet-type flow in the core region surrounded by donut-shaped vertical flow, were identified in the upstream of the second-stage orifice [6].

Insufficient understanding of the complex flow pattern (including cavitation) inside the multi-stage orifice, important to safe NPP operation, makes it difficult for the NPP operator or regulator to predict the pressure drop, cavitation, and erosion characteristics depending on either the operation condition or orifice geometry. To solve this difficulty, the main contents and scope of this study were composed as follows:

- To verify whether the numerical modeling available in ANSYS CFX R19.1 can predict reliably and accurately the complex flow inside the multi-stage orifice, the numerical analysis was performed on the six-stage orifice test facility, and the simulation results were compared with the measured data.
- To assess the adequacy of the changed operating flow rate proposed by a domestic NPP operator as the corrective measure about the flow leakage in the AFW pump recirculation line, the pattern of the cavitation flow in the eight-stage orifice and the connecting pipe depending on the magnitude of the operating flow rate was predicted by using ANSYS CFX R19.1.
- Using ANSYS Mechanical, the structural analysis was conducted for the eight-stage orifice and the connecting pipe under the same operating flow rate condition used for the flow analysis, and the structural integrity was evaluated for the allowable stress.

Additionally, modal analysis was performed to predict the possible occurrence of cavitation due to a pressure fluctuation.

- The geometric dimensions and arrangement of orifice hole position at the individual stage of the multi-stage orifice may have a significant influence on the characteristics of pressure drop and flow patterns (including cavitation). Therefore, these effects were examined by simulating an analysis model in which the hole diameter of the eighth stage orifice was changed under the design flow condition.

2. Validation of the Numerical Modeling

To validate whether the numerical modeling available in ANSYS CFX R19.1, predicts reliably and accurately the complex flow inside the multi-stage orifice, the numerical simulation was performed on the six-stage orifice and then the calculated pressure drop between inlet and outlet sections of the multi-stage orifice depending on the operating flow rate was compared with the measured data. For reference, the numerical modeling in this section was successfully validated for the single-stage orifice flowmeter [7].

2.1. Analysis Model

The analysis model in this section is based on a single-phase pressure drop test in a multi-stage letdown orifice pipe, performed by Wang et al. [3] of Shanghai University of Science and Technology. For reference, the multi-stage orifice installed in the NPP is generally operated at high pressure and the local flow velocity passing through the orifice hole may exceed 100 m/s. Therefore, it is difficult to find detailed experimental data to validate the simulation results for multiphase flow including cavitation. Figure 3 shows the schematic diagram of the multi-stage orifice used as the test apparatus. The corresponding multi-stage orifice consisted of six stages and the total length was 700 mm (For reference, the total length of the present analysis model was extended to 990 mm to guarantee no reverse flow at the outlet boundary). The length (L_s) between the five orifice disks (or plates) located on the upstream side was equal to 101.6 mm respectively. The sixth stage orifice disk had a hole in the pipe centerline and was connected to the remaining upstream five orifice stages by two flanges. In addition, the sixth stage orifice disk can be easily replaced by a disk having a different orifice hole diameter [3]. Particularly, the orifice holes from the second to the fifth stage were alternately and eccentrically arranged in the opposite direction from the pipe centerline.

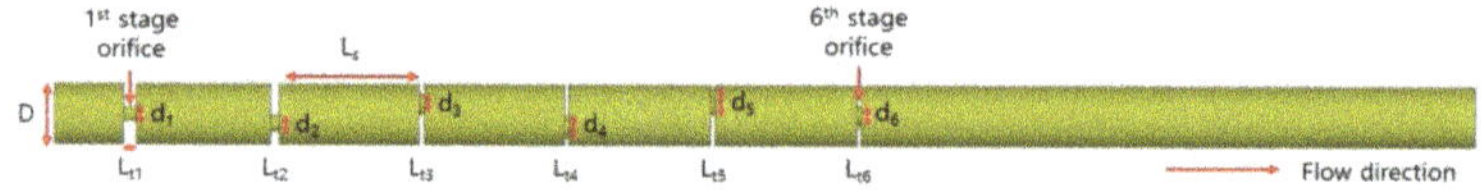

Figure 3. Schematic diagram of the six-stage orifice.

The main geometrical dimensions of the multi-stage orifice were summarized in Table 1. Here, $\mathbf{L_t}$ is the thickness of the orifice disk, and β is the ratio of the orifice hole diameter (d) to the inner diameter of the connecting pipe (D). The diameter ratio increased from the first to fifth stage orifice, while the sixth stage orifice showed a smaller magnitude than the fourth stage orifice.

Table 1. The geometrical information for six-stage orifice [3].

Stage No.	L_t (mm)	d (mm)	D (mm)	β = d/D
1	9.0	10.31		0.241
2	6.0	12.14		0.284
3	4.0	14.27	42.82	0.333
4	3.0	16.69		0.390
5	3.0	19.30		0.451
6	3.0	15.00		0.350

The flow rate was supplied using a ten-stage centrifugal pump with a rated flow rate of 5.0 m^3/h and maximum discharge pressure of 1.9 MPa [3]. The test flow rate was adjusted using a butterfly valve installed downstream of the connecting pipe [3]. The pressure drop between the orifice inlet and outlet was calculated using the measured data obtained through a series of repeated tests [3]. Water at 54.4 °C was used as the working fluid.

2.2. Numerical Modeling

In this study, the turbulent flow inside the six-stage orifice was calculated under steady, single-phase, and incompressible flow conditions using ANSYS CFX R19.1. The spatial discretization error may be caused by the accuracy order of the difference scheme and grid spacing. For either flow that is not parallel to the grid line or complex flows, it is recommended not to use the discretization scheme with the first-order accuracy [8]. Therefore, in the present study, the convective terms of the momentum and turbulence transport equations were calculated by applying the high-resolution scheme equivalent to the second-order accuracy. For reference, in the high-resolution scheme, the values of the blend factor are determined using the local solution for the entire computational domain. For flow regions where the gradient of the variable is small, the blend factor has a value close to 1 and as a result, this scheme has the second-order accuracy. In the case of flow regions where the gradient of the variable changes rapidly, the blend factor has a value close to 0 and consequently this scheme has the first-order accuracy to prevent distortion of the simulation result and maintain robustness. When the root mean square residual of the individual equations was 10^{-5} or less and the change of the main variables was very small, the calculation was judged to be converged.

In CFD simulations, the types of errors can be divided into numerical errors and model errors. The turbulence model is one of the main causes of model error. In general, the flow patterns are quite complex in the issues related to IST, but no turbulence model can accurately simulate them throughout the whole computational domain. In this study, the turbulent flow inside the multi-stage orifice was calculated applying the standard k-ε model widely used in the industry among the turbulence models based on the Reynolds-averaged Navier–Stokes (RANS) equation available in ANSYS CFX R19.1. Differential transport equations for the turbulence kinetic energy (k) and turbulence dissipation rate (ε) are as follows:

$$\frac{\partial(\rho k)}{\partial t} + \frac{\partial}{\partial x_i}(\rho k U_i) = \frac{\partial}{\partial x_j}\left[\left(\mu + \frac{\mu_t}{\sigma_k}\right)\frac{\partial k}{\partial x_j}\right] + P_k - \rho\varepsilon \tag{1}$$

$$\frac{\partial(\rho \varepsilon)}{\partial t} + \frac{\partial}{\partial x_i}(\rho \varepsilon U_i) = \frac{\partial}{\partial x_j}\left[\left(\mu + \frac{\mu_t}{\sigma_\varepsilon}\right)\frac{\partial \varepsilon}{\partial x_j}\right] + \frac{\varepsilon}{k}(C_{\varepsilon 1} P_k - C_{\varepsilon 2}\rho\varepsilon) \tag{2}$$

$$\mu_t = \rho C_\mu \frac{k^2}{\epsilon} \tag{3}$$

where U_i is the mean velocity components, ρ is the fluid density, μ is the molecular (or dynamic) viscosity, $C_\mu = 0.09$, $C_{\varepsilon 1} = 1.44$, $C_{\varepsilon 2} = 1.92$, $\sigma_k = 1.0$, and $\sigma_\varepsilon = 1.3$ are turbulence constants and μ_t is the turbulent (or eddy) viscosity. P_k is the production of turbulence kinetic energy due to the mean velocity gradient. The standard k-ε model is numerically stable and has a well-established flow regime with good predictive performance, so it has been adopted by most general-purpose CFD software and used as a representative turbulence model in the related industries. However, this model may have limitations in accurately predicting boundary layer separation, flow accompanying rapid changes in the average strain rate, rotational flow, and flow over the curved surfaces [9].

As shown in Figure 4, the computational grid in the form of an unstructured hexahedron was generated using the ICEM-CFD, grid generation software, for the same sized computational domain as the test facility. The total number of computational nodes used in the calculation was about 7.58 × 10^6. For reference, detailed information on the grid system used in the calculation was summarized in Table 2. The full geometry of the six-stage orifice was considered in case the flow could not maintain the symmetrical pattern when passing through the orifice hole. Additionally, in order to properly predict the complex turbulent flow inside the orifice, a dense grid distribution was applied near the wall and the orifice hole.

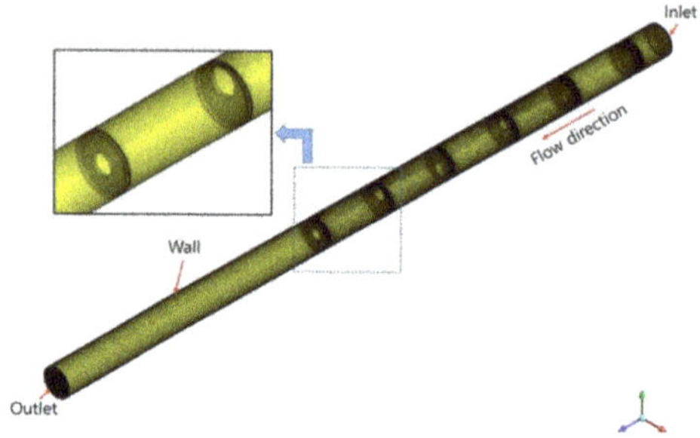

Figure 4. Grid system for six-stage orifice.

Table 2. Grid information for six-stage orifice.

Items	Values	Items	Values
Total number of nodes	7.58 × 10^6	Max. face angle	135.0°
Total number of elements	7.48 × 10^6	Max. edge length ratio	771.7
Min. face angle	45.0°	Max. element volume ratio	18.8

As the inlet boundary condition, the velocity profile for the fully developed flow obtained through the separate flow analysis applying the corresponding flow conditions for the orifice-free pipe with the same pipe diameter and turbulence intensity of 5% were applied. Average static pressure was used as the outlet condition. The walls were assumed to be smooth with zero surface roughness and a no-slip condition was applied there. A scalable wall function was applied to calculate the flow near the wall. This wall function can solve one of the main problems of the standard wall function in that it can be applied to any dense grid without resulting in the erroneous modeling of the laminar and buffer regions of the boundary layer.

2.3. Validation Results

Figure 5a shows the pressure sampling points at seven locations of the multi-stage orifice. P_0 and P_6 represent pressure sampling points located at the inlet and outlet of the multi-stage orifice, respectively, and their positions were approximately 30 mm away from the adjacent orifice disk. The remaining five pressure sampling points, P_1 to P_5, were located downstream from the adjoining orifice disk by a distance of 3 d_1~3 d_5, respectively. Figure 5b shows the result of comparing the calculated pressure drop ($\Delta p = P_0 - P_6$) between the inlet and outlet sections of the multi-stage orifice with the measured data depending on the inlet flow rate. As the inlet flow rate increased, the magnitude of the pressure drop also increased, and the predicted static pressure drop in the range of the inlet flow rate 5.24~6.21 m^3/h was consistent within a maximum deviation of 5% compared to the measurement results. However, the gradient between the measured inlet flow rate and pressure drop decreased as the flow rate became smaller, while the predicted result maintained an almost constant gradient. In this regard, it was confirmed that the Reynolds number based on both pipe diameter and the mean inlet velocity at an inlet flow rate of 5.24 m^3/h was 8.68 × 10^4 (Reynolds number for the other inlet flow rates can be found in Table 3), which corresponded to a completely turbulent flow, but the reason for the decrease in the gradient between the measured inlet flow rate and pressure drop could not be found.

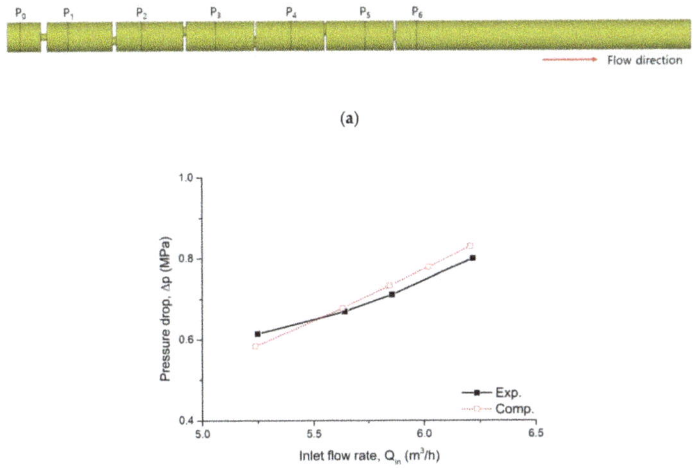

(b)

Figure 5. Comparison of measured and predicted static pressure drop versus inlet flow rate: (**a**) pressure sampling points and (**b**) static pressure drop versus inlet flow rate.

Table 3. Pressure drop between each orifice stage.

Flowrate (m^3/h)	Reynolds Number	P_0–P_1 (MPa)	P_1–P_2 (MPa)	P_2–P_3 (MPa)	P_3–P_4 (MPa)	P_4–P_5 (MPa)	P_5–P_6 (MPa)
5.24	8.68×10^4	0.25	0.11	0.08	0.046	0.027	0.067
5.64	9.32×10^4	0.29	0.13	0.09	0.055	0.031	0.08
5.85	9.67×10^4	0.31	0.14	0.1	0.06	0.034	0.086
6.02	9.96×10^4	0.33	0.15	0.11	0.062	0.035	0.09
6.21	1.03×10^5	0.35	0.16	0.12	0.065	0.038	0.096

On the other hand, as shown in Table 3, the smaller the orifice hole diameter, the higher the flow velocity, so the magnitude of the pressure drop at the individual orifice stage was correspondingly in the order of 1st > 2nd > 3rd > 6th > 4th > 5th stage orifice.

Besides the validation for the calculation results, the distribution of the predicted flow velocity and streamline inside the multi-stage orifice can be found in the author's research report [10].

3. Effect of the Operating Flow Rate

As previously shown in Figures 1 and 2, flow leakage due to cavitation erosion happened at the socket welding part downstream of the multi-stage orifice installed in the AFW pump recirculation line. As the corrective measure, the domestic NPP operator changed the operating flow rate to prevent the occurrence of cavitation flow. To assess the appropriateness of this corrective action, the audit calculation from the regulatory perspective was conducted by using ANSYS CFX R19.1, and simulation results were explained in this section.

3.1. Analysis Model

The analysis model is a multi-stage orifice installed in the recirculation line of the AFW pump to limit the flow rate at the high discharge pressure of the AFW pump and to decompress the corresponding system. Figure 6 shows the schematic diagram of the multi-stage orifice and the connecting pipe. The multi-stage orifice consisted of eight stages. The first and eighth stage orifice had holes in the center of the pipe, while the second to seventh stage orifice were alternately and eccentrically arranged in the opposite direction from the pipe centerline. Hole diameter (d), the spacing between the orifice disks (L_s), and thickness of orifice disk (L_t) for each orifice stage were the same. Since the quantitative dimensions of the multi-stage orifice were related to the intellectual property rights of the manufacturer, specific information could not be provided. Water at 40 °C was used as the working fluid.

Figure 6. Schematic diagram of the eight-stage orifice and the connecting pipe.

As previously shown in Figure 1, the recirculation line is opened when the NPP is in the mode of start/stop transient or the AFW pump is operated while the AFW pump outlet valve in the MFW line is closed. As the pressure in the steam generator side decreases, the recirculation flow rate is determined by the system resistance-pump performance matching characteristics. The minimum (or design) flow rate for the safe operation of the AFW in the recirculation line is about 19.3 m^3/h, and a maximum operation time of a quarter-hour is allowed [11]. On the other hand, the minimum flow rate for continuous safe operation without stopping the AFW pump was about 34.1 m^3/h.

3.2. Numerical Modeling

3.2.1. Flow Analysis

In this study, the turbulent flow inside an eight-stage orifice was calculated using ANSYS CFX R19.1 under steady, multiphase, and incompressible flow conditions. The discretization accuracy for the convective terms of the momentum equation and turbulence transport equation, turbulence model, grid type, and so on were the same as those described in Section 2.2. The mixture model was used to consider an interphase transfer. This model solves the continuity, momentum, energy equation for the mixture, and the volume fraction equation for the secondary phase (vapor). For the cavitation flow simulation, liquid–vapor mass transfer is governed by the following the vapor volume fraction equation:

$$\frac{\partial(\rho_v \alpha_v)}{\partial t} + \frac{\partial}{\partial x_i}(\rho_v \alpha_v v_i) = R_{vap} - R_{cond} \tag{4}$$

where ρ_v is the vapor density, α_v is the vapor volume fraction, and v_i is the directional velocity component, R_{vap} and R_{cond} are the mass transfer rates correspond to the vaporization and condensation during the cavitation process respectively. The Rayleigh–Plesset equation describing the bubble growth in the liquid is given by:

$$R_{vap} = F_{vap} \frac{3\alpha_n (1 - \alpha_v) \rho_v}{R_B} \sqrt{\frac{2}{3} \frac{(p_v - p)}{\rho_l}}, p < p_v \tag{5}$$

$$R_{cond} = F_{cond} \frac{3\alpha_v \rho_v}{R_B} \sqrt{\frac{2}{3} \frac{(p - p_v)}{\rho_l}}, p > p_v \tag{6}$$

where $R_{nuc} = 10^{-6}$ m is the nucleation site radius, p_v is the vapor pressure at the liquid temperature, p is the liquid pressure, ρ_l is the liquid density, $\alpha_n = 5 \times 10^{-4}$ is the volume fraction of the nucleation sites, and F_{vap} = 50 and F_{cond} = 0.01 are an empirical factor depend on vaporization and condensation designed for the different rates (vaporization is generally much faster than condensation). Saturation pressure set to about 7.36 kPa. For reference, the numerical method used in this study was summarized in Table 4. According to the author's previous study [12,13], the numerical modeling explained in this section reliably predicted the cavitation flow occurring inside the IST-related components (e.g., cavitating Venturi and sharp-edged orifice).

Table 4. Numerical method.

Items		Contents
Discretization accuracy for the convective terms	Momentum eqn.	High resolution
	Turbulence transport eqn.	High resolution
Interphase transfer model		Mixture
Cavitation model		Rayleigh–Plesset
Turbulence model		Standard k-ε
Near-wall treatment method		Scalable wall function
Convergence criterion		$<10^{-5}$

On the other hand, to obtain accurate simulation results when calculating cavitation flow using CFD software, it is essential to consider an appropriate grid type, especially at the location where cavitation may occur. In this regard, the authors confirmed through a previous study [14] that the grid shape had a significant effect on the cavitation analysis results (cavitation inception condition, discharge coefficient magnitude, etc.) inside a sharp-edged orifice. Based on the results of the previous study, to properly predict the cavitation flow inside the multi-stage orifice, a dense grid distribution was applied near the wall and the orifice hole as shown in Figure 7.

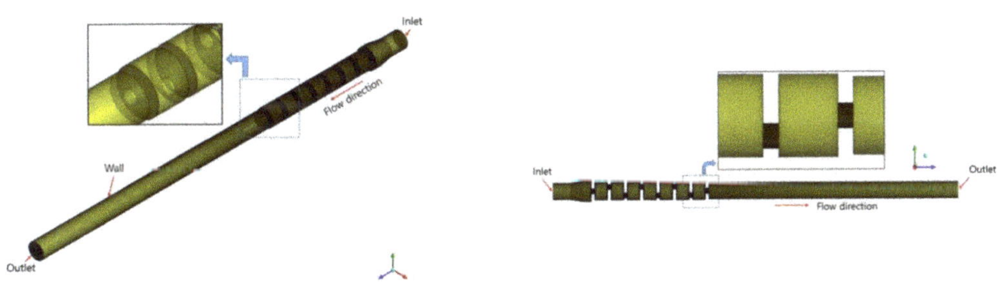

(a) (b)

Figure 7. Grid system for eight-stage orifice: (**a**) isometric-view and (**b**) x-axis view.

In addition, as shown in Table 5, a grid sensitivity study was performed for three types of grid systems. Type1 was the coarsest grid, and Type2 and 3 had dense grids near walls and orifice holes. Overall, the difference in the analysis results depending on the grid size was not large, and to understand in more detail the complex flow field (including cavitation flow) inside the multi-stage orifice with the reasonable computation cost, the prediction results for an intermediate grid (Type2, total number of nodes: 5.02×10^6) were explained in this paper.

Operating flow rate (Q_{in} = 34.1, 37.0, 39.0, and 41.5 m^3/h), turbulence intensity of 5%, and eddy viscosity ratio of 10 were applied as inlet conditions. The volume fraction of the liquid phase (water) at the inlet was assumed to be 1. As an outlet condition, the measured gauge pressure of about 310 kPa was used. The walls were assumed to be smooth and no-slip condition was applied there. A scalable wall function was applied to calculate the flow near the wall.

Table 5. Grid information.

Items	Type1	Type2	Type3
Total number of nodes	3.21×10^6	5.02×10^6	7.04×10^6
Total number of elements	3.15×10^6	4.94×10^6	6.94×10^6
Min. face angle	45.3°	45.2°	45.2°
Max. face angle	134.9°	135.0°	135.0°
Max. edge length ratio	611.6	61.3	61.3
Max. element volume ratio	9	9	9

3.2.2. Structural Analysis

Figure 8 shows the geometry modeling for the structural analysis of an eight-stage orifice made by ANSYS SpaceClaim. In addition to the geometry modeling for flow analysis shown in Figure 6, the actual thickness and the welding parts between the multi-stage orifice and the connecting pipes were considered. However, the thinning of the welding part due to cavitation erosion was not considered in this study.

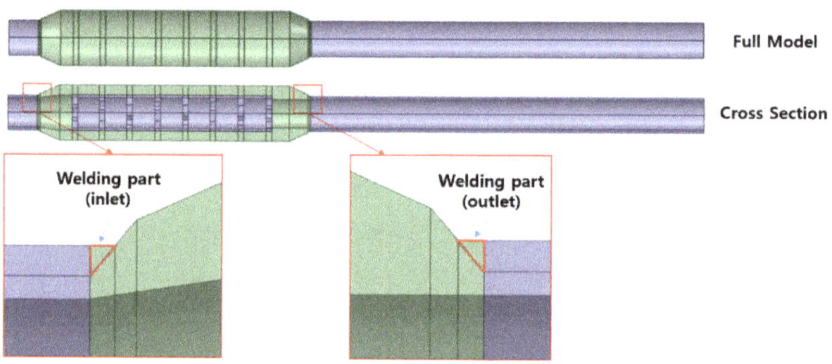

Figure 8. Geometry modeling for the structural analysis of the eight-stage orifice.

Figure 9 shows the grid shape for the structural analysis of an eight-stage orifice generated using ANSYS Meshing. To improve the accuracy of the analysis model, a 3D SOLID element representing quadratic displacement behavior was applied. The element has three degrees of freedom per node and provides plasticity, hyperelasticity, creep, stress stiffening, large deflection, and strain [15]. To assess the sensitivity depending on the element size, the structural analysis was performed by changing the default element size to 1.5, 2.5, and 3.0 mm.

As shown in Figure 10, the predicted stress intensity at the same location was compared. As the total number of nodes increased, the rate of change in the magnitude of stress intensity was reduced, and the tendency to converge to a constant value was shown. Based on the results of the grid sensitivity study as described above, the default element size for the structural analysis of an eight-stage orifice was determined to be 1.5 mm.

As shown in Figure 11, the material property values were applied by referring to ASME Code Section-II, Part-D. In the case of the welding part, stainless steel property was used.

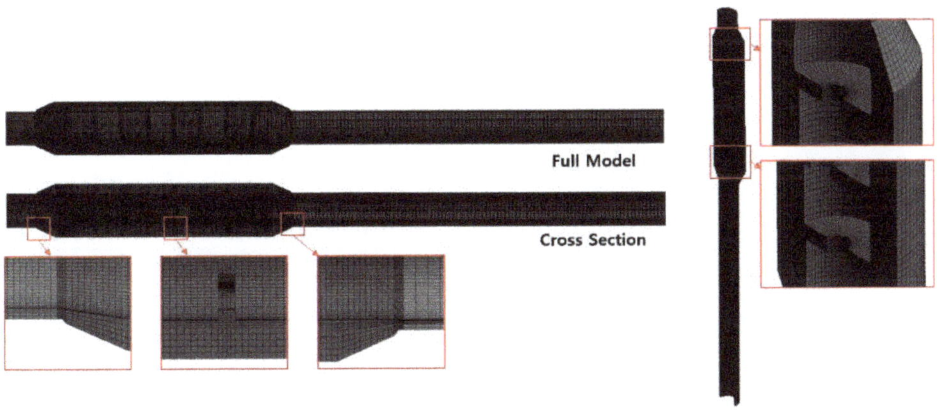

Figure 9. Grid system for the structural analysis of the eight-stage orifice.

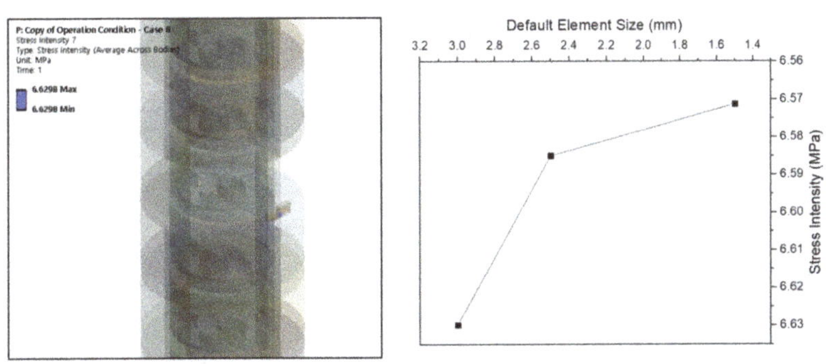

(a) (b)

Figure 10. Results of the grid sensitivity study for the structural analysis: (**a**) location for the assessment of stress intensity and (**b**) default element size versus stress intensity.

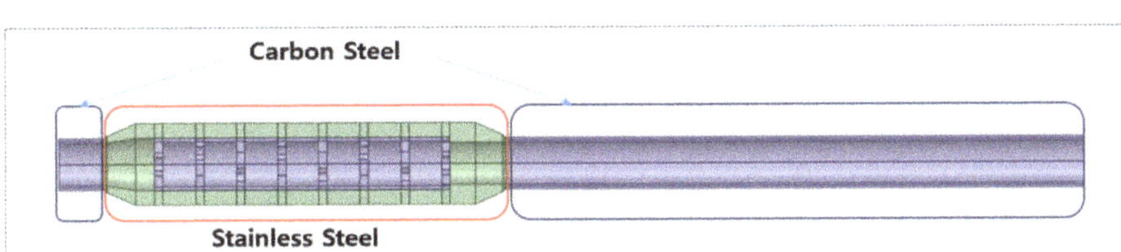

(a)

Material	Elastic Modulus (MPa)	Poisson's Ratio	Allowable Stress Intensity (MPa)
Stainless Steel	195,000	0.31	115
Carbon Steel	200,000	0.30	128

(b)

Figure 11. Material properties values per zone: (**a**) type of material properties applied per zone and (**b**) material properties values.

As shown in Figure 12, constraints were applied to the cross-sections of the upstream and downstream connecting pipes of the multi-stage orifice. Specifically, in the upstream pipe, the degrees of freedom in the hoop and axial directions were fixed, but the degree of freedom in the radial direction was free. On the other hand, in the downstream pipe, the degree of freedom in the hoop direction was fixed, but the degrees of freedom in the radial and axial directions were free. The flow analysis results obtained using ANSYS CFX R19.1 were applied as the pressure boundary condition at the inner wall of the multi-stage orifice and the connecting pipe. To be more specific, the Program Controlled Mapping option was selected to transfer the wall pressure data across a dissimilar mesh interface, that is, from ANSYS CFX to ANSYS Mechanical. This option can determine the appropriate settings based on the source and target mesh, and the data transfer type.

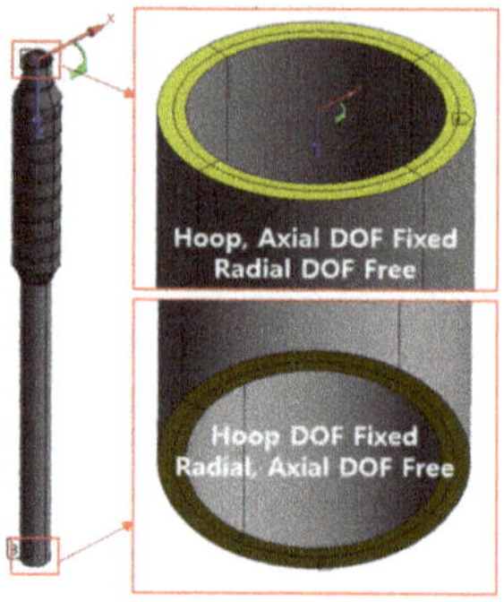

Figure 12. Constraint conditions.

3.3. *The Computational Results*

3.3.1. Flow Analysis

Figure 13a shows the distribution of the vapor volume fraction in the second half of the multi-stage orifice depending on the operating flow rate.

As the operating flow rate increased, the region with a vapor volume fraction of 0.9 or more expanded to the downstream of the eighth stage orifice. In fact, at 41.5 m^3/h, a flow rate condition in which leakage in the welding part due to cavitation erosion really occurred, a cavitation flow region was widely formed downstream of the eighth stage orifice. Besides, cavitation flow still appeared around the hole of the eighth stage orifice even at 34.1 m^3/h, corresponded to not only the operating flow rate that the NPP operator suggested as the corrective measure but also the minimum flow rate required for continuous safe operation without stopping the AFW pump.

Figure 14 shows the distribution of flow velocity and streamlines inside the eight-stage orifice and the connecting pipe (for the symmetric y-z plane) depending on the operating flow rate. The flow patterns between the sixth to seventh and seventh to eighth stage orifices were similar regardless of the magnitude of the operating flow rate. However, as the operating flow rate increased, the peak value of flow velocity also increased, and the high-speed jet flow region passing through the hole of the eighth stage orifice tended to expand. In addition, since the flow passing through the hole of the eighth stage orifice was directed to the upper part of the connecting pipe, the primary recirculation flow was formed at the lower part of the jet flow.

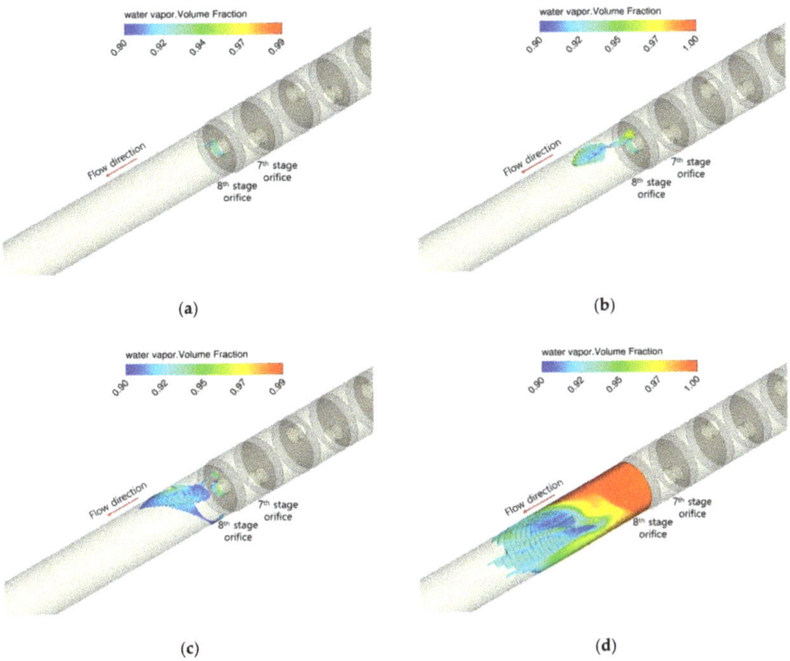

Figure 13. Iso-volume of vapor volume fraction: (**a**) Q_{in} = 34.1 m^3/h (Re = 2.74 × 10^5); (**b**) Q_{in} = 37.0 m^3/h (Re = 2.97 × 10^5); (**c**) Q_{in} = 39.0 m^3/h (Re = 3.13 × 10^5); and (**d**) Q_{in} = 41.5 m^3/h (Re = 3.33 × 10^5).

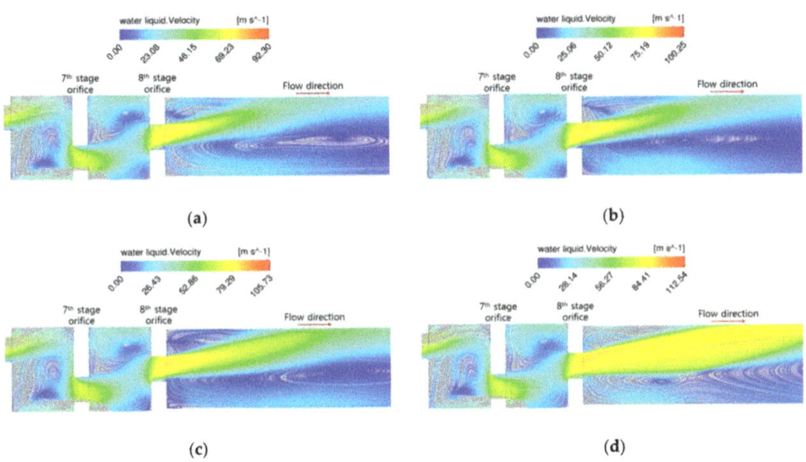

Figure 14. Distribution of flow velocity and streamlines inside the eight-stage orifice and the connecting pipe (for the symmetric y-z plane) depending on the operating flow rate: (**a**) Q_{in} = 34.1 m^3/h (Re = 2.74 × 10^5); (**b**) Q_{in} = 37.0 m^3/h (Re = 2.97 × 10^5); (**c**) Q_{in} = 39.0 m^3/h (Re = 3.13 × 10^5); and (**d**) Q_{in} = 41.5 m^3/h (Re = 3.33 × 10^5).

Figure 15 shows the mass-averaged vapor volume fraction in the seventeen axial cross-sections (6.35 mm spacing between cross-sections) downstream of the eighth stage orifice depending on the operating flow rate. The mass-averaged vapor volume fraction of the corresponding cross-sections also increased as the operating flow rate increased except for the cross-section number from 1 to 4 in the case of the operating flow rate of 37.0 m^3/h condition. On the other hand, at 41.5 m^3/h, which is a flow condition in which leakage occurred at the welding part of the multi-stage orifice and the connecting pipe due to cavitation erosion, the mass-averaged vapor volume fraction from cross-section number 13 to 14 showed the maximum value. These cross-sections corresponded to the location where flow leakage appeared. Therefore, it may be judged that the simulation results obtained by applying the numerical modeling described in Section 3.2.1 can properly reflect the actual situation where the leakage occurred.

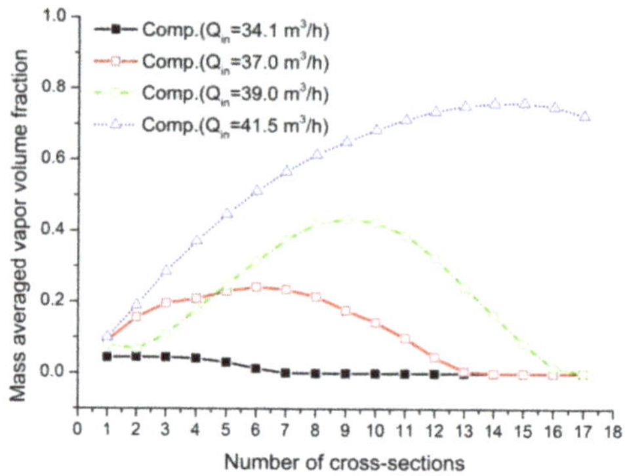

Figure 15. The mass-averaged vapor volume fraction in the axial cross-sections downstream of the eighth stage orifice depending on the operating flow rate.

Based on the above-mentioned flow analysis results, it was confirmed that the operating flow rate was the main factor to influence the cavitation behavior inside the multi-stage orifice. Therefore, the reduction in the operating flow rate suggested by the NPP operator is judged to be an appropriate corrective action. However, it was found that cavitation still happened in the vicinity of the changed operating flow rate of 34.1 m^3/h, i.e., minimum flow rate required for the continuous safe operation of the AFW pump. Therefore, as an additional measure, it is necessary to shorten the operating time of the AFW pump and reinforce periodic monitoring (vibration measurement during the AFW pump operation) [1]. Separately, it is required to review the design appropriateness of the multi-stage orifice as a fundamental solution to prevent flow leakage from the welding part between the multi-stage orifice and the connecting pipe caused by cavitation erosion.

3.3.2. Structural Analysis

Table 6 and Figure 16 show the predicted results of the stress intensity distribution in the multi-stage orifice and the connecting pipes depending on the operating flow rate by applying the numerical modeling for the structural analysis, described in Section 3.2.2, to ANSYS Mechanical. As the operating flow rate increased, the maximum value of the stress intensity also increased. The maximum value of the stress intensity occurred in the upstream connecting pipe of the multi-stage orifice, which is believed to be due to the depressurization occurring as the flow entered into the multi-stage orifice passed through the orifice hole. On the other hand, it was not evident whether the high-stress intensity caused by cavitation flow occurred in the welding part between the multi-stage orifice and the connecting pipe.

Table 6. The predicted results of stress components depending on the operating flow rate.

Case	Flowrate (m^3/h)	Diameter Ratio (d_8/d)	Stress Intensity (MPa)	Membrane Stress (MPa)	Membrane + Bending Stress (MPa)	Allowable Stress (MPa)	Location of Maximum Stress
OP1	34.1	1	123.6	105.9	121.8	207	Upstream connecting pipe
OP2	37		145.8	125	143.7		
OP3	39		162	138.8	159.7		
OP4	41.5		183.8	157.5	181.2		

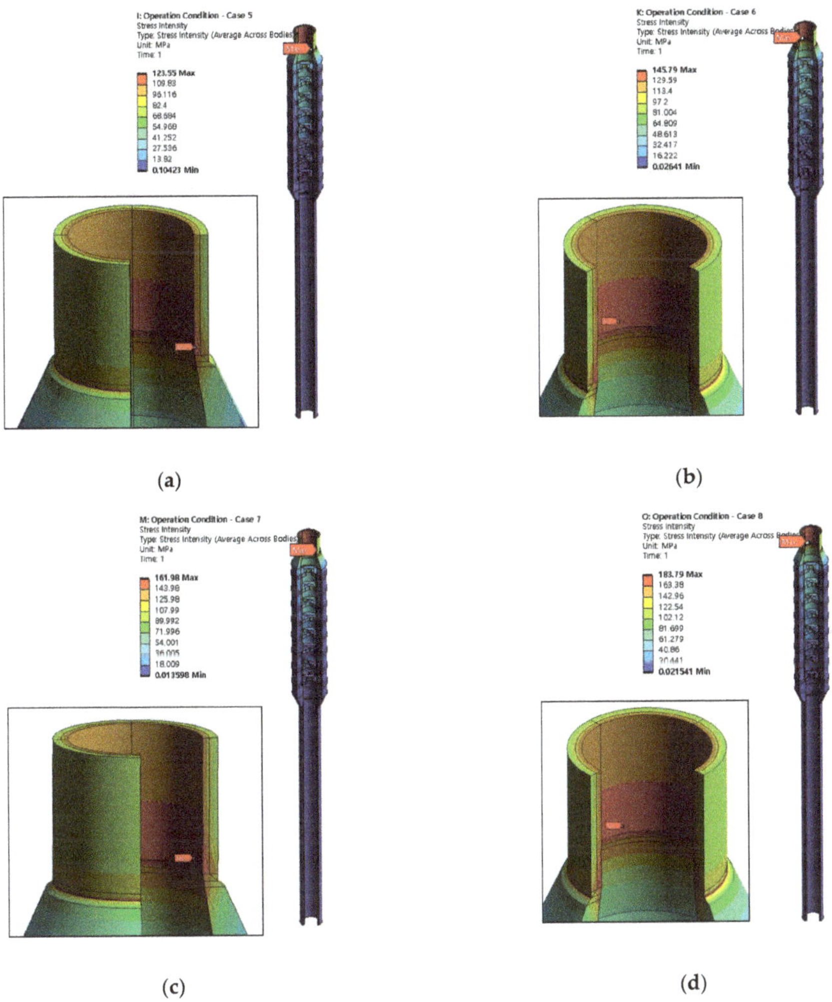

Figure 16. Distribution of the predicted stress intensity depending on the operating flow rate: (**a**) Q_{in} = 34.1 m^3/h (Re = 2.74 × 10^5); (**b**) Q_{in} = 37.0 m^3/h (Re = 2.97 × 10^5); (**c**) Q_{in} = 39.0 m^3/h (Re = 3.13 × 10^5); and (**d**) Q_{in} = 41.5 m^3/h (Re = 3.33 × 10^5).

In summary, the analysis result of the stress intensity distribution in the multi-stage orifice and the connecting pipe depending on the operating flow rate showed a value less than the allowable stress of 207 MPa. Therefore, the decrease in the operating flow rate suggested by the NPP operator is considered to be reasonable in that it can secure more margins for the structural integrity of the corresponding facility, including the welding part between the multi-stage orifice and the connecting pipe. As described above, since the maximum value of the stress intensity occurred in the connecting pipe upstream of the multi-stage orifice, it will be expected that the above-mentioned analysis result will not change significantly even if the additional structural analysis is performed by considering the thinning of the welding part between the multi-stage orifice and the connecting pipe due to cavitation erosion.

Table 7 shows the predicted deformation results for each direction in the multi-stage orifice and the connecting pipe depending on the operating flow rate. The deformation size in each direction also increased as the operating flow rate increased, except for the deformation in the hoop direction at Q_{in} = 37.0 m^3/h. At the same operating flow rate, the amount of deformation for each direction was in the order of axial > radial > hoop. In the case of radial deformation, the maximum deformation value was shown in the upstream connecting pipe of the multi-stage orifice where high stress acted. For axial deformation, the maximum deformation value was shown at the seventh stage orifice. In the case of the deformation in the hoop direction, the maximum deformation value was shown either at the third stage orifice or in the downstream connecting pipe of the multi-stage orifice depending on the operating flow rate.

Table 7. The predicted results of deformation components depending on the operating flow rate.

Case	Flowrate (m^3/h)	Diameter Ratio (d_8/d)	Max. Radial Deformation (mm)	Location of Max. Deformation	Max. Hoop Deformation (mm)	Location of Max. Deformation	Max. Axial Deformation (mm)	Location of Max. Deformation
OP1	34.1	1	0.013	Upstream connecting pipe	0.0017	3rd stage orifice	0.0154	7th stage orifice
OP2	37		0.0155		0.0044	Downstream connecting pipe	0.0183	
OP3	39		0.0171		0.0023	3rd stage orifice	0.0203	
OP4	41.5		0.0194		0.0037	Downstream connecting pipe	0.0231	

Additionally, a modal analysis was performed to generate frequencies for the finite element model and the boundary conditions applied to the connecting pipe inlet and outlet ends. The first 20 modes were summarized in Figure 17, and some of the corresponding mode shapes were shown in Figure 18. From the mode shapes, it was found that the first seven beam modes ranged from 137 to 4321 Hz. Specifically, corresponding values of 1061 Hz and 1612 Hz were found for the third and fourth beam modes, between 1000 and 2000 Hz, which showed several peak values in the frequency domain resulting from cavitation at the specific flow rate [2]. This may be one of the reasons for a possible resonance effect between the structural frequency and the forcing frequency. Four radial modes were shown at 863 Hz, 2286 Hz, 3754 Hz, and 4661 Hz. Moreover, two axial modes were found at 1252 Hz and 2227 Hz for orifice modes with forward and backward motions of the flow direction, respectively. The modal characteristics obtained by the modal analysis give the possible solution to escape the peak values of pressure fluctuation between 1000 and 2000 Hz. Decreasing flow rate may be one of the options to eliminate resonance between the structural frequency and the cavitation frequency.

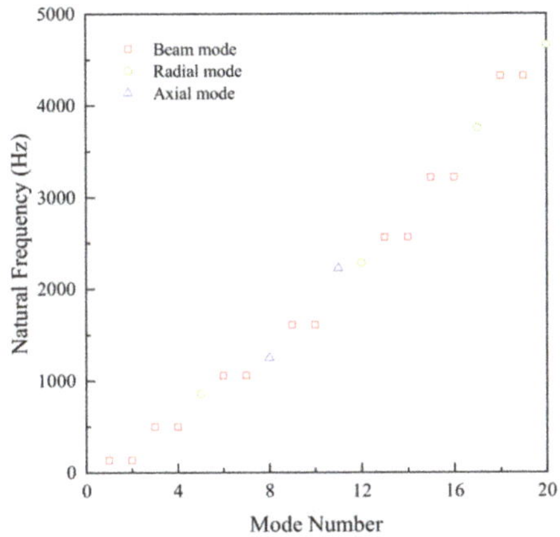

Figure 17. Natural frequencies of the multi-stage orifice and the connecting pipe.

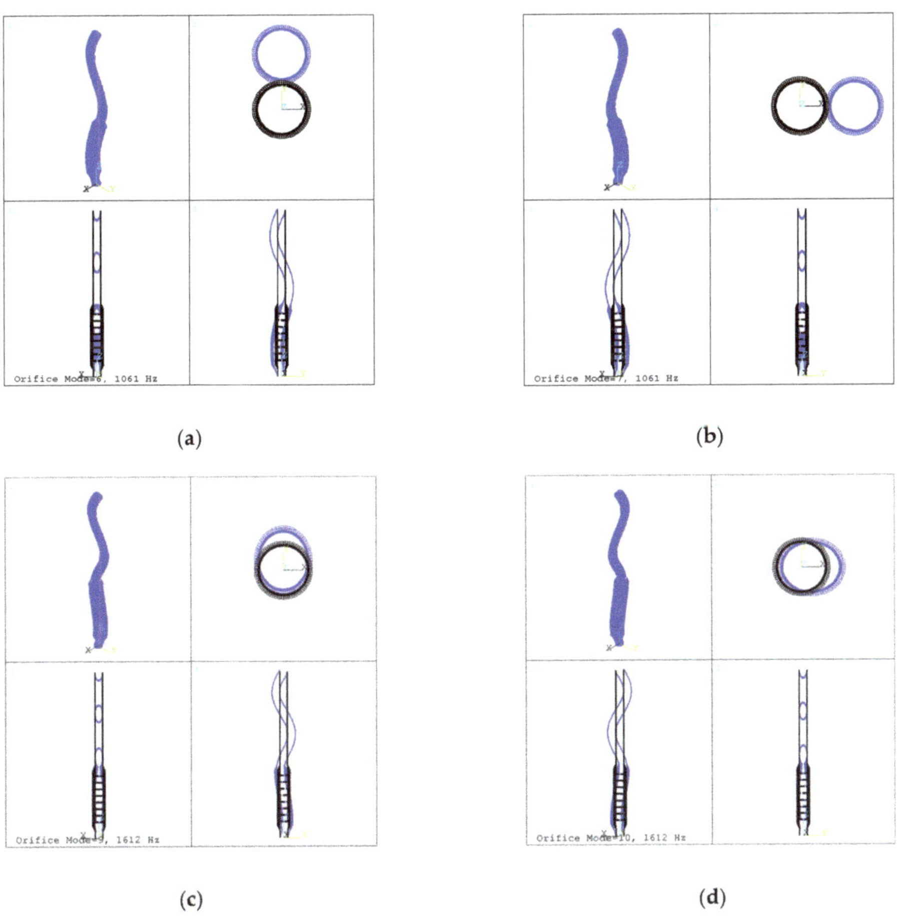

Figure 18. Mode shapes of the multi-stage orifice and the connecting pipe: (**a**) $d_8/d = 0.84$; (**b**) $d_8/d = 1.0$ (original case); (**c**) $d_8/d = 1.24$; and (**d**) $d_8/d = 1.49$.

4. Effect of the Orifice Hole Diameter Size

Apart from the variables related to the performance of the multi-stage orifice, an important design consideration is to ensure that cavitation does not occur inside the multi-stage orifice [5]. In this regard, the dimensions and arrangement positions of the orifice holes in individual stages of the multi-stage orifice can have a significant influence on the pressure drop characteristics and flow patterns (including cavitation). Therefore, in this section, the effect of the change in the hole diameter (d_8) of the eighth stage orifice (shown in Figure 6) on the pressure drop characteristics and flow patterns was investigated.

4.1. Analysis Model

The analysis model in this section additionally considered the cases that the ratio (d_8/d) of the changed hole diameter (d_8) to the existing hole diameter (d) of the eighth stage orifice was 0.84, 1.24, and 1.49, respectively. In this case, the ratio (d_8/D) of the hole diameter (d_8) of the eighth stage orifice to the inner diameter (D) of the upstream connecting pipe was 0.22, 0.32, and 0.38, respectively. Except for the hole diameter of the eighth stage orifice, the remaining geometric specifications were maintained. On the other hand, the simulation was performed on the design flow rate of about 19.3 m^3/h ($Re = 1.53 \times 10^5$), which allows the safe operation of the AFW pump for a maximum of 15 min. The corresponding flow rate was the reduced amount up to 54% compared to the operating flow rate (Q_{in} = 34.1, 37.0, 39.0, and 41.5 m^3/h) described in Section 3.2.1.

4.2. Numerical Modeling

4.2.1. Flow Analysis

Except for applying the design flow rate (Q_{in} = 19.3 m^3/h) as the inlet condition, the discretization scheme for the convective terms of the momentum and turbulence transport equation, turbulence model, multiphase flow model, grid shape, and boundary conditions were the same as those described in Section 3.2.1.

4.2.2. Structural Analysis

Except that additional flow analysis result obtained using ANSYS CFX R19.1 for the change of both the hole diameter of the eighth stage orifice and the flow rate was applied as the pressure boundary condition on the inner wall of the multi-stage orifice and the connecting pipes, the grid shape, material properties, and constraint conditions were the same as those described in Section 3.2.2.

4.3. The Computational Results

4.3.1. Flow Analysis

Figure 19 shows the distribution of the vapor volume fraction in the second half of the multi-stage orifice depending on the size of the hole diameter (d_8) in the eighth stage orifice. As the size of the hole diameter (d_8) in the eighth stage orifice increased, the main cavitation region moved from the hole of the eighth stage orifice to that of the seventh stage orifice [16]. For $d_8/d = 1.24$, both the size of the cavitation region and the peak value of the vapor volume fraction were the smallest [16]. For $d_8/d = 0.84$, the cavitation region was extended to the downstream of the eighth stage orifice, and the maximum value of the vapor volume fraction was the largest [16]. In all cases, no cavitation region was formed upstream of the seventh stage orifice (i.e., from the inlet of the upstream connecting pipe to the inlet of the seventh stage orifice) [16].

On the other hand, in the case of the original multi-stage orifice ($d_8/d = 1.0$) installed in the AFW pump recirculation line of the domestic NPP, a cavitation region was formed inside the hole of the eighth stage orifice under the design flow condition as shown in Figure 19b. Therefore, based on the simulation results of this study, it is necessary to review the adequacy of the multi-stage orifice design.

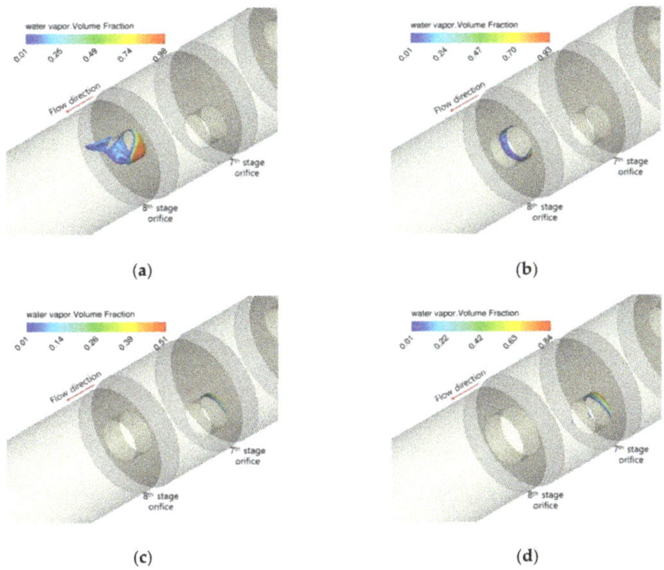

Figure 19. Iso-volume of vapor volume fraction depending on the size of the hole diameter (d_8) in the eighth orifice stage: (**a**) $d_8/d = 0.84$; (**b**) $d_8/d = 1.0$ (original case); (**c**) $d_8/d = 1.24$; and (**d**) $d_8/d = 1.49$.

Figure 20 shows the distribution of flow velocity and streamlines in the second half of a multi-stage orifice (for the symmetric y–z plane) depending on the size of the hole diameter (d_8) in the eighth orifice stage. In this study, only the hole diameter of the eighth stage orifice was changed and therefore the streamline pattern in the upstream of the eighth stage orifice was similar to each other [16]. On the other hand, since the location of the eighth stage orifice hole was higher than that of the seventh stage orifice hole, the flow passed through the eighth stage orifice hole upward [16]. The angle of this upward flow also increased as the size of the hole diameter in the eighth stage orifice increased [16]. As the size of the eighth stage orifice hole diameter was smaller, water (liquid phase) velocity passing through the eighth stage orifice hole was much faster [16]. Therefore, the peak value of the water (liquid phase) velocity for $d_8/d = 0.84$ was found near the hole of the eighth stage orifice. For other d_8/d cases (for example, $d_8/d = 1.0$, 1.24, and 1.49), the maximum value of the water (liquid phase) velocity was shown near the hole entrance of the second stage orifice [16].

Figure 21 shows the pressure drop (Δp) depending on the size of the hole diameter in the eighth stage orifice. The pressure drop (Δp) is the difference in static pressure between the upstream and downstream cross-sections of the multi-stage orifice [16]. The corresponding cross-sections were located at 19 mm from the first and eighth stage orifice [16]. As shown in Figure 21, it was found that reducing the hole diameter in the eighth stage orifice resulted in increasing the pressure drop [16]. This trend in the static pressure drop can be also found in the experimental results of Wang et al. [3].

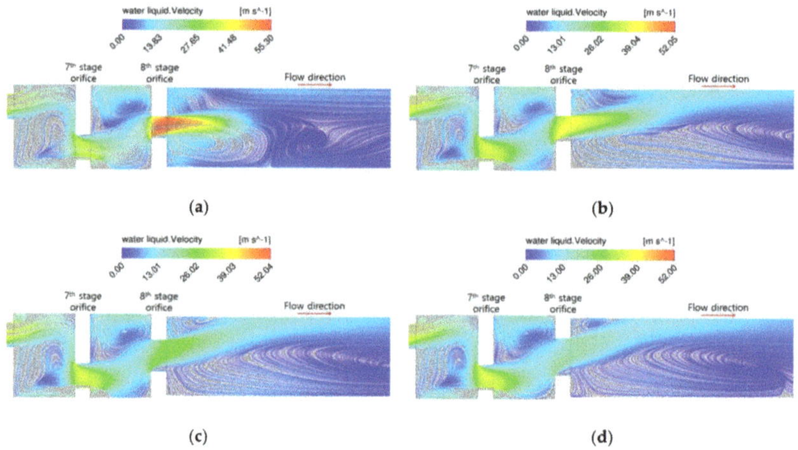

Figure 20. Distribution of flow velocity and streamlines in the second half of the multi-stage orifice (for the symmetric y-z plane) depending on the size of the hole diameter (d_8) in the eighth orifice stage: (**a**) $d_8/d = 0.84$; (**b**) $d_8/d = 1.0$ (original case); (**c**) $d_8/d = 1.24$; and (**d**) $d_8/d = 1.49$.

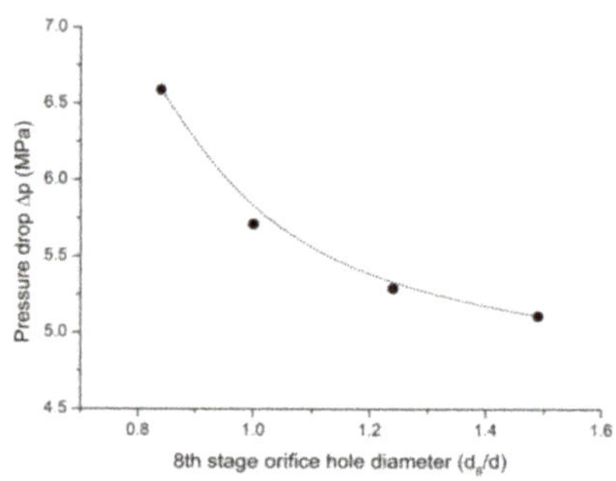

Figure 21. Pressure drop depending on the size of the hole diameter in the eighth stage orifice.

4.3.2. Structural Analysis

Figure 22 and Table 8 show the predicted results of the stress intensity distribution in the multi-stage orifice and the connecting pipes depending on the hole diameter size of the eighth stage orifice under the design flow rate condition. For reference, numerical modeling for the structural analysis described in Section 4.2.2 was applied to ANSYS Mechanical. As the hole diameter of the eighth stage orifice decreased, the maximum value of the stress intensity increased. Similar to the analysis results in Section 3.2.2, because the flow entering into the multi-stage orifice experienced decompression in the process of passing through the orifice hole, it was judged that the maximum value of the stress intensity occurred in the connecting pipe upstream of the multi-stage orifice. However, it was confirmed that the hole diameter size of the eighth stage orifice did not significantly affect the stress intensity distribution in the multi-stage orifice and the connecting pipe. The predicted stress intensity distribution in the multi-stage orifice and the connecting pipe depending on the hole diameter size of the eighth orifice stage showed more margin for the allowable stress of 207 MPa than that for the operating flow rate (Q_{in} = 34.1, 37.0, 39.0, and 41.5 m^3/h).

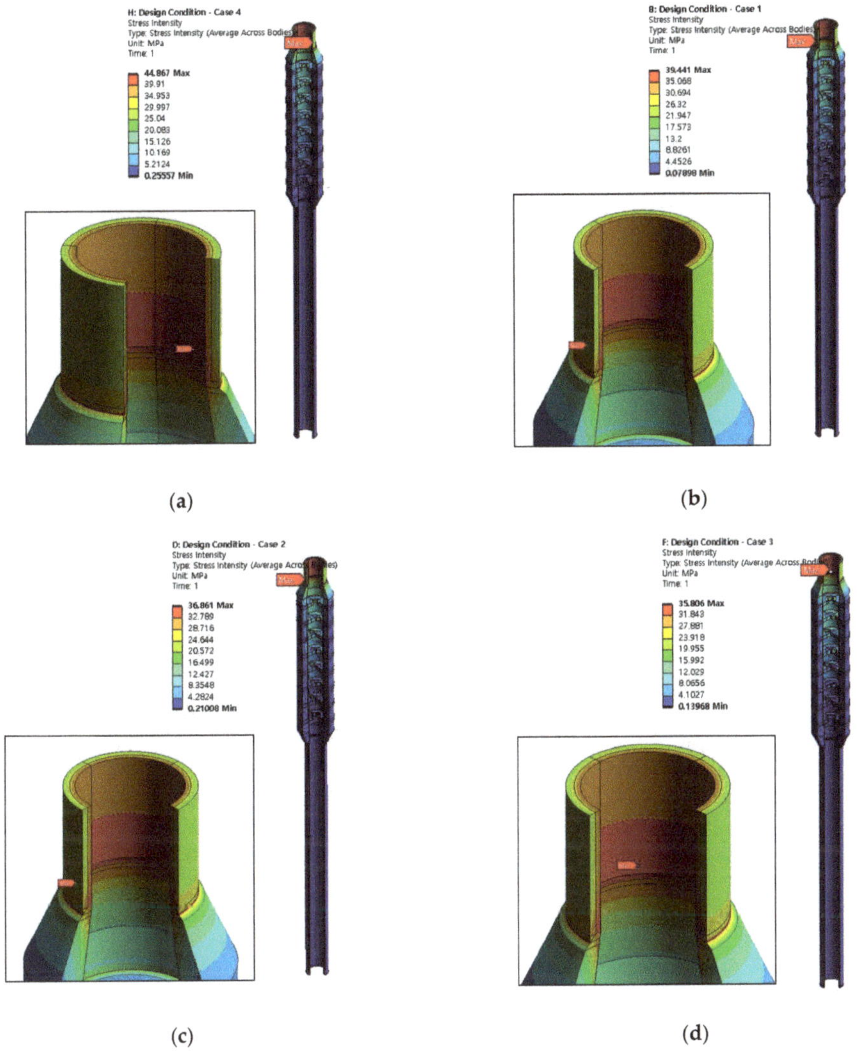

Figure 22. Distribution of stress intensity in the multi-stage orifice and the connecting pipes depending on the size of the hole diameter (d_8) in the eighth orifice stage: (**a**) $d_8/d = 0.84$; (**b**) $d_8/d = 1.0$ (original case); (**c**) $d_8/d = 1.24$; and (**d**) $d_8/d = 1.49$.

Table 8. The predicted results of stress components depending on the size of the hole diameter (d_8) in the eighth orifice stage.

Case	Flowrate (m^3/h)	Diameter Ratio (d_8/d)	Stress Intensity (MPa)	Membrane Stress (MPa)	Membrane + Bending Stress (MPa)	Allowable Stress (MPa)	Location of Maximum Stress
DS1	19.3	0.84	44.87	38.43	44.23	207	Upstream connecting pipe
DS2		1	39.44	33.77	38.88		
DS3		1.24	36.86	31.56	36.34		
DS4		1.49	35.81	30.65	35.3		

Table 9 shows the predicted deformation results for each direction in the multi-stage orifice and the connecting pipe depending on the hole diameter (d_8) size of the eighth stage orifice. In general, the smaller the hole diameter (d_8) size of the eighth stage orifice, the greater the amount of deformation in each direction. At the same hole diameter (d_8) size of the eighth stage orifice, the deformation size for each direction was in the order of axial > radial > hoop. For radial deformation, the maximum deformation value was shown in the upstream connecting pipe of the multi-stage orifice subjected to high stress. In the case of axial deformation, the maximum deformation value was indicated at the seventh or eighth stage orifice. For the deformation in the hoop direction, the maximum deformation value was shown either at the third stage orifice or in the downstream connecting pipe of the multi-stage orifice depending on the hole diameter (d_8) size of the eighth stage orifice.

Table 9. The predicted results of deformation components depending on the size of the hole diameter (d_8) in the eighth orifice stage.

Case	Flowrate (m^3/h)	Diameter Ratio (d_8/d)	Max. Radial Deformation (mm)	Location of Max. Deformation	Max. Hoop Deformation (mm)	Location of Max. Deformation	Max. Axial Deformation (mm)	Location of Max. Deformation
DS1	19.3	0.84	0.0046	Upstream connecting pipe	0.0007	Downstream connecting pipe	0.0063	8th stage orifice
DS2		1	0.0041		0.0006	3rd stage orifice	0.0049	7th stage orifice
DS3		1.24	0.0038		0.0005		0.0045	
DS4		1.49	0.0037		0.0005		0.0044	

5. Conclusions

In this study, CFD simulation was performed for a six-stage orifice test facility to validate whether the numerical modeling available in ANSYS CFX R19.1 predicted reliably and accurately the complex flow inside the multi-stage orifice. In addition, to assess the adequacy of the changed operating flow rate proposed by the domestic NPP operator as a corrective measure for the flow leakage in the AFW pump recirculation line, the cavitation flow pattern in the multi-stage orifice and the connecting pipe depending on the operating flow rate was simulated. Additionally, using ANSYS Mechanical, the structural analysis was performed for the multi-stage orifice and the connecting pipe under the same operating flow rate condition used for the flow analysis, and the structural integrity was evaluated for the allowable stress. Finally, the effect of the change in the size of the hole diameter at the eighth-stage orifice on the pressure drop characteristics and flow patterns (including cavitation) under the design flow rate condition was evaluated. The main conclusions are as follows:

- For the six-stage orifice, selected as the benchmark analysis model, the difference in the static pressure drop between the orifice inlet and outlet depending on the operating flow rate was consistent within a maximum error of 5% compared to the measured data. Therefore, the numerical modeling applied in this study may be valid, and the calculation results may be judged to be reliable to a certain level. However, it is necessary to additionally provide detailed measurement data (velocity vector, vapor volume fraction, turbulence quantities, etc.) to validate the CFD software for complex flow patterns that can occur in the multi-stage orifice, including cavitation flow.
- As the operating flow rate increased, the cavitation region expanded, and through this, it was confirmed that the operating flow rate was a key factor to influence the cavitation behavior inside the multi-stage orifice. Therefore, the reduction of the operating flow rate proposed by the domestic NPP operator concerning the flow leakage in the AFW pump recirculation line is considered to be an appropriate corrective action. However, it was found that cavitation flow still happened even in the vicinity of the corrected operating flow rate, so it is necessary to conduct the operating time reduction of the AFW pump and periodic monitoring reinforcement (e.g., vibration measurement during the AFW pump operation, etc.) suggested as additional measures by the domestic NPP operator. Furthermore, it should be necessary to fundamentally review the adequacy of the multi-stage orifice design.

- As the operating flow rate increased, the maximum value of the stress intensity acting on the multi-stage orifice and the connecting pipe also increased but was predicted to be less than the allowable stress for the connecting pipe made of carbon steel in which the actual leakage occurred. Therefore, it is judged that the integrity of the multi-stage orifice and the connecting pipe can be maintained under the condition that there is no pipe thinning due to cavitation erosion.
- One of the important design considerations for the multi-stage orifice is to ensure that no cavitation occurs there. However, in the case of the original multi-stage orifice ($d_8/d = 1.0$) installed in the AFW pump recirculation line of the domestic NPP, it was found that the cavitation region was formed inside the hole of the eighth stage orifice even when the corresponding flow rate was the reduced amount up to 54% compared to the operating flow rate. Therefore, it is necessary to review the appropriateness of the multi-stage orifice design.
- As the hole diameter size of the eighth stage orifice decreased, the maximum value of the stress intensity that occurred in the upstream connecting pipe of the multi-stage orifice increased. However, it was found that the change in the size of the hole diameter of the eighth stage orifice under the design flow rate condition did not significantly affect the stress intensity distribution in the multi-stage orifice and the connecting pipe.

Author Contributions: Conceptualization, G.L., J.B. and S.K.; methodology, G.L.; software, G.L.; validation, G.L.; formal analysis, G.L. and M.J.; investigation, G.L., J.B. and S.K.; resources, S.K. and J.B.; data curation, G.L.; writing—original draft preparation, G.L. and M.J.; writing—review and editing, G.L., M.J. and J.B.; visualization, G.L.; supervision, G.L.; project administration, G.L.; funding acquisition, G.L. All authors have read and agreed to the published version of the manuscript.

Funding: This work was supported by the Nuclear Safety Research Program through the Korea Foundation Of Nuclear Safety (KOFONS) using the financial resource granted by the Nuclear Safety and Security Commission (NSSC) of the Republic of Korea (No. 1805007).

Acknowledgments: This work was supported by the Supercomputing Center/Korea Institute of Science and Technology Information with supercomputing resources including technical support (project number: KSC-2019-CRE-0236). The author gratefully thanks Choi in the Central Research Center of Korea Hydro & Nuclear Power, Lee and Chang in the Tae Sung S&E for giving the valuable technical comments, and Lee in the Korea Institute of Nuclear Safety for providing the schematic diagram of the AFW system.

Conflicts of Interest: The authors declare no conflict of interest. The opinions expressed in this paper are those of the author and not necessarily those of the Korea Institute of Nuclear Safety (KINS). Any information presented here should not be interpreted as official KINS policy or guidance.

References

1. Korea Hydro & Nuclear Power. *Comprehensive Action Plane for Leakage of the Welding Part, Reporting Material*; Korea Hydro & Nuclear Power: Gyeongju, Korea, 2018.
2. Lee, H.S.; Choi, M.H. Review of High Vibration of Multistage Orifice Piping by Cavitation. In Proceedings of the KSFM Annual Meeting, Pyeongchang, Korea, 3–5 July 2019; pp. 214–217.
3. Wang, H.; Xie, S.; Sai, Q.; Zhou, C.; Lin, H.; Chen, E. Experiment study on pressure drop of a multistage letdown orifice tube. *Nucl. Eng. Des.* **2013**, *265*, 633–638.
4. Bai, W.; Duan, Q.; Zhang, Z. Numerical investigation on cavitation within letdown orifice of PWR nuclear power plant. *Nucl. Eng. Des.* **2002**, *305*, 230–245. [CrossRef]
5. Niyogi, K.K.; Anton, S.; Majumdar, D.M. Prediction of Performance of Multi-stage Orifice Assembly using CFD Code. In Proceedings of the 26th International Conference on Nuclear Engineering, ICONE26-81186, London, UK, 22–26 July 2018.
6. Araoye, A.A.; Badr, H.M.; Ahmed, W.H. Investigation of flow through multi-stage restricting orifices. *Ann. Nucl. Energy* **2017**, *104*, 75–90. [CrossRef]
7. Lee, G.H.; Bae, J.H. Numerical study for the effect of diameter ratio magnitudes on the flow pattern around a flat orifice flowmeter. *J. Comput. Fluids Eng.* **2019**, *24*, 42–49. [CrossRef]
8. Menter, F. *CFD Best Practice Guidelines for CFD Code Validation for Reactor Safety Applications*; ECORA CONTRACT N°FIKS-CT-2001-00154; European Commission: Brussels, Belgium, 2002; p. 8.

9. ANSYS. Turbulence and Near-wall Modeling. In *ANSYS CFX-Solver Modeling Guide*; ANSYS: Canonsburg, PA, USA, 2018.
10. Lee, G.H.; Bae, J.H.; Kang, S.H. *Preliminary Assessment for the Welding Part Integrity of Multistage Orifice in the Auxiliary Feedwater System, NSTAR-20NS13-129*; Korea Foundation of Nuclear Safety: Seongnam, Korea, 2020.
11. Korea Hydro & Nuclear Power. Auxiliary Feedwater. In *The Final Safety Analysis Report*; Korea Hydro & Nuclear Power: Gyeongju, Korea, 2020.
12. Lee, G.H.; Bae, J.H. CFD Simulation of Cavitation Flow inside a Cavitating Venturi using ANSYS CFX. In Proceedings of the Transactions of the Korean Nuclear Society Spring Meeting, Jeju, Korea, 9–10 July 2020.
13. Lee, G.H.; Bae, J.H. Comparative study for the prediction of cavitating flow inside a square-edged orifice using different commercial CFD software. *E3S Web Conf.* **2019**, *128*, 1–3. [CrossRef]
14. Lee, G.H.; Bae, J.H. Assessment of Mesh Topology Effect on the Analysis Result of Cavitation Phenomenon inside an Orifice. In Proceedings of the KSCFE Spring Conference, Jeju, Korea, 3–5 July 2019.
15. ANSYS. Element Library. In *ANSYS Element Reference*; ANSYS: Canonsburg, PA, USA, 2020.
16. Lee, G.H.; Bae, J.H.; Kang, S.H. Numerical Study on the Flow Characteristics inside a Multi-stage Orifice depending on the Orifice Hole Diameter Sizes at the Design Condition. In Proceedings of the 31th International Symposium on Transport Phenomena, Honolulu, HI, USA, 13–16 October 2020.

energies

Article

Simulation of a Fast-Charging Porous Thermal Energy Storage System Saturated with a Nano-Enhanced Phase Change Material

Mohammad Ghalambaz [1,2], S.A.M. Mehryan [3], Hassan Shirivand [4], Farshid Shalbafi [5], Obai Younis [6,7], Kiao Inthavong [8], Goodarz Ahmadi [9] and Pouyan Talebizadehsardari [1,2,*]

1 Metamaterials for Mechanical, Biomechanical and Multiphysical Applications Research Group, Ton Duc Thang University, Ho Chi Minh City 758307, Vietnam; mohammad.ghalambaz@tdtu.edu.vn
2 Faculty of Applied Sciences, Ton Duc Thang University, Ho Chi Minh City 758307, Vietnam
3 Young Researchers and Elite Club, Yasooj Branch, Islamic Azad University, Yasooj 7591493686, Iran; alal171366244@gmail.com
4 Faculty of Mechanical and Energy Engineering, Shahid Beheshti University, Tehran 1983969411, Iran; hassan.shirivand@yahoo.com
5 Department of Mechanical Engineering, Shahid Chamran University of Ahvaz, Ahvaz 61355, Iran; farshid.shalbafi@gmail.com
6 Department of Mechanical Engineering, College of Engineering at Wadi Addwaser, Prince Sattam Bin Abdulaziz University, Wadi Addwaser 11991, Saudi Arabia; oubeytaha@hotmail.com
7 Department of Mechanical Engineering, Faculty of Engineering, University of Khartoum, Khartoum 11111, Sudan
8 Mechanical & Automotive, School of Engineering, RMIT University, Bundoora, VIC 3083, Australia; kiao.inthavong@rmit.edu.au
9 Department of Mechanical and Aeronautical Engineering, Clarkson University, Potsdam, NY 13699, USA; gahmadi@clarkson.edu
* Correspondence: ptsardari@tdtu.edu.vn

Citation: Ghalambaz, M.; Mehryan, S.A.M.; Shirivand, H.; Shalbafi, F.; Younis, O.; Inthavong, K.; Ahmadi, G.; Talebizadehsardari, P. Simulation of a Fast-Charging Porous Thermal Energy Storage System Saturated with a Nano-Enhanced Phase Change Material. *Energies* **2021**, *14*, 1575. https://doi.org/10.3390/en14061575

Academic Editor: Alessandro Mauro

Received: 27 January 2021
Accepted: 6 March 2021
Published: 12 March 2021

Abstract: The melting of a coconut oil–CuO phase change material (PCM) embedded in an engineered nonuniform copper foam was theoretically analyzed to reduce the charging time of a thermal energy storage unit. A nonuniform metal foam could improve the effective thermal conductivity of a porous medium at regions with dominant conduction heat transfer by increasing local porosity. Moreover, the increase in porosity contributes to flow circulation in the natural convection-dominant regimes and adds a positive impact to the heat transfer rate, but it reduces the conduction heat transfer and overall heat transfer. The Taguchi optimization method was used to minimize the charging time of a shell-and-tube thermal energy storage (TES) unit by optimizing the porosity gradient, volume fractions of nanoparticles, average porosity, and porous pore sizes. The results showed that porosity is the most significant factor and lower porosity has a faster charging rate. A nonuniform porosity reduces the charging time of TES. The size of porous pores induces a negligible impact on the charging time. Lastly, the increase in volume fractions of nanoparticles reduces the charging time, but it has a minimal impact on the TES unit's charging power.

Keywords: nonuniform metal foam; melting heat transfer; thermal energy storage

1. Introduction

The design of thermal energy storage (TES) systems has improved significantly in recent years. Phase change materials (PCMs) have been implemented in various practical applications such as buildings materials [1], air heating and cooling in buildings [2], cooling of electronic components [3], recovering low-temperature industrial waste heat [4], and automotive applications [5].

Most of the advances in TES system design have been around improving the response time of storage units during the charging and discharging process and the synthesis of new composite PCM materials. For instance, using expanded graphite additives [6], tree-like fins [7], stepped fins [8], and heat pipes [9] are some of the recent techniques. Moreover, thermal conductive structures such as three-dimensional (3D) porous diamond foams [10] and continuous diamond–carbon nanotube foams [11] have also shown promising performance.

Many researchers attempted using a frame of thermal conductive metals and porous metal foams, saturating them with phase change material. The thermal conductive structure of the metal foam carries out the heat to and from PCM inside the pores. Thus, the composite PCM–metal foam channels the heat from the heat sources and improves the thermal conductivity. Open metal foams allow some degree of natural convection heat transfer in molten areas of the TES unit. The natural convection heat transfer is an important phenomenon that transfers thermal energy by advection in a molten region. Considering open metal foams and natural convection effects, Talebizadehsardari [12] utilized a composite of metal foam and PCMs and designed an air heater for domestic application. The results indicated that the geometrical shape of air passages induces dominant effects on the PCM unit's thermal behavior. Sardari et al. [13] used an aluminum foam–PCM composite TES unit to absorb a wall-mounted radiator's heat during off-peak loads. Later, the TES unit releases the absorbed heat when the primary heating system turns off. The authors showed that using the metal foam shortens the charging time by 95% compared to a simple PCM. Interestingly, the increase in metal foam porosity (97%) produces a positive influence and reduces the TES unit's charging/discharging time.

Zhao et al. [14] compared the advantages of using fins with those of using metal foams in reducing the melting time of a PCM in a shell-and-tube shape TES unit. They found an optimum design for the fins. They then compared the melting time of using the optimum fin with a case made using the same amount of metal foam instead of the fins. They reported that properly designed fins could be as advantageous as metal foam.

Using nanoparticles is another approach to synthesize nano-enhanced phase change materials (NePCMs). Zhang et al. [15] dispersed copper oxide nanoparticles in RT28 PCM to improve the charging heat transfer rate in a wavy-channel TES unit. The presence of nanoparticles improved the heat transfer rate and reduced the charging time. Bondareva et al. [16] examined the melting of an alumina–paraffin NePCM inside a copper radiator. This investigation showed that the nanoparticle presence increases the viscosity of molten PCM, limiting the mobility of liquid PCM.

On the other hand, nano-additives enhance the thermal conductivity of the PCM. Thus, using NePCM was most advantageous at an early melting heat transfer stage, where the conduction regime was dominant. It can be concluded that the advantage of using nanoparticles mainly depends on the internal structure of the TES enclosure and dominant mechanisms of heat transfer.

All of the above studies used a uniform metal foam to improve the thermal conductivity of composite PCM. Mahdi et al. [17] employed a cascade (multiple segments) of PCM–metal foam composites to enhance a TES unit's discharging time. They also employed 5% nanoparticles to further enhance the thermal conductivity of the system. Using the multiple segments and nanoparticles, they reduced the discharging time by 94%.

The literature studies showed that the natural convection effects and the thermal conductivity of composite PCMs could significantly influence a TES unit's phase change behavior. A dense porous structure (low porosity) increases the composite thermal conductivity but suppresses convection flows. This is while a high porosity metal foam allows convection flows. Thus, a nonuniform porous metal foam could be advantageous from both points of view. The present study investigates the melting process of a PCM embedded in a nonuniform metal foam. The influence of using various concentrations of nanoparticles, porous densities, pores sizes, and porous gradients on the full charging time of the TES unit are addressed. The Taguchi optimization approach was employed to systematically find an optimum design of nonuniform porosity. In particular, the study addressed the following research questions: (1) Does a nonuniform porosity reduce the melting time for a fixed amount of metal foam? (2) What is the impact of nanoparticles and pore size on the melting time of TES?

2. Governing Equations and Boundary Conditions

A schematic view of the physical and computational domains with dimensions of the latent heat thermal energy storage unit is depicted in Figure 1a,b. The TES unit is a cylinder with a longitudinal inner tube containing heat transfer fluid (HTF). The cylinder is filled with an inhomogeneous porous medium in which porosity and permeability vary along the z-direction. The porosity in the direction of the z-axis could be increasing, constant, or decreasing. Biobased coconut oil containing the CuO nanoparticles fills the pores. Table 1 lists the properties of coconut oil, CuO, copper metal foam, and water.

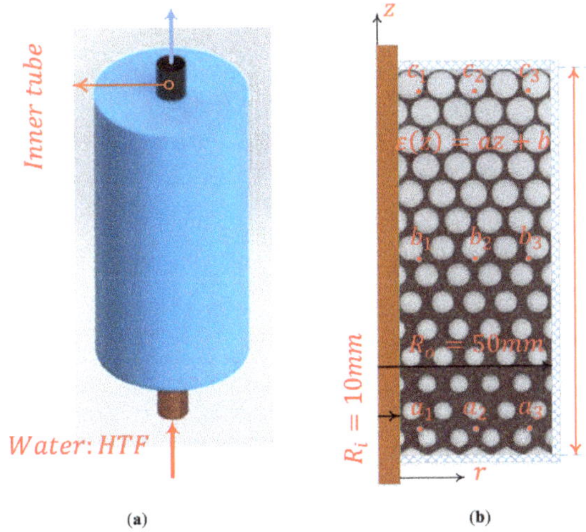

Figure 1. The physcial model of the storage unit: (**a**) cylindrical thermal energy storage (TES) system filled with phase change material (PCM) and metal foam, along with a heat transfer fluid tube in the center; (**b**) the computational domain.

Table 1. Thermophysical properties of the coconut oil and the nanoadditives [18,19].

Properties	Coconut Oil (Measured)		Heat Transfer Fluid	CuO Nanoadditives	Metal Foam
	Solid		Liquid		
$\theta_{melting}$ (K)	297	Not applicable	Not applicable	Not applicable	Not applicable
θ_{window} (K)	2	Not applicable	Not applicable	Not applicable	Not applicable
l_{ppcm} (J·kg^{-1})	-	Not applicable	1.03×10^4	Not applicable	Not applicable
ρ (kg·m^{-3})	920	993.73	914	6500	8900
μ_{lppcm} (N·s·m^{-2})	Not applicable	7.05×10^{-4}	3.26×10^{-2}	Not applicable	Not applicable
C_p (J·kg·K^{-1})	3750	4178	2010	540	386
κ (W·m^{-1}·K^{-1})	2.28×10^{-1}	6.23×10^{-1}	1.66×10^{-1}	18	380

The modeling was done under the following assumptions: (1) the melted NePCM behaves as an incompressible Newtonian fluid, and the flow in the pores is laminar; (2) nanoparticles do not settle, and they are homogeneously dispersed in the PCM; (3) expansion of the NePCM is neglected; (4) the properties of the HTF in tube are considered to be constant and Newtonian. Here, a linear porosity profile is assumed as

$$\varepsilon(z) = az + b, \tag{1}$$

where Equation (1) shows the linear variation of porosity which is proposed in the present study. As seen, the porosity varies linearly in the vertical direction. The values of porosity at the lower and upper walls are

$$\varepsilon(z=0) = \varepsilon_0 = b, \tag{2}$$

$$\varepsilon(z=L) = \varepsilon_L. \tag{3}$$

According to the values of porosity on the lower and upper walls, and the average porosity of the domain, i.e., ε_{avg}, the b parameter is calculated as follows:

$$b = \varepsilon_{avg} - \frac{aL}{2}. \tag{4}$$

The equations for the transient melting process include mass conservation, momentum balance, and energy conservation equations [20–22].

Mass conservation:

$$\nabla \cdot \vec{U} = 0; \tag{5}$$

Momentum equations:

$$\begin{aligned} &\frac{1}{\varepsilon}\frac{\partial\left(\rho_{lnpcm}\vec{U}\right)}{\partial t} + \tfrac{1}{\varepsilon}\rho_{lnpcm}\left(\vec{U}\cdot\nabla\right)\vec{U}\tfrac{1}{\varepsilon} + \nabla\cdot pI = \nabla\cdot\left(\mu_{lnpcm}\tfrac{1}{\varepsilon}\nabla\vec{U}\right) \\ &-\rho_{lnpcm}\beta_{lnpcm}\vec{g}\left(\theta_{melting}-\theta\right) + \left(M(\theta) - \tfrac{\mu_{lnpcm}}{\sigma}\right)\vec{U} \end{aligned} \tag{6}$$

where $\vec{U}$, p, and θ are the velocity vector, pressure, and temperature, respectively. Here, ε, β, μ, and g are the porosity, volume expansion coefficient, viscosity, and gravity acceleration, respectively. The subscript $lnpcm$ denotes the molten PCM, and $\theta_{melting}$ is the phase change temperature. The linear Boussinesq model is employed to apply the buoyancy force, and the Forchheimer term was neglected since the natural convection velocities are very small. The source term $M(\theta)$ is a function of temperature (θ) and is zero in liquid PCM, but it rises to large values in the solid PCM. The large values of $M(\theta)$ induce significant resistance forces to the fluid motion and force the velocities to zero in a solid PCM. The source term $M(\theta)$ is defined as

$$\begin{aligned} &M(\theta) = B_{vc}\frac{\xi^2(\theta)-2\xi(\theta)+1}{\xi^3(\theta)+10^{-3}} \\ &\xi(\theta) = \begin{cases} 0 & \theta < \theta_{melting} - \theta_{window}/2 \\ 0.5 - \frac{\theta_{melting}-\theta}{\theta_{window}} & \theta_{melting} - 0.5\theta_{window} < \theta < \theta_{melting} + 0.5\theta_{window} \\ 1 & \theta > \theta_{melting} + 0.5\theta_{window} \end{cases} \end{aligned} \tag{7}$$

where B_{vc} is a large value, which intensifies the magnitude of the temperature-dependent term. The porous permeability, σ, as a function of porosity, is calculated as follows [23,24]:

$$\begin{aligned} &\sigma = \frac{0.73\times10^{-3}}{(1-\varepsilon)^{0.224}R_m^{-2}\left(R_zR_m^{-1}\right)^{1.11}} \\ &R_zR_m^{-1} = \sqrt{\frac{1.3924(1-\varepsilon)}{3\pi}}\frac{1}{1-e^{25(\varepsilon-1)}} \\ &R_m = 0.254\times10^{-7}\chi^{-1}\,(PPI) \end{aligned} \tag{8}$$

where R_z and R_m are the pore characteristics, as introduced in [23,24], and χ is the pore density in pore size per inch (*PPI*).

Energy conservation [25]:

$$\begin{aligned} &\left[\xi(\theta)\left[(\rho c_p)_{eff\_lnpcm} - (\rho c_p)_{eff\_snpcm}\right] + (\rho c_p)_{eff\_snpcm}\right]\frac{\partial\theta}{\partial t} + (\rho c_p)_{lnpcm}\vec{U}.\nabla\theta = \\ &\nabla.\left(\kappa_{eff,nepcm}\nabla\theta\right) + (VF_{na}-1)\rho_{ppcm}l_{ppcm}\varepsilon\frac{\partial\xi(\theta)}{\partial t} \end{aligned} \tag{9}$$

where

$$\left(\rho c_p\right)_{eff,lnpcm(snpcm)} = (1-\varepsilon)\left(\rho c_p\right)_{sm} + \varepsilon\left(\rho c_p\right)_{lnepcm(snepcm)}. \tag{10}$$

The subscript *snepcm* denotes the characteristics of the NePCM in solid state, and *sm* refers to the solid matrix of metal foam. The term $\xi(\theta)$ denotes the liquid volume fraction and can be changed in each element. The variation of the liquid fraction represents the variation of the stored/released latent heat. The thermal conductivity of the porous medium saturated with the NePCM is considered as the weighted average of the two media and is computed as

$$\kappa_{eff,nepcm} = \xi(\theta)\kappa_{eff,lnpcm} + (1-\xi(\theta))\xi_{eff,snpcm}, \tag{11}$$

in which,

$$\kappa_{eff,lnpcm(snpcm)} = \frac{\Xi_1}{\Xi_1}, \tag{12}$$

$$\Xi_1 = \left[\kappa_{lnpcm(snpcm)} + \left(\left(\tfrac{\pi-\pi\varepsilon}{3}\right)^{0.5} - \tfrac{1-\varepsilon}{3}\right)\left(\kappa_{sm} - \kappa_{lnpcm(snpcm)}\right)\right] \times\left[\kappa_{lnpcm(snpcm)} + \left(\tfrac{1-\varepsilon}{3}\right)\left(\kappa_{mf} - \kappa_{lnpcm(snpcm)}\right)\right] \tag{13}$$

$$\Xi_2 = \kappa_{lnpcm(snpcm)} + \left[\frac{4}{3}\left(\frac{1-\varepsilon}{3\pi}\right)^{0.5}(1-\varepsilon) + \left(\frac{\pi-\pi\varepsilon}{3}\right)^{0.5} - (1-\varepsilon)\right]\left(\kappa_{mf} - \kappa_{lnpcm(snpcm)}\right), \tag{14}$$

The governing equations for flow and heat transfer of HTF passing the tube are laminar convection heat transfer.

Mass conservation:

$$\nabla\cdot\vec{u} = 0; \tag{15}$$

Momentum Balance:

$$\rho_{HTF}\frac{\partial\vec{u}}{\partial t} + \rho_{HTF}\left(\vec{u}\cdot\nabla\right)\vec{u} + \nabla\cdot pI = \nabla\cdot\left(\mu_{HTF}\nabla\vec{u}\right); \tag{16}$$

Energy conservation:

$$\left(\rho c_p\right)_{HTF}\frac{\partial\theta}{\partial t} + \vec{u}.\nabla\left(\left(\rho c_p\right)_{HTF}\theta\right) = \nabla.\left(\kappa_{HTF}\nabla\theta\right), \tag{17}$$

where subscript HTF shows the heat transfer fluid.

The thermophysical properties of the nano-enhanced phase change material can be defined as weighted functions of the pure PCM and nanoparticle properties. These properties are tabulated in Table 2. The initial and boundary conditions are listed in Table 3.

Table 2. Properties of the nano-enhanced phase change material.

Density	$\rho_{nepcm} = \rho_{ppcm} - VF_{na}\rho_{ppcm} + VF_{na}\rho_{na}$ $\rho_{ppcm}(\theta) = \xi(\theta)\rho_{lppcm} - \xi(\theta)\rho_{sppcm} + \rho_{sppcm}$	(18)
Dynamic viscosity	$\mu_{lnpcm} = \left(\mu_{lppcm}^{-0.4} - \mu_{lppcm}^{-0.4}VF_{na}\right)^{-2.5}$	(19)
Coefficient of thermal expansion	$\rho_{lnpcm}\beta_{npcm} = \rho_{lppcm}\beta_{lppcm} - VF_{na}\rho_{lppcm}\beta_{lppcm} + VF_{na}\rho_{na}\beta_{na}$	(20)
Thermal conductivity	$\kappa_{lnpcm(snpcm)} = \kappa_{lppcm(sppcm)}\frac{\Pi_1}{\Pi_2}$ $\Pi_1 = \left(\kappa_{na} + 2\kappa_{lppcm(sppcm)}\right) - 2VF_{na}\left(\kappa_{lppcm(sppcm)} - \kappa_{na}\right)$ $\Pi_2 = \left(\kappa_{na} + 2\kappa_{lppcm(sppcm)}\right) + VF_{na}\left(\kappa_{lppcm(sppcm)} - \kappa_{na}\right)$	(21)
Heat capacity	$\left(\rho c_p\right)_{lnepcm(snepcm)} = \left(\rho c_p\right)_{lppcm(sppcm)} + VF_{na}\left(\rho c_p\right)_{,na} - \left(\rho c_p\right)_{lppcm(sppcm)}$	(22)

Table 3. Initial and boundary conditions.

Initial condition	$U_{r,nepcm} = U_{z,nepcm} = 0,\ \theta_{nepcm} = 293K$	(23)
Interface of the tube wall and NePCM domain	$\left(\kappa\frac{\partial\theta}{\partial r}\right)_{HTF} = \left(\kappa\frac{\partial\theta}{\partial r}\right)_{eff,nepcm},\ \theta_{HTF} = \theta_{nepcm}$	(24)
Entrance of the tube	$\theta_{HTF} = 293K,\ U_{r,HTF} = 0,\ U_{z,HTF} = 0.01m/s$	(25)
Outlet of the tube	$U_{r,HTF} = 0,\ \left(\frac{\partial\theta}{\partial z}\right)_{HTF} = \left(\frac{\partial U_z}{\partial z}\right)_{HTF} = 0$	(26)
The right side surface of the NePCM domain	$U_{r,nepcm} = U_{z,nepcm} = 0,\ \left(\frac{\partial\theta}{\partial r}\right)_{nepcm} = 0$	(27)
Top and bottom surface of the NePCM domain	$U_{r,nepcm} = U_{z,nepcm} = 0,\ \left(\frac{\partial\theta}{\partial z}\right)_{nepcm} = 0$	(28)

Sensible energy, latent energy, and total energy stored in the unit are

$$ES(t) = \int_V \left(\underbrace{\underbrace{(1 - VF_{na})\rho_{ppcm}\varepsilon(z) l_{ppcm}}_{Latent\ energy} + \underbrace{(\rho C_p)_{eff\_nepcm}(\theta - \theta_{initial})}_{Sensible\ energy}}_{Total\ energy\ stored} \right) dV, \quad (29)$$

where V is the total volume of the NePCM domain. The melt volume fraction is

$$MVF(t) = \frac{S_{lnepcm}}{S_{lnepcm} + S_{snepcm}}. \quad (30)$$

The charging power of the TES unit is defined as

$$CP = \frac{ES|_{MVF=1}}{t|_{MVF=1}}. \quad (31)$$

Moreover, the uniformity of the temperature field can be measured using

$$TU^2(t) = \left(\int_V (\theta - \theta_{mean})^2 dV\right) \Big/ \left(\int_V dV\right), \quad (32)$$

where θ_{mean} is the mean temperature of the NePCM domain.

3. Numerical Approach and Grid Dependency

The model equations described above were solved by employing the finite element approach. A mesh independence test was performed to determine a balance between the precision and performance of the numerical simulation. Structural meshes with rectangular elements were employed for both computational domains. Initially, a mesh with 75 × 75 elements was considered. Then, five levels of finer meshes were used to test the solution accuracy and grid optimization. The simulation results for different grids are depicted in Figure 2. Figure 2a,b show, respectively, the mesh size's influence on the melting volume fraction (*MVF*) and local temperature at a specified point. It is seen that the temperature field is more sensitive to the mesh size. On the basis of Figure 2b, a mesh size with 125 × 125 elements was selected to satisfy the balance between the solution accuracy and simulation efficiency.

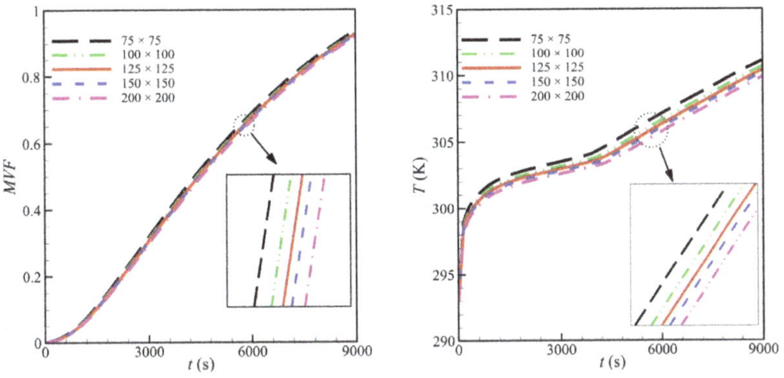

(**a**) (**b**)

Figure 2. The variation of (**a**) melting volume fraction (MVF), and (**b**) the temperature at selected points with coordinates ($5R_i$, $0.5L$) for various cases of grid size for $\varepsilon_{avg} = 0.84$, $VF_{na} = 0.04$, $a = 0.6$, and $PPI = 30$.

To guarantee the reliability of the implemented model and the employed numerical solution, we studied the liquid fraction field of biobased coconut oil contained in a uniform porous rectangular enclosure by conducting the two-dimensional (2D) numerical simulation. The predicted liquid fraction field was compared with the experimental observations of Al-Jethelah et al. [18]. In this validation test, a net heat flux was imposed on the cubic cavity's left side wall, and the insulated boundary conditions were set at the other surfaces. The porosity of copper metal foam used in the container was 0.92, and the corresponding permeability was 3.3142×10^{-7} m^2. The liquid fraction field obtained from the present model is in good agreement with the experimental field observation of Al-Jethelah et al. [18], as shown in Figure 3.

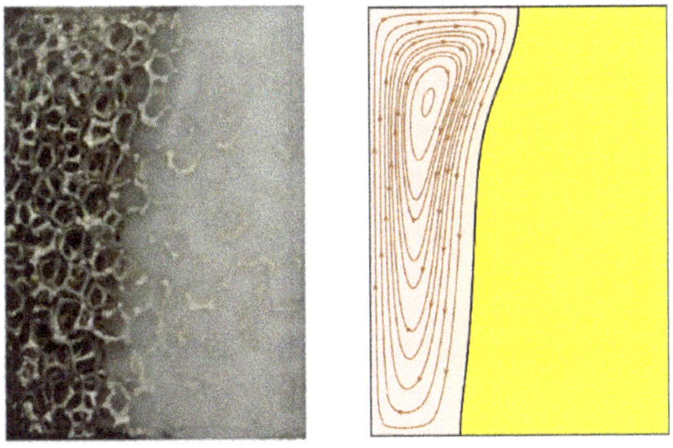

Experimental work [24] Present work

Figure 3. A comparison of melting fields simulated in the present simulations and the experimental study conducted by Al-Jethelah et al. [18].

To verify the computational model for application to an inhomogeneous porous domain, a square cavity occupied with an inhomogeneous metal foam utilized in [26] was studied. Here, the insulation boundary conditions were assigned to the cavity's upper and lower surfaces. However, the left and right vertical side boundaries were kept at high and low temperatures. The isotherms and streamlines of the present study are in excellent agreement with those of Xiong et al. [26], as illustrated in Figure 4. In this analysis, porosity was a linear function of the y-coordinate such that ε_{avg}=0.7. In addition, the nondimensional parameters were as follows: $Ra = 10^6$ and $Da = 10^{-1}$.

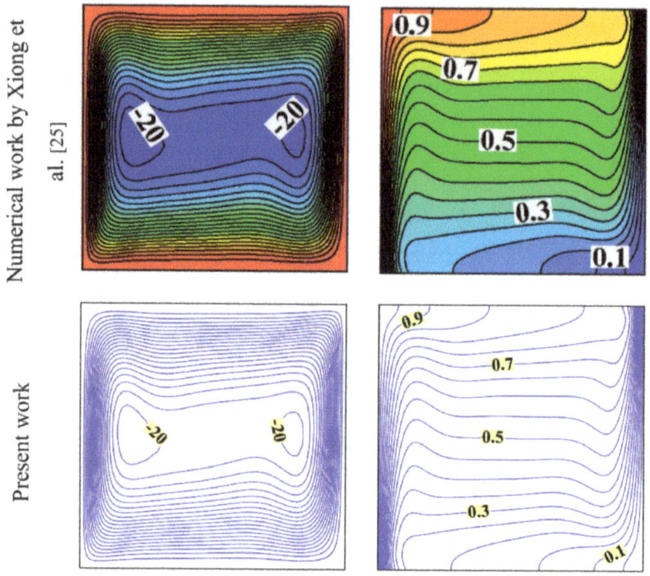

Figure 4. Comparisons of the predicted streamlines and isotherms with the simulations of Xiong et al. [26].

4. Results and Discussion

Here, the Taguchi optimization method was adopted to systematically investigate the impact of design parameters on the melting rate of PCM in the TES unit. The volume fraction of nanoparticles (VF_{na}), average porosity of porous medium (ε_{avg}), intensity of porosity gradient (a), and pore size per inch (PPI) were selected as design parameters. These parameters are known as control factors in the context of the Taguchi method. The aim was to minimize the full melting time (charging time). Thus, the charging time was selected as the target parameter. Since the time should be minimized, the target function was selected as "the lower, the better" for the Taguchi method. Following the Taguchi method, the control factors were divided into several levels. Here, five levels were selected for each factor. The details of the levels and factors are summarized in Table 4. The porosity gradient was selected in the range of −0.6 to +0.6, which covered a negative and positive gradient distribution. The zero gradient case represents uniform porosity.

Table 4. The range and levels of control factors.

Factors	Description	Level 1	Level 2	Level 3	Level 4	Level 5
A	VF_{na} (Volume fraction)	0.00	0.01	0.02	0.03	0.04
B	ε_{avg} (Average of porosity)	0.800	0.825	0.850	0.875	0.900
C	a (Gradient of porosity)	−0.6	−0.3	0.0	0.3	0.6
D	PPI (Pore per inch of the metal matrix)	10	15	20	25	30

The total combination of four factors and five levels resulted in 5^4 combination variables. A full melting process should be computed for each combination, which is computationally impractical. The Taguchi method uses an orthogonal table to probe the possible solution space efficiently. Here, we adopted the L25 design. The L25 design selects only 25 unique combinations of possible designs out of 5^4 possibilities. These selected combinations are summarized in Table 5. The simulations were carried out for all 25 cases listed in Table 5, and the corresponding values for the time of full melting ($t|_{MVF=1}$), stored energy (ES), and TES unit power (CP) at full charge are reported in Table 5. The Taguchi method was then used to compute the signal to noise (S/N) ratio values. The S/N ratio indicates the robustness of a factor level to possible noises. Thus, a factor with the highest S/N value is a promising candidate to produce an optimum design (lowest charging time). Using the values of $t|_{MVF=1}$ from Table 5, the S/N of the Taguchi method was computed, and the results are summarized in Table 5.

Table 5. Taguchi L25 orthogonal table corresponding to range and levels of control parameters. PPI, pore size per inch; ES, stored energy; CP, unit power.

Experiment Number	Control Parameters				Full Melting Characteristics			S/N Ratio
	VF_{na}	ε_{avg}	a	PPI	$t\|_{MVF=1}$	ES (kJ)	CP (kW)	
1	0.00	0.800	−0.6	10	10,200	933.745	0.09154	−80.1720
2	0.00	0.825	−0.3	15	10,500	965.132	0.09192	−80.4238
3	0.00	0.850	0.0	20	11,100	1011.959	0.09117	−80.9065
4	0.00	0.875	0.3	25	12,150	1071.543	0.08819	−81.6915
5	0.00	0.900	0.6	30	15,900	1192.082	0.07497	−84.0279
6	0.01	0.800	−0.3	20	9900	943.927	0.09535	−79.9127
7	0.01	0.825	0.0	25	10,350	984.372	0.09511	−80.2988
8	0.01	0.850	0.3	30	10,950	1025.580	0.09366	−80.7883
9	0.01	0.875	0.6	10	12,750	1109.630	0.08703	−82.1102
10	0.01	0.900	−0.6	15	14,850	1072.519	0.07222	−83.4345
11	0.02	0.800	0.0	30	9600	950.417	0.09900	−79.6454
12	0.02	0.825	0.3	10	10,050	987.757	0.09828	−80.0433
13	0.02	0.850	0.6	15	11,100	1050.637	0.09465	−80.9065
14	0.02	0.875	−0.6	20	12,150	975.115	0.08026	−81.6915
15	0.02	0.900	−0.3	25	12,900	1023.790	0.07936	−82.2118
16	0.03	0.800	0.3	15	9300	953.180	0.10249	−79.3697
17	0.03	0.825	0.6	20	10,050	1007.866	0.10029	−80.0433
18	0.03	0.850	−0.6	25	11,250	952.128	0.08463	−81.0231
19	0.03	0.875	−0.3	30	11,850	997.453	0.08417	−81.4744
20	0.03	0.900	0.0	10	12,900	1054.967	0.08178	−82.2118
21	0.04	0.800	0.6	25	9150	963.774	0.10533	−79.2284
22	0.04	0.825	−0.6	30	10,500	928.251	0.08840	−80.4238
23	0.04	0.850	−0.3	10	10,950	967.676	0.08837	−80.7883
24	0.04	0.875	0.0	15	11,700	1018.223	0.08703	−81.3637
25	0.04	0.900	0.3	20	13,200	1085.484	0.08223	−82.4115

Following the standard Taguchi method, Table 6 shows the S/N and rank values of each level and control factor. Figure 5 shows a graphical representation of S/N values in Table 6. The highest value of S/N for each control parameter shows the promising level for the minimum charging time. Table 6 also shows the rank of each factor with respect to the variation of melting time. As seen, the average porosity (ε_{avg}) was the most effective parameter influencing the melting time, followed by the volume fraction of nanoparticles. The porosity gradient (a) and pore size (PPI) were ranked third and fourth. Here, δ indicates the maximum difference between the computed S/N values of each factor. A higher δ value denotes a more influential factor.

Table 6. The S/N and rank values of the control factors.

Levels	VF_{na}	ε_{avg}	a	PPI
1	−81.44	−79.67	−81.35	−81.07
2	−81.31	−80.25	−80.96	−81.10
3	−80.90	−80.88	−80.89	−80.99
4	−80.82	−81.67	−80.86	−80.89
5	−80.84	−82.86	−81.26	−81.27
δ	0.62	3.19	0.49	0.38
Rank	2	1	3	4

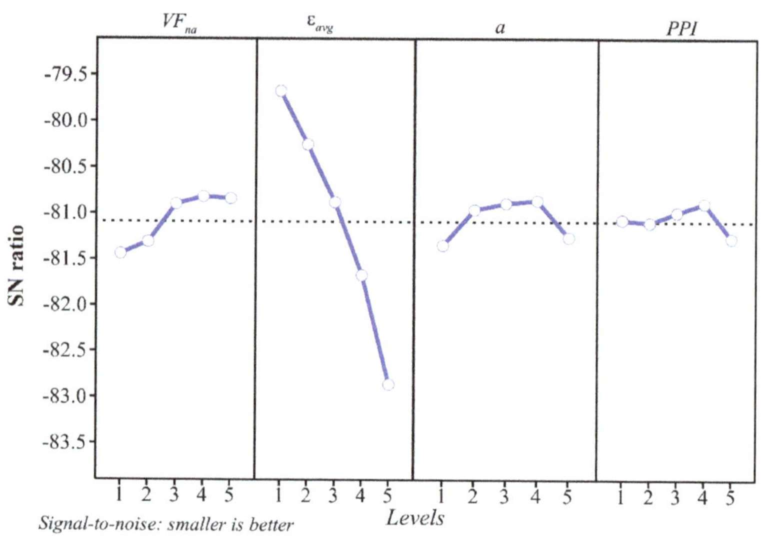

Figure 5. Mean values of the signal-to-noise (S/N) ratios for all the levels of the controlling parameters. Optimum case: $A = 5$, $B = 1$, $C = 4$, and $D = 4$.

Figure 5 shows that Levels 4, 1, 4, and 4 corresponded to the highest S/N ratios for nanoparticle volume fraction (VF_{na}), average porosity (ε_{avg}), porosity gradient (a), and pore size (PPI), respectively. The approach for computing the S/N ratio was "smaller is better"; thus, a case with a lower melting time produced a larger S/N ratio value. Using Table 4, these levels can be read as $VF_{na} = 0.04$, $\varepsilon_{avg} = 0.8$, $a = 0.3$, and PPI = 25. The Taguchi method estimated a melting time of 8784 s for this design parameter combination. A simulation was performed for this specific combination to confirm the reduction in the TES unit's charging time. The simulation outcomes showed a full melting (charging) time of 9097 s, which is slightly higher than the estimated time of 8784 s estimated using the Taguchi method. However, a comparison between the computed melting time and the 25 cases of Table 5 confirms that this melting time was the smallest. Thus, the proposed design according to the Taguchi method was selected as the optimum design. The details of the optimum design are shown in Table 7.

Table 7. The optimum values of the controlling parameters.

Factors				Time for MVF = 1	Estimated Time Fir MVF = 1	ES (kJ)	CP (kW)
VF_{na}	ε_{avg}	a	PPI				
0.04	0.8	0.3	25	9097	8784s	931.508	0.10240

Table 5 also shows that case 5 resulted in the highest melting time (15,900 s). Considering the investigated range of control parameters, the optimum melting time of 9097s was 74% better than the highest melting time. It should be noted that case 5 contained no nanoparticles and had the maximum average porosity of 0.9. Moreover, case 21 with a full melting time of 9150 s provided a melting time that was close to the optimum case. The only difference between these two cases was the porosity gradient.

The results of Table 5 were also used to develop a linear relationship for design parameters and the charging time of the TES. That is,

$$\text{Time} + 8892 - 243.0\, VF_{na} + 1047.0\, \varepsilon_{avg} - 9.0\, a + 45.0\, PPI. \quad (33)$$

Here, 16 more cases were adopted to further explore the impact of the variation of the design parameters on the optimum design. The details of selected cases are presented in Table 8. Figure 6a shows the time evaluations of MVF during the charging process for various values of nanoparticle volume fraction. It is seen that the increase in the concentration of nanoparticles slightly increased the MVF. The differences were only visible in the final stages of charging.

Table 8. Table of 16 cases for further analysis around the optimum design.

Experiment No.	Case	Parameter Study	Control Parameters				$t\|_{MVF=1}$	ES (kJ)	CP (kW)
			VF_{na}	ε_{avg}	a	PPI			
1	1	VF_{na}	0.00	0.800	0.3	25	9473	978.696	0.10195
2	2		0.01	0.800	0.3	25	9400	966.340	0.10226
3	3		0.02	0.800	0.3	25	9327	965.680	0.10219
4	4		0.03	0.800	0.3	25	9254	953.160	0.10249
5	5	ε_{avg}	0.04	0.825	0.3	25	9861	975.162	0.09850
6	6		0.04	0.850	0.3	25	10669	1012.875	0.09378
7	7		0.04	0.875	0.3	25	11687	1039.620	0.08886
8	8		0.04	0.900	0.3	25	13102	1085.454	0.08223
9	9	a	0.04	0.800	−0.6	25	9796	909.159	0.09183
10	10		0.04	0.800	−0.3	25	9530	919.070	0.09574
11	11		0.04	0.800	0.0	25	9317	937.623	0.09922
12	12		0.04	0.800	0.6	25	9150	963.774	0.10533
13	13	PPI	0.04	0.800	0.3	10	9180	952.508	0.10242
14	14		0.04	0.800	0.3	15	9180	952.519	0.10242
15	15		0.04	0.800	0.3	20	9180	952.599	0.10243
16	16		0.04	0.800	0.3	30	9180	952.577	0.10243

Figure 6b depicts the predicted temperature of point A in the enclosure. The temperature in the solid region increased sharply until it reached the fusion temperature. Then, the temperature slightly increased until the entire PCM changed to liquid at this point. The temperature then increased again. The increase in nanoparticle concentration slightly increased the temperature of point A in the liquid region due to the improved thermal conductivity of NePCM and better heat transfer between the hot liquid in the tube and the PCM inside the enclosure. The melting interfaces during the melting process were very close; thus, they were not plotted.

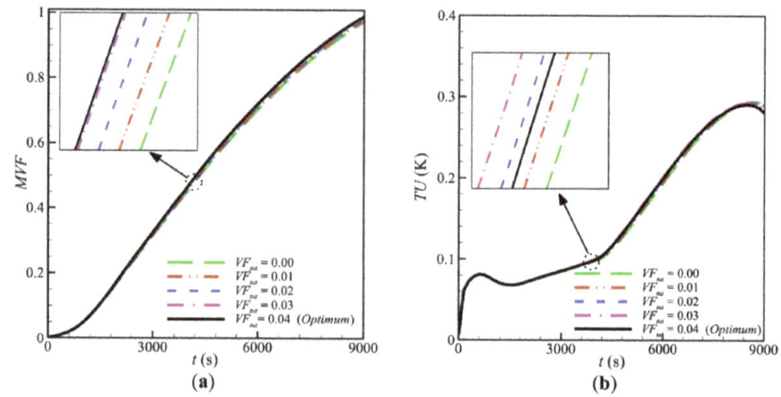

Figure 6. The time variation of (**a**) *MVF*, and (**b**) temperature distribution for various nano-additive volume fractions for R_o = 0.1 m, R_i = 0.01 m, L = 0.3 m, ε_{avg} = 0.800, a = 0.3, and PPI = 25.

Figure 7 shows the phase change interfaces (ℓ = 0.5) for various average porosity (ε_{avg}) values and different melting times. It is seen that the melting interface advanced in the enclosure toward the right wall with time.

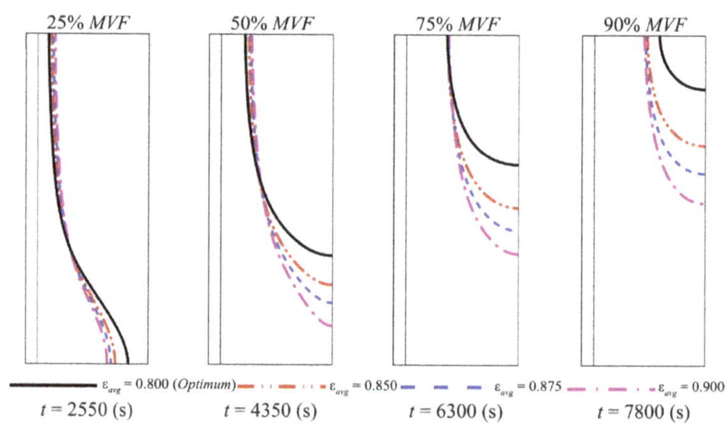

Figure 7. Melting interface at four steps of melting progress for various average porosities for VF_{na} = 0.04, a = 0.3, and PPI = 25.

Interestingly, the melting started at the bottom since the HTF fluid at the entry was hot, and the convection heat transfer coefficients at the undeveloped region of the HTF tube was also larger. Thus, the heat transfer between HTF liquid and NePCM at the bottom was more intense. The advancement of the melting interface accelerated as the porosity decreased (mass of metal foam increased). This was due to the increase in thermal conductivity of metal foam, which enhanced the heat transfer rate. It should be noted that the inlet temperature of HTF fluid was constant; hence, a higher composite thermal conductivity at the PCM side increased the heat transfer rate. Thus, the amount of stored heat in an enclosure increased for a higher composite thermal conductivity. Therefore, as seen in Figure 7 at t = 7800 s, there was a clear difference between the melting interfaces. Figure 8a illustrates the time history of MVF for various average porosities. The average porosity induced notable changes in the values of MVF. In agreement with the melting interfaces of Figure 7, this figure shows that the increase in average porosity raised the MVF.

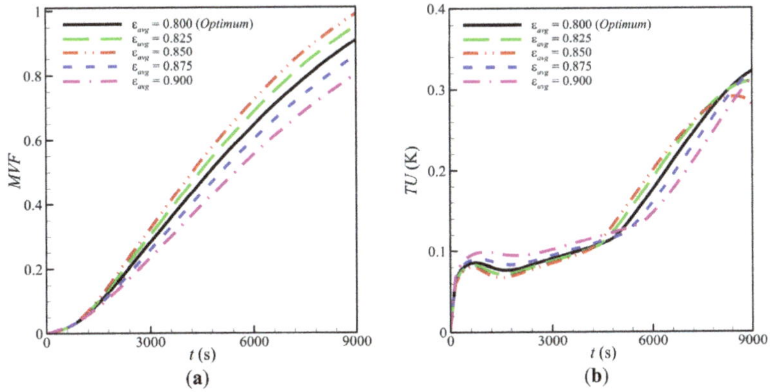

Figure 8. The variation of (**a**) MVF, and (**b**) temperature uniformity for the various average porosities as a function of time for $VF_{na} = 0.04$, $a = 0.3$, and $PPI = 25$.

Figure 8b depicts the temperature uniformity inside the PCM domain. The enclosure temperature distribution experienced a slightly higher temperature gradient during the melting process when average porosity was maximum. The rise in temperature was due to the lower composite thermal conductivity and large temperature gradients. At the early stages of heat transfer, the NePCM was in a solid state, and there was no practical molten region or natural convection circulations. As the MVF increased and the molten region grew, natural convection flows occurred. During the phase change, the temperature uniformity stayed almost constant. When the molten region grew significantly, the natural convection circulation started, and the temperature gradients also increased.

Figure 9 is plotted to show the impact of pore sizes on MVF and temperature evaluations of point A. It is seen that the variation of pore size induced a negligible impact on the MVF. There was a slight temperature variation for point A around 6000 s. Half of the enclosure was in a molten state, and there was a dominant natural convection heat transfer flow. In a microscopic view, the overall heat, which diffused to the composite PCM (foam and PCM), first channeled through the pore walls, and then the PCM inside the pores could absorb it. A decrease in the pore size could provide a larger contact surface and a better melting rate. However, since the amount of foam was low (80% or higher porosity), the heat transfer rate was limited by the foam's capacity to conduct heat into the solid PCM region. Since the amount of foam was limited (constant porosity), the condition heat transfer through the foam was also limited. Thus, the change in pore size could impact the heat transfer, but it did not play a dominant role. Moreover, a smaller pore size led to a smaller permeability, which reduced the natural convection effects.

The variation of pore sizes could affect the permeability of the medium, according to Equation (4); thus, the impact of pore size on the flow and heat transfer can be boosted in convective dominant heat transfer regimes. Since the influence of the pore size (PPI) on the MVF was minimal, the melting interfaces were not plotted for the sake of brevity.

Figure 10 display the dependency of the porosity gradient on the evaluation of melting interfaces at various time steps. Interestingly, the increase in a shifted the melting interface toward the left wall and resulted in an increased melting rate. According to Equation (1), the increase in a meant more metal foam at the bottom and less metal foam at the top. Thus, the PCM's thermal conductivity at the bottom of the enclosure was higher than that at the top. At the bottom and in the initial stages of phase change, the dominant mechanism of heat transfer was conduction. Thus, the increase in thermal conductivity enhanced the heat transfer. As the phase change continued, the molten region grew, and convection heat transfer occurred in the enclosure's top regions. In natural convection flows, the increase in porosity of metal foams increased the medium's permeability. Consequently, the liquid PCM could circulate more easily in the top area, contributing to convection heat transfer.

The relative difference in full charging of TES for $a = 0.3$ (optimum case) and $a = 0.6$ (case 12, Table 8) was only 0.5%, and Figure 10 shows the close competition of these two cases at 7800 s. Case $a = 0.3$, which was the adopted optimum design, led to a better melting at the final stage of thermal energy storage (above 90% MVF) compared to case 12 in Table 8.

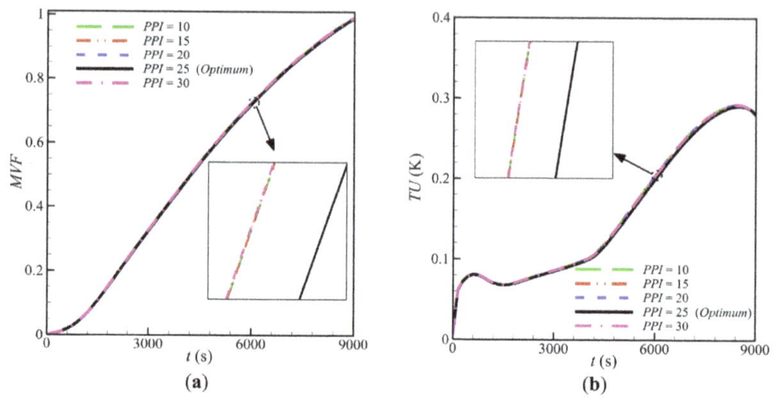

Figure 9. The variation of (**a**) MVF, and (**b**) temperature distribution at the various cases of pore per inch of the metal matrix as a function of time for $VF_{na} = 0.04$, $a = 0.3$, and $\varepsilon_{avg} = 0.800$.

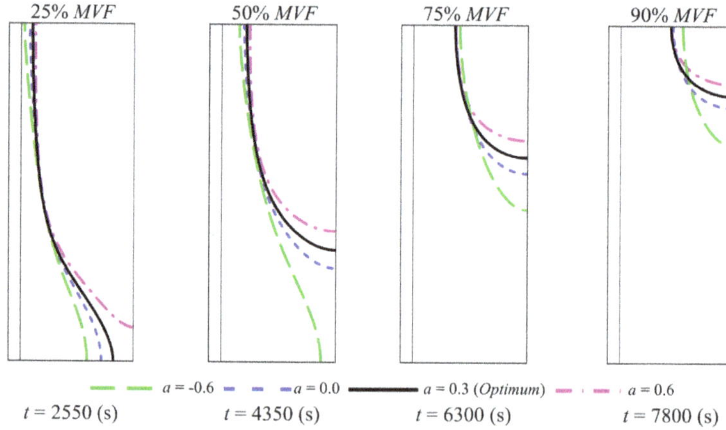

Figure 10. Melting interface at four steps of melting for various porosity gradients for $VF_{na} = 0.04$, $PPI = 25$, and $\varepsilon_{avg} = 0.800$.

Figure 11 illustrates the time history of MVF and temperature uniformity for various values of porosity gradients. Figure 11a shows that an increase in a improved the MVF during the middle stages of phase change. This finding is in agreement with the advancement of the melting interfaces, as observed in Figure 10. However, in the final stages of charging, the case $a = 0.3$ provided better performance than the case of $a = 0.6$. The temperature uniformity for the initial stages of the melting process was almost independent of the a parameter. However, in the middle stages, where the molten region grew, the increase in a parameter promoted the temperature nonuniformity. In the middle stages, the increase in a promoted temperature nonuniformities. A high value of a led to a stronger heat transfer flow and natural convection flows.

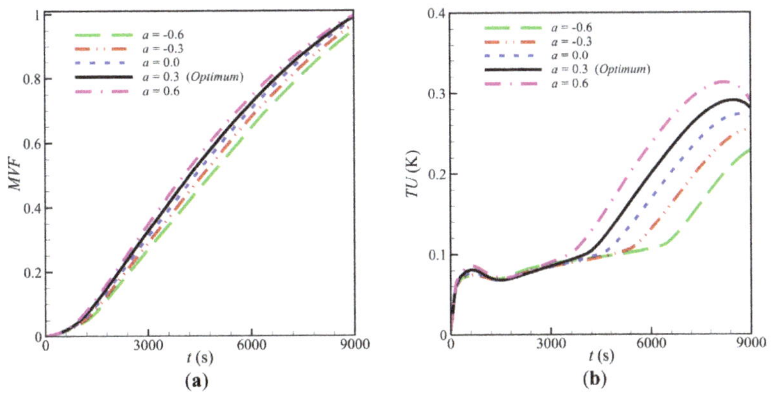

Figure 11. The variation of (**a**) *MVF*, and the (**b**) temperature distribution at various gradients of porosity as a function of time for $VF_{na} = 0.04$, $PPI = 25$, and $\varepsilon_{avg} = 0.800$.

Nine points in the NePCM–metal foam region were selected to investigate the temperature variations during the melting process. These points were located at $a_1(2R_i, 0.1L/3)$, $a_2(2R_i, L/2)$, $a_3(2R_i, 2.9L/3)$, $b_1(5R_i, 0.1L/3)$, $b_2(5R_i, L/2)$, $b_3(5R_i, 2.9L/3)$, $c_1(9R_i, 0.1L/3)$, $c_2(9R_i, L/2)$, and $c_3(9R_i, 2.9L/3)$. Figure 12 shows the variation of temperatures at these nine points in the PCM domain during the melting process. All points started with an initial temperature of 293 K followed by a linear increase in temperature. The linear segment showed pure conduction in a super-cold solid PCM before it reached a melting temperature. Then, there was a flattened segment, in which the temperature slightly increased. This was the phase change stage, where the PCM absorbed the energy in the form of latent heat. Then, there was a semi-linear increase in the temperature, which corresponded to the heat transfer in the molten region with no phase change. The time variations of temperatures for all cases showed that the impact of the porosity gradient (*a*) on temperature profiles was minimal in the solid state and during melting phase change. However, after the phase change, the molten NePCM circulated in the enclosure, enhancing the effect of the porosity gradient.

Here, a_1, b_1, and c_1 were the points placed at the bottom of the enclosure. Point a_1 was next to the tube wall, point b_1 was in the middle, and c_1 was next to the insulated shell wall. Since a_1 was the closest point next to the tube, it experienced a sharp temperature rise at the beginning of the charging process. In contrast, c_1 showed a smooth temperature variation since it was far from the tube. Point c_3 was placed near the top-right corner of the enclosure, which was the last place to be melted. Thus, during the melting process, this point remained at the fusion temperature.

Figure 13 compares the full charging power of TES units for various values of design factors. Figure 13a shows that the variation of nanoparticle concentration slightly changed the unit power. This is because the presence of nanoparticles improved the heat transfer rate; they also reduced the heat capacity of the storage unit. Figure 13b depicts an increasing trend of power (*CP*) as the porosity gradient (*a*) increased. Both cases of *a* = 0.3 and a = 0.6 showed almost similar charging powers. The increase in average porosity (Figure 13c) reduced the charging power. As the average porosity increased, the heat capacity of the TES also increased. However, a reduction in the composite PCM's effective thermal conductivity weakened the heat transfer rate and consequently increased the charging time. The variation of pore sizes had a negligible impact on the TES power since PPI variation did not induce notable changes in either the unit's heat capacity or the melting time.

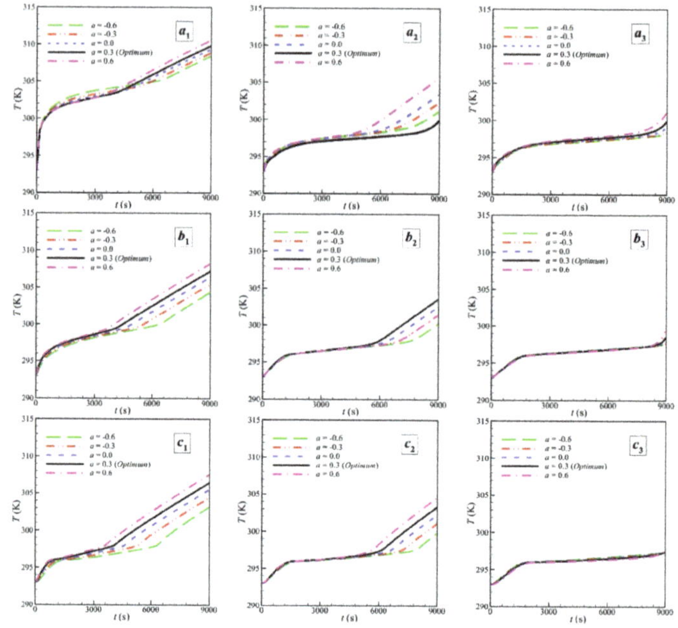

Figure 12. Variation of temperature at selected points during the charging process for optimum condition.

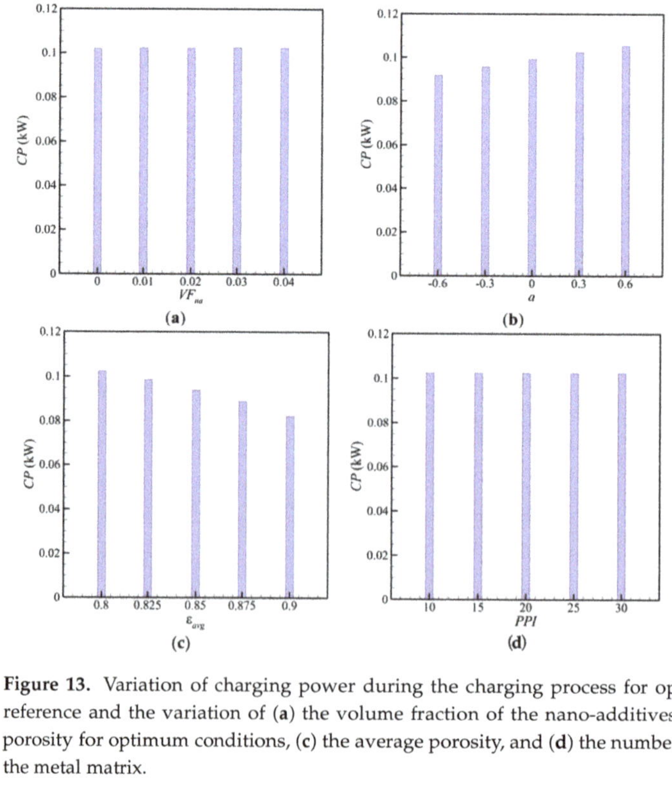

Figure 13. Variation of charging power during the charging process for optimum designs as a reference and the variation of (**a**) the volume fraction of the nano-additives, (**b**) the gradient of porosity for optimum conditions, (**c**) the average porosity, and (**d**) the number of pores per inch of the metal matrix.

5. Conclusions

The charging process of a TES unit filled with a nonuniform metal foam was studied. The metal foam was saturated with coconut oil–CuO NEPCM, and the porosity was linearly changed from bottom to top. Hot water flowed in the heat transfer tube. Finite element method FEM was applied to solve the governing partial differential equations for momentum and phase change energy. The Taguchi optimization method was employed to find an optimum design for the TES unit. The results led to the following conclusions:

- The average porosity was the dominant factor influencing the charging time and power. Thus, in the design of TES units improved by metal foams, the first step should be selecting the average porosity of the metal foam. The uniformity of the porous medium contributed to heat transfer, and a positive value of porosity gradient ($a > 0$) could reduce the charging time of the TES unit. A positive porosity gradient corresponded to low porosity at the bottom and high porosity at the top of the PCM enclosure.
- The volume fraction of nanoparticles was the second most important parameter affecting the TES unit's phase change behavior. The increase in nanoparticle volume fraction smoothly decreased the charging time.
- The porous medium's average pore size had a negligible influence on the charging time. This design factor was the least critical parameter among design variables.

Author Contributions: Conceptualization, M.G., S.A.M.M., and P.T.; methodology, M.G., S.A.M.M., H.S., F.S., O.Y., K.I., G.A., and P.T.; software, S.A.M.M., H.S., and F.S.; validation, S.A.M.M. and M.G.; formal analysis, S.A.M.M., H.S., and F.S.; investigation, M.G., S.A.M.M., H.S., F.S., O.Y., K.I., G.A., and P.T.; resources, M.G. and O.Y.; writing—original draft preparation, M.G., S.A.M.M., H.S., F.S., O.Y., K.I., G.A., and P.T.; writing—review and editing, M.G., S.A.M.M., H.S., F.S., O.Y., K.I., G.A., and P.T.; visualization, S.A.M.M., H.S., and F.S.; supervision, M.G. Authorship must be limited to those who have contributed substantially to the work reported. All authors have read and agreed to the published version of the manuscript.

Funding: This research received no external funding.

Institutional Review Board Statement: Not applicable.

Informed Consent Statement: Not applicable.

Data Availability Statement: Data is contained within the article.

Conflicts of Interest: The authors declare no conflict of interest.

References

1. Pomianowski, M.; Heiselberg, P.; Zhang, Y. Review of thermal energy storage technologies based on PCM application in buildings. *Energy Build.* **2013**, *67*, 56–69. [CrossRef]
2. Iten, M.; Liu, S.; Shukla, A. A review on the air-PCM-TES application for free cooling and heating in the buildings. *Renew. Sustain. Energy Rev.* **2016**, *61*, 175–186. [CrossRef]
3. Sahoo, S.K.; Das, M.K.; Rath, P. Application of TCE-PCM based heat sinks for cooling of electronic components: A review. *Renew. Sustain. Energy Rev.* **2016**, *59*, 550–582. [CrossRef]
4. Du, K.; Calautit, J.; Eames, P.; Wu, Y. A state-of-the-art review of the application of phase change materials (PCM) in Mobilized-Thermal Energy Storage (M-TES) for recovering low-temperature industrial waste heat (IWH) for distributed heat supply. *Renew. Energy* **2021**, *168*, 1040–1057. [CrossRef]
5. Jaguemont, J.; Omar, N.; Bossche, P.V.D.; Mierlo, J. Phase-change materials (PCM) for automotive applications: A review. *Appl. Therm. Eng.* **2018**, *132*, 308–320. [CrossRef]
6. Bahraseman, H.G.; Languri, E.M.; East, J. Fast charging of thermal energy storage systems enabled by phase change materials mixed with expanded graphite. *Int. J. Heat Mass Transf.* **2017**, *109*, 1052–1058. [CrossRef]
7. Liu, X.; Huang, Y.; Zhang, X.; Zhang, C.; Zhou, B. Investigation on charging enhancement of a latent thermal energy storage device with uneven tree-like fins. *Appl. Therm. Eng.* **2020**, *179*, 115749. [CrossRef]
8. Nakhchi, M.; Esfahani, J. Improving the melting performance of PCM thermal energy storage with novel stepped fins. *J. Energy Storage* **2020**, *30*, 101424. [CrossRef]
9. Tiari, S.; Qiu, S.; Mahdavi, M. Discharging process of a finned heat pipe–assisted thermal energy storage system with high temperature phase change material. *Energy Convers. Manag.* **2016**, *118*, 426–437. [CrossRef]

10. Zhang, L.; Zhou, K.; Wei, Q.; Ma, L.; Ye, W.; Li, H.; Zhou, B.; Yu, Z.; Lin, C.-T.; Luo, J.; et al. Thermal conductivity enhancement of phase change materials with 3D porous diamond foam for thermal energy storage. *Appl. Energy* **2019**, *233–234*, 208–219. [CrossRef]
11. Hu, N.; Li, H.; Wei, Q.; Zhou, K.; Zhu, W.; Zhang, L.; Li, S.; Ye, W.; Jiao, Z.; Luo, J.; et al. Continuous diamond-carbon nanotube foams as rapid heat conduction channels in composite phase change materials based on the stable hierarchical structure. *Compos. Part B Eng.* **2020**, *200*, 108293. [CrossRef]
12. Talebizadehsardari, P.; Mohammed, H.I.; Mahdi, J.M.; Gillott, M.; Walker, G.S.; Grant, D.; Giddings, D. Effect of airflow channel arrangement on the discharge of a composite metal foam-phase change material heat exchanger. *Int. J. Energy Res.* **2021**, *45*, 2593–2609. [CrossRef]
13. Sardari, P.T.; Babaei-Mahani, R.; Giddings, D.; Yasseri, S.; A Moghimi, M.; Bahai, H. Energy recovery from domestic radiators using a compact composite metal Foam/PCM latent heat storage. *J. Clean. Prod.* **2020**, *257*, 120504. [CrossRef]
14. Zhao, C.; Opolot, M.; Liu, M.; Bruno, F.; Mancin, S.; Hooman, K. Numerical study of melting performance enhancement for PCM in an annular enclosure with internal-external fins and metal foams. *Int. J. Heat Mass Transf.* **2020**, *150*, 119348. [CrossRef]
15. Zhang, X.; Sheikholeslami, M.; Jafaryar, M.; Sheremet, M.A.; Shafee, A.; Babazadeh, H. Simulation for melting of paraffin for saving energy with utilize of nanoparticles. *J. Mol. Liq.* **2020**, *313*, 113574. [CrossRef]
16. Bondareva, N.S.; Gibanov, N.S.; A Sheremet, M. Melting of nano-enhanced PCM inside finned radiator. *J. Phys. Conf. Ser.* **2018**, *1105*, 012023. [CrossRef]
17. Mahdi, J.M.; Mohammed, H.I.; Hashim, E.T.; Talebizadehsardari, P.; Nsofor, E.C. Solidification enhancement with multiple PCMs, cascaded metal foam and nanoparticles in the shell-and-tube energy storage system. *Appl. Energy* **2020**, *257*, 113993. [CrossRef]
18. Al-Jethelah, M.; Ebadi, S.; Venkateshwar, K.; Tasnim, S.; Mahmud, S.; Dutta, A. Charging nanoparticle enhanced bio-based PCM in open cell metallic foams: An experimental investigation. *Appl. Therm. Eng.* **2019**, *148*, 1029–1042. [CrossRef]
19. Choi, S.-K.; Kim, S.-O.; Lee, T.-H. Computation of the Natural Convection of Nanofluid in a Square Cavity with Homogeneous and Nonhomogeneous Models. *Numer. Heat Transf. Part A Appl.* **2013**, *65*, 287–301. [CrossRef]
20. Nield, D.A.; Bejan, A. *Convection in Porous Media*; Springer: New York, NY, USA, 2013; Volume 3.
21. Shahsavar, A.; Ali, H.M.; Mahani, R.B.; Talebizadehsardari, P. Numerical study of melting and solidification in a wavy double-pipe latent heat thermal energy storage system. *J. Therm. Anal. Calorimet* **2020**, *141*, 1785–1799. [CrossRef]
22. Yang, X.; Yu, J.; Xiao, T.; Hu, Z.; He, Y.-L. Design and operating evaluation of a finned shell-and-tube thermal energy storage unit filled with metal foam. *Appl. Energy* **2020**, *261*, 114385. [CrossRef]
23. Mahdi, J.M.; Nsofor, E.C. Melting enhancement in triplex-tube latent heat energy storage system using nanoparticles-metal foam combination. *Appl. Energy* **2017**, *191*, 22–34. [CrossRef]
24. Sardari, P.T.; Mohammed, H.I.; Giddings, D.; Walker, G.S.; Gillott, M.; Grant, D. Numerical study of a multiple-segment metal foam-PCM latent heat storage unit: Effect of porosity, pore density and location of heat source. *Energy* **2019**, *189*, 116108. [CrossRef]
25. Xu, Y.; Li, M.-J.; Zheng, Z.-J.; Xue, X.-D. Melting performance enhancement of phase change material by a limited amount of metal foam: Configurational optimization and economic assessment. *Appl. Energy* **2018**, *212*, 868–880. [CrossRef]
26. Xiong, Q.; Poor, H.Z.; Izadi, M.; Assareh, E. Natural heat exchange in inhomogeneous porous medium using linear and quadratic porosity distribution. *Int. J. Therm. Sci.* **2021**, *161*, 106731. [CrossRef]

energies

MDPI

Article

Conical Two-Phase Swirl Flow Atomizers—Numerical and Experimental Study

Marek Ochowiak [1,*], Daniel Janecki [2], Andżelika Krupińska [1], Sylwia Włodarczak [1], Tomasz Wilk [3] and Radosław Olszewski [3]

[1] Department of Chemical Engineering and Equipment, Poznan University of Technology, 60-965 Poznan, Poland; andzelika.krupinska@put.poznan.pl (A.K.); sylwia.wlodarczak@put.poznan.pl (S.W.)

[2] Institute of Environmental Engineering and Biotechnology, University of Opole, Kominka 6a, 45-032 Opole, Poland; zecjan@uni.opole.pl

[3] Faculty of Chemistry, Adam Mickiewicz University, 61-614 Poznan, Poland; tomasz.wilk@adob.com.pl (T.W.); radoslaw.olszewski@adob.com.pl (R.O.)

* Correspondence: marek.ochowiak@put.poznan.pl

Abstract: This paper presents the results of numerical simulations for the developed and discussed conical two-phase atomizers with swirl flow, differing in the ratio of the height of the swirl chamber to its diameter. Experiments were carried out for SAN-1 with $H_S/D_S = 1$ and SAN-2 with $H_S/D_S = 4$ atomizers. The study was conducted over a range of Reynolds number for liquid Re_L = (1400; 5650) and for gas Re_G = (2970; 9900). Numerical calculations were performed with the use of computational fluid dynamics (CFD), which were verified on the basis of experimental data. Based on the analysis of experimental studies and simulations results the influence of operational parameters and changes of the atomizer geometry on the generated spray was demonstrated. As the gas flow rate increased and the swirl chamber height decreased, the spray angle increased. Higher velocity values of the liquid and greater turbulence occur in the center of the spray. The flow inside the atomizer determines the nature of the spray obtained. The geometry of the swirl chamber influences the air core formed inside the atomizer, and this determines the atomization effect. The results of numerical simulations not only confirm the results of experimental studies, but also provide additional information on internal and external fluid flow.

Keywords: conical swirl atomizer; atomization; CFD; Eulerian model

Citation: Ochowiak, M.; Janecki, D.; Krupińska, A.; Włodarczak, S.; Wilk, T.; Olszewski, R. Conical Two-Phase Swirl Flow Atomizers—Numerical and Experimental Study. *Energies* **2021**, *14*, 1745. https://doi.org/10.3390/en14061745

Academic Editor: Pouyan Talebizadeh Sardari

Received: 18 February 2021
Accepted: 18 March 2021
Published: 21 March 2021

Publisher's Note: MDPI stays neutral with regard to jurisdictional claims in published maps and institutional affiliations.

1. Introduction

Spraying liquids is a phenomenon that can be observed both naturally occurring in nature and in the form of an effect induced by human activities [1–5]. This is the process of transforming bulk liquid into a large number of droplets as a result of internal and external forces acting. The obtained atomization effect depends on many input variables, e.g., properties of the sprayed liquid, geometry of atomizer, or operating conditions [1,6].

The geometry of the atomizer, especially the orifice, affects the atomization characteristics. It can be determined on the basis among others sheet thickness, breakup length, spray angle, Sauter mean diameter or droplet size distribution. There may be cavitation in the atomizer. This phenomenon is often achieved deliberately to obtain a liquid atomization. However, when the intensity of cavitation is too high, it chokes the flow and reduces the flow rate [7,8].

The theory of the atomization process includes both the aspect of the decomposition of streams and membranes into droplets, as well as the so-called secondary drop decay. Spraying can take place in various ways, it is possible to generate both conical and flat membranes as well as liquid streams, which then disintegrate [1]. The nature of the spray decay depends primarily on the speed of liquid outflow from the sprayer and on the type of sprayed liquid [9]. Depending on the ratio of gas and liquid densities, different mechanisms of stream formation are observed. For higher gas-to-liquid density ratios, it is very important to take into account the Kelvin–Helmholtz (K–H) instability, because the structure of the vortex is crucial for the spray effect (insights into the dynamics of spray—swirl interactions) [10].

Due to the scope and commonness of this process, it seems reasonable to explore the secrets of its theoretical and practical foundations, to learn about its possibilities and limitations. As a result, there are numerous scientific publications and subsequent modifications to already existing commercially available atomizers. One of the main limitations is related to the measuring apparatus [11,12].

One of the most developing branches of fluid mechanics is multiphase flows [13]. This is the result of their ever-growing practical importance, among others in the chemical, agrifood, energy, transport and environmental protection industries. The two-phase flow also deserves special attention when spraying liquids. Two-phase swirl flow atomizers are extremely interesting solutions [13–15]. They are characterized by high efficiency and reliability, while maintaining a simple structure. They make it possible to obtain a spray characterized by small droplets and a better quality of atomization compared to standard single-phase atomizers. They show low sensitivity to the rheological properties of the sprayed liquid and the range of operating parameters. This allows for a wider range of their applications, while maintaining the economic aspect. The introduction of the second phase (air) into the swirl chamber allows the streams or films of the liquid to be broken down satisfactorily into droplets [16].

The development of science and the progress of technology make it necessary to implement information technology tools that support the modeling of unit operations. In the literature, more and more publications on spraying in the context of numerical methods can be found. Some of them are comparing the data obtained experimentally with the simulation report. On the basis of the convergence of information, the correctness of the calculations is then determined and analyses are performed that were impossible to perform based only on the experimental results [17–21].

For dispersed flows, two types of models are most often used: Euler-Euler (E-E) [22,23] and Eulerian-Lagrangian (E-L) [24–27]. In both approaches, the flow of the continuous phase is described by means of generalized transport equations. In the E-E model, the dispersed phase together with the continuous phase are treated as a mixture.

The interfacial momentum exchange coefficient F_{jk} is calculated from the formula:

$$F_{jk} = \frac{3}{4}\rho_k \alpha_k \alpha_j \frac{C_D}{d_p}\left|\vec{u}_j - \vec{u}_k\right| \tag{1}$$

where C_D is the resistance coefficient calculated from the classical Schiller–Naumann model:

$$C_D = \begin{cases} \frac{24\left(1+0,15Re^{0.687}\right)}{Re} & dla\ Re \leq 1000 \\ 0.44 & dla\ Re > 1000 \end{cases} \tag{2}$$

where Re is the Reynolds number [28]:

$$Re = \frac{\rho_{k|\vec{u}_j - \vec{u}_k|d_p}}{\eta_k} \tag{3}$$

where:
u—actual velocity (m/s); ρ—density (kg/m^3); d_p—diameter of inlet port (m), η_k—liquid viscosity (Pa·s).

Obtaining a correlation between the atomizer geometry, process operating conditions and the spraying effect was the goal of many studies based on numerical methods [24,29].

Vashahi et al. [29] presented in their work a hybrid swirl atomizer aimed at agricultural use. The goal was to create a design that would allow the transition from spraying in the form of a hollow liquid cone to a full cone. This effect was achieved by creating the possibility to extend the length of the swirl chamber depending on preferences. A numerical computational fluid dynamics (CFD) simulation was performed based on the VOF (volume of fluid) model. The obtained results allowed to determine the internal structure of the flow in the atomizer, describe the interaction of air and water in the swirl chamber and outside the atomizer, and determine the spray angle. On the basis of the experimental tests, the obtained spray angles were measured. This parameter was adopted as a comparative criterion. The discrepancy between the experimental and numerical results was small, hence the simulation was assessed as carried out correctly.

Belhadef et al. [23] conducted a number of experimental studies using phase-Doppler anemometry (PDA), allowing to determine the axial velocity of the droplets and the Sauter mean diameter. These data were then compared with the results obtained by means of numerical simulations. In order to perform the calculations for the spraying process, the Euler model was used (the calculations concerned the turbulent range—large values of Re_L and We_L numbers), which was improved to achieve the best convergence of the results obtained with both methods. Conducting a CFD simulation allowed to obtain additional information on, among others: the nature of the generated aerosol/spray (stream shape, recirculation zones, droplet size distribution) or velocity and pressure gradients.

Tonini et al. [30] determined by means of numerical simulations the influence of the atomizer geometry and operating conditions on the atomization process in water mist atomizers. For this purpose, three-dimensional large eddy simulations based on the VOF methodology were implemented. It was assumed that the flow is incompressible and runs under isothermal conditions. The differences of the spraying effects obtained with the use of atomizers with conical and cylindrical swirl chambers of different dimensions were analyzed. In addition, the modification of the geometry also concerned the slope of the vortex channels. Thanks to the performed calculations, it was possible to fully characterize the liquid flow in the atomizer, quantify the number of swirls in a given cross-section, the momentum flux distribution and determine the thickness of the liquid ring forming. The performed numerical simulations also allowed to determine the characteristic flow regimes and link them with the preset liquid injection pressures.

The aim of this paper is to present the performed numerical simulations for the proposed two phase atomizers using swirl motion. Experimental tests for the considered constructions are discussed in detail in [14]. In the literature there is a lack of data about internal flow for such atomizers. The analyzed atomizers differ in the geometry of the swirl chamber—the ratio of the height of the swirl chamber to its diameter. This invariant is one of the most important geometric parameters determining the quality of the spray obtained [31,32]. Moreover, the influence of the selected operational parameters (velocities of liquid and gas) of the process on the obtained spraying effect was examined. In the works [14,15] the considered input quantities were correlated, i.e., the flow rate of both media and the geometry of the atomizer with the spray effect achieved. Based on the research, it was possible to determine the obtained discharge coefficient, spray angle, Sauter mean diameter and diameter distribution. The characteristics of the two-phase internal and external flow are complemented by the determination of the fluid velocity values obtained thanks to numerical simulations. It has been shown experimentally that there is a critical value of the ratio $H_S/D_S = 3$, above which the resulting spray shows a different characteristic compared to the atomizers with the same shape of the swirl chamber, but the values $H_S/D_S \leq 3$. On the basis of experimental results interpretation, it was decided to perform numerical simulations to obtain phase velocity values and to visualize the flow inside the atomizer.

2. Materials and Methods

Two out of nine pretested atomizers, i.e., SAN-1 with $H_S/D_S = 1$ and SAN-2 with $H_S/D_S = 4$ atomizers, were selected for numerical simulations. Figure 1 shows the geometry of the analyzed atomizers.

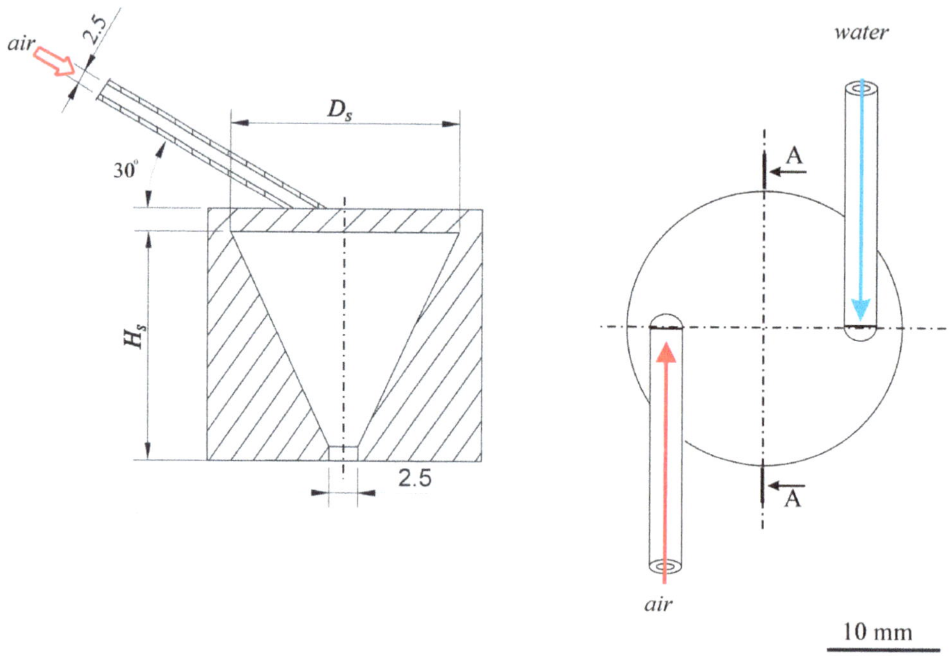

Figure 1. The geometry of the analyzed atomizers.

Table 1 presents the detailed characteristics of selected atomizers and the designations used in the further part of the work.

Table 1. The dimensions of the atomizers.

Designation in the Article	Diameter of Chamber D_S (mm)	Height of Chamber H_S (mm)	H_S/D_S
SAN-1	20 ± 0.1	20 ± 0.1	1
SAN-2	20 ± 0.1	80 ± 0.1	4

Numerical calculations were performed with the use of computational fluid dynamics CFD using the ANSYS Fluent R18.1 software. Because the purpose of the CFD simulation was to visualize the nature of the flow inside the atomizer (phase velocities, volumetric fractions) and to determine the spraying angles, the Euler-Euler model was used. The Euler-Euler approach (Euler model) includes an averaged mass and momentum balance for each phase present in the system. The Euler model treats all phases as continuous fluids with different velocities, volume fractions, and physicochemical properties. The differential equations of mass balance (continuity equation) and momentum for turbulent flow must be solved along with the equations describing the turbulence model. The most popular and simplest turbulence model was used for the calculations, the k-ε model, which is characterized by both quick convergence and good stability of the calculation process [23,33–38]. During the preliminary calculations, the k-ω and Reynold stress (RSM) models were tested, but their application resulted in slower convergence and lower stability of the calculation process. Based on that, it was decided to use the k-ε model.

ANSYS Fluent offers three options for the k-ε turbulence model for multiphase flows: mixture turbulence model (default), distributed turbulence model or per phase turbulence model. The calculations used a distributed turbulence model, applied when secondary phases are dispersed in a continuous phase. Turbulent predictions for the continuous phase q are obtained with the standard k-ε model supplemented with additional conditions that include interfacial turbulent momentum transfer. Predictions of the size of turbulence for dispersed phases are obtained using Tchen's theory on the dispersion of discrete particles by homogeneous turbulence [31].

The equations describing the model for the continuous phase q are as follows:

$$\frac{\partial}{\partial t}(\alpha_q\rho_q k_q) + \nabla\left(\alpha_q\rho_q k_q\vec{U}_q\right) = \nabla\left(\alpha_q\left(\mu_q + \frac{\mu_{t,q}}{\sigma_k}\right)\nabla k_q\right) + \alpha_q G_{k,q} - \alpha_q\rho_q\varepsilon_q + \alpha_q\rho_q\Pi_{k_q} \tag{4}$$

$$\frac{\partial}{\partial t}(\alpha_q\rho_q\varepsilon_q) + \nabla\left(\alpha_q\rho_q\vec{U}_k\varepsilon_q\right) = \nabla\left(\alpha_q\left(\mu_q + \frac{\mu_{t,q}}{\sigma_\varepsilon}\right)\nabla\varepsilon_q\right) + \alpha_q\frac{\varepsilon_q}{k_q}\left(C_{1\varepsilon}G_{k,q} - C_{2\varepsilon}\rho_q\varepsilon_q\right) + \alpha_q\rho_q\Pi_{\varepsilon_q} \tag{5}$$

The turbulence quantities for the dispersed phase are not obtained from the transport equations. The time and length scales that characterize motion are used to evaluate dispersion coefficients, correlation functions and turbulent kinetic energy of the dispersed phase.

$\vec{U}_q$, α_i, ρ_i, μ_i, are the phase-weighted velocity, volume fraction, phase density and viscosity, respectively.

The turbulent viscosity is calculated from the formula:

$$\mu_{t,q} = \rho_q C_\mu \frac{k_q^2}{e_q} \tag{6}$$

while the production of turbulence energy from the formula:

$$G_{k,q} = \mu_{t,q} S^2 \tag{7}$$

where:

$$S = \sqrt{2S_{ij}S_{ij}} \tag{8}$$

S_{ij} is a tensor of averaged deformation coefficients, terms Π_{kq} and $\Pi_{\varepsilon q}$ are source terms that can be taken into account to model a turbulent interaction between the dispersed phase and continuous phases. The values of the constants C and the turbulent Prandtl numbers σ in the k-ε turbulence model are as follows $C_{1\varepsilon}$ = 1.44; $C_{2\varepsilon}$ = 1.92; C_μ = 0.09; σ_k= 1; σ_ε = 1.3 [31].

The geometries of the atomizers used in the calculations were created on the basis of the commercial DesignModeler program in accordance with the real dimensions presented in work [14]. In order to obtain reliable results of numerical calculations, a number of simulations were carried out, enabling the selection of an appropriate grid and time step for these calculations. Four types of meshes were generated to check the sensitivity of the mesh. For SAN-1 atomizer: 134,169, 118,704, 57,595, 42,048 cells, and for SAN-2 atomizer: 143,825, 89,301, 59,475, 52,736 cells. Calculations were performed for the generated meshes. Figure 2 shows the results of the calculated mean liquid volume fractions at the outlet of the SAN-2 atomizer for each of these meshes.

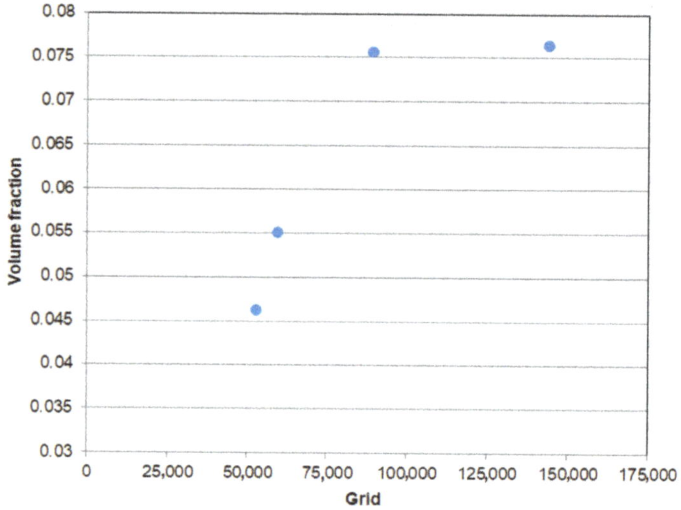

Figure 2. Compare the calculated mean liquid volume fractions at the outlet of the SAN-2 atomizer for different numbers of mesh elements.

Taking into account both the accuracy and the computational cost, the simulations used an intermediate mesh size, i.e., 89,301 cells. For a grid with 143,825 cells, with increased calculation time, the result was similar to that for a grid with 89,301 cells. By analyzing the calculation results for the SAN-1 atomizer identically, it was found that the mesh with 118,704 cells was the most optimal (Figure 3). The quality of the computational mesh was determined using the skewness value, which for our meshes was: 0.233 for the SAN-1 and 0.231 for the SAN-2.

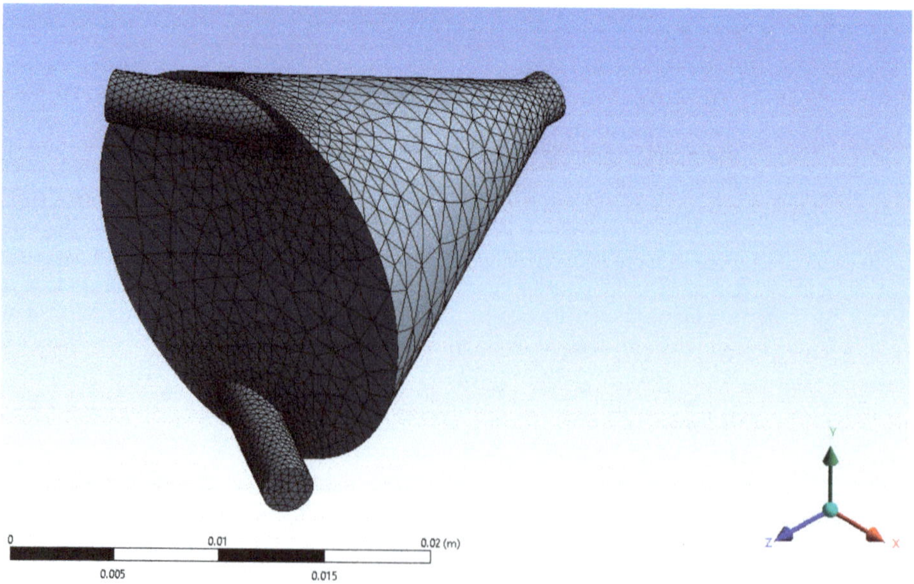

Figure 3. Geometry of the swirl chamber of the analyzed atomizer—computational grid.

The boundary conditions: inlets with a specified velocity, an outlet with a specified pressure and no slip on the walls were also introduced to the program. The simulation for the transient state with a time step of 0.001 s was used for the calculations. The operating parameters ranges are shown in the Table 2.

Table 2. The operating parameters ranges.

Re (-)	*u* (m/s)	$\dot{V}$ (m^3/s)	$\dot{G}$ (kg/s)
Re_L = (1400; 5650)	u_L = (0.57; 2.27)	$\dot{V_L} = (2.78 \times 10^{-6};\ 1.11 \times 10^{-5})$	$\dot{G_L} = (2.78 \times 10^{-3}; 1.11 \times 10^{-2})$
Re_G = (2970; 9900)	u_G = (17; 57)	$\dot{V_G} = (8.75 \times 10^{-5}; 2.91 \times 10^{-4})$	$\dot{G_G} = (1.05 \times 10^{-4};\ 3.50 \times 10^{-4})$

The numerical model was validated on the basis of the experimental results available in the works [14,15]. The basis for determining the correctness of the calculations is a comparison of the compliance of the obtained spray angle values for individual test points. An exemplary summary of data concerning the spray angle obtained experimentally and using numerical simulations for the SAN-1 and SAN-2 atomizers and the methodology of comparison are shown in Figures 4 and 5. The mean difference between CFD numerical calculations and experimental studies is around 6%. The maximum difference between numerical results and experimental data does not exceed 20%. The analysis of the obtained data showed that, with the increase of the Reynolds number for the gas, the spray angle value also increases.

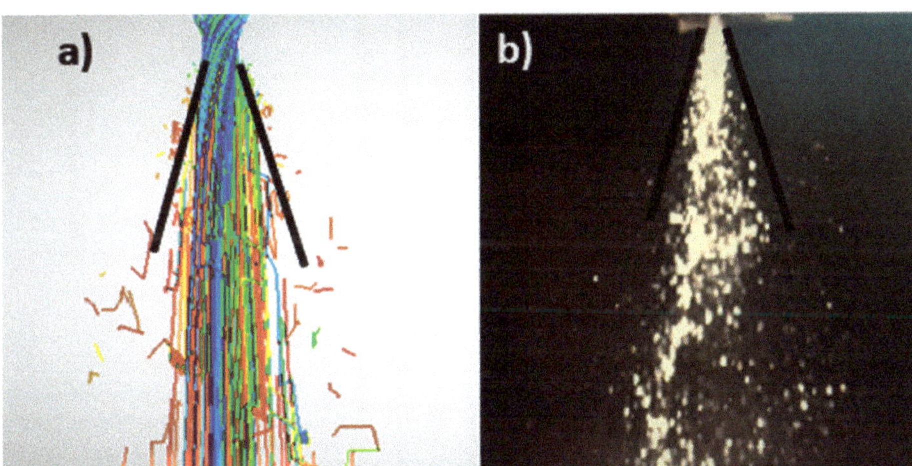

Figure 4. Comparison methodology measurement a spray angle: (**a**) numerical simulations, (**b**) experimental studies.

In order to obtain reliable results of numerical calculations, a number of simulations were carried out, enabling the selection of an appropriate grid and time step for these calculations. Tetrahedral numerical grids composed of 89,301 to 118,704 elements were plotted on the defined structure (Figure 1). The boundary conditions: inlets with a specified velocity, an outlet with a specified pressure and no slip on the walls were also introduced to the program. The simulation for the transient state with a time step of 0.001 s was used for the calculations.

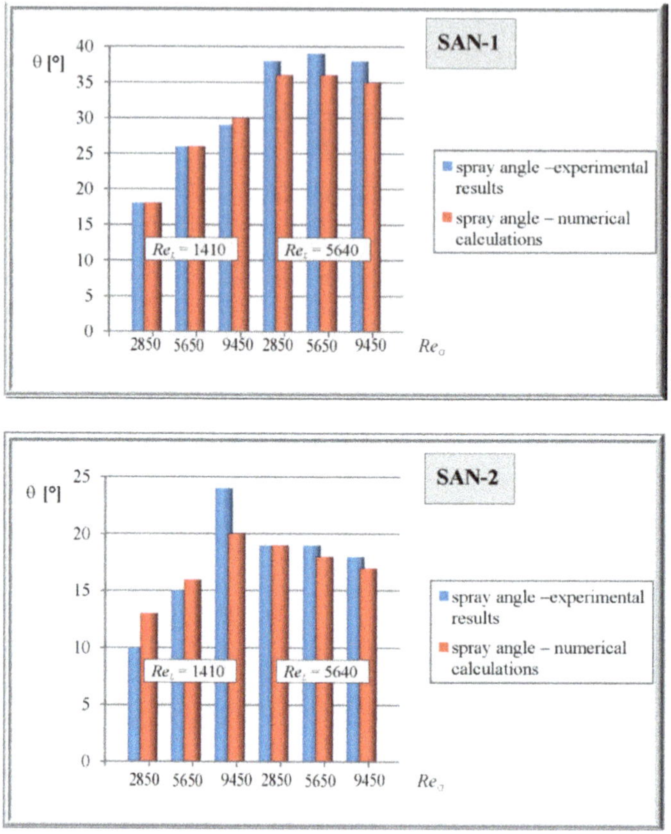

Figure 5. Verification of the correctness of computational fluid dynamics (CFD) numerical calculations—results.

3. Results

Figure 6 shows an example of the fluid flow velocity field inside the selected atomizer (SAN-1) and after leaving the system with the preset operating parameters Re_L = 4250 and Re_G = 3960.

It can be observed that higher liquid velocity and greater turbulence occur in the central part of the spray. Three velocity ranges can then be distinguished: the widest area of low velocity, the area of medium velocity and the narrowest area of high velocity. The maximum velocity is noticeable at the outlet of the atomizer. Then, as the spray becomes conical in shape, the velocity of the liquid slows down. This is the result of resistance forces (dependent on kinetic energy and aerodynamic drag) and the interaction of the droplets with the induced air movement, which is confirmed by the literature data presented by Sun et al. [37] and Shi and Kleinstreuer [39]. The expansion of the spray is facilitated by the radial velocity component. It also intensifies diffusion into the air. Figures 7 and 8 illustrate the velocity field distribution in selected cross-sections for the SAN-1 atomizer at the media flow defined as Re_L = 4250 and Re_G = 5650.

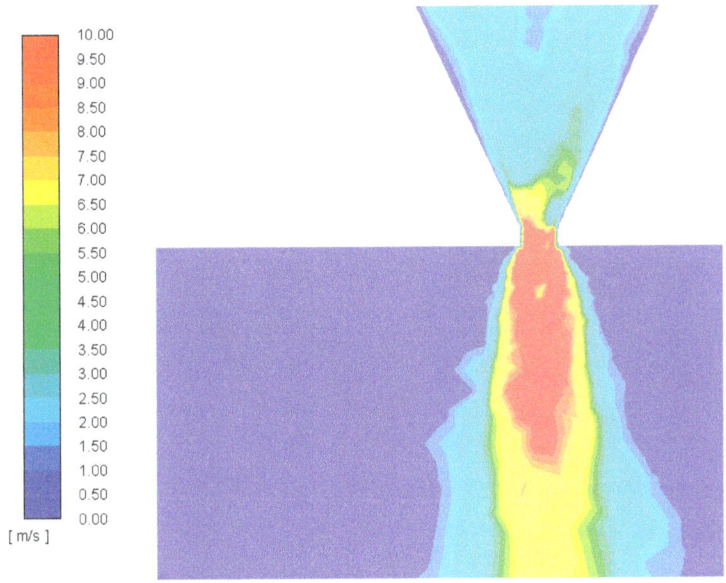

Figure 6. Distribution of liquid velocity inside and outside the atomizer SAN 1.

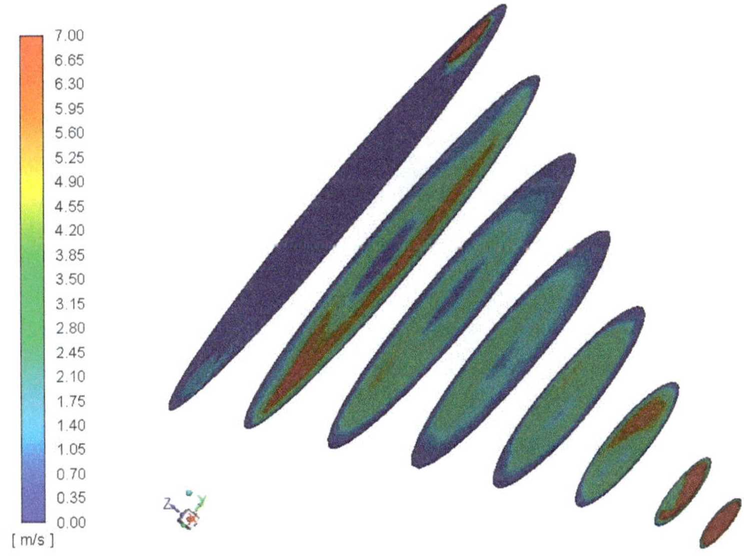

Figure 7. Velocity fields map.

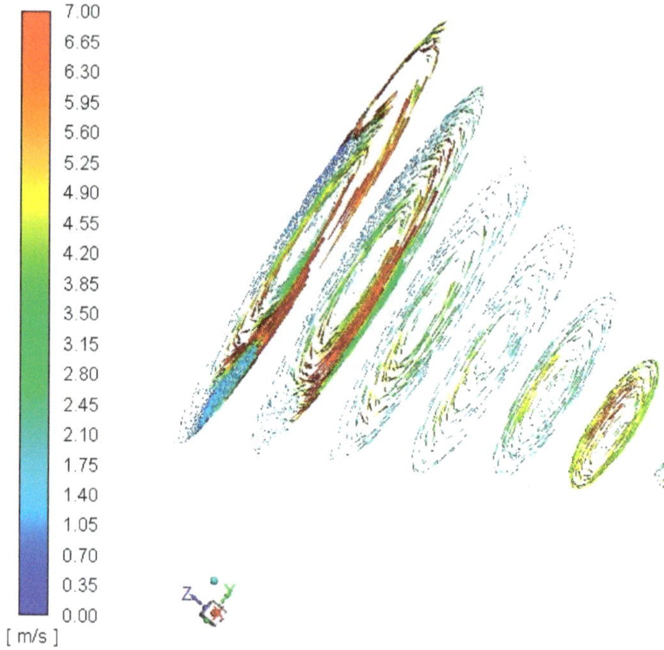

Figure 8. Spatial maps of the distribution of velocity vectors of liquids inside the SAN-1 atomizer.

The obtained images confirm the formation of swirl motion inside the mixing chamber and the increase in the velocity of the liquid flow towards the nozzle outlet of the atomizer. The formation of the air core inside the atomizer is due to the centrifugal forces induced by the tangential introduction of both media. The flow of the two-phase mixture in the outlet becomes fully annular. The liquid flows as a thin film over the wall of the outlet orifice, and the gas flows at high velocity in the center of the outlet. This effect is desirable due to the accumulation of surface energy, increased stream instability, and thus increased susceptibility of the liquid to decay, i.e., improved atomization quality.

Figures 9 and 10 show the effect of changing the flow conditions of individual media on the obtained atomization effect and liquid velocity. With the increase of the Reynolds number of the liquid, with the constant Reynolds number of the gas, a larger spray angle was observed. On the other hand, the increase in the Reynolds number of the gas with the constant Reynolds number of the liquid resulted in a significant (about threefold) increase in the velocity of the liquid, especially in the central area of the aerosol. Moreover, in all the cases, the maximum speed of the droplet is observed in the central region of the generated spray.

The geometry of the atomizer also influences the obtained liquid velocity distribution. Figure 11 shows the results obtained for the SAN-1 atomizer (H_S/D_S = 1) and the SAN-2 atomizer (H_S/D_S = 4).

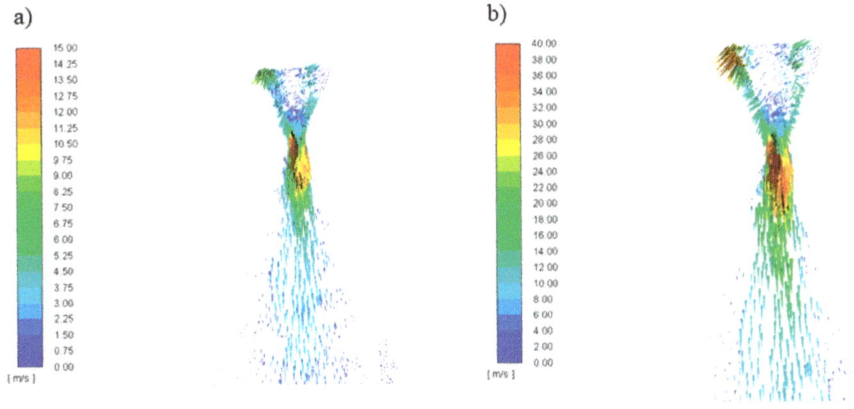

Figure 9. Vector velocity field for SAN-1 atomizer: (**a**) $Re_L = 1410$, $Re_G = 2850$, (**b**) $Re_L = 5640$, $Re_G = 2850$.

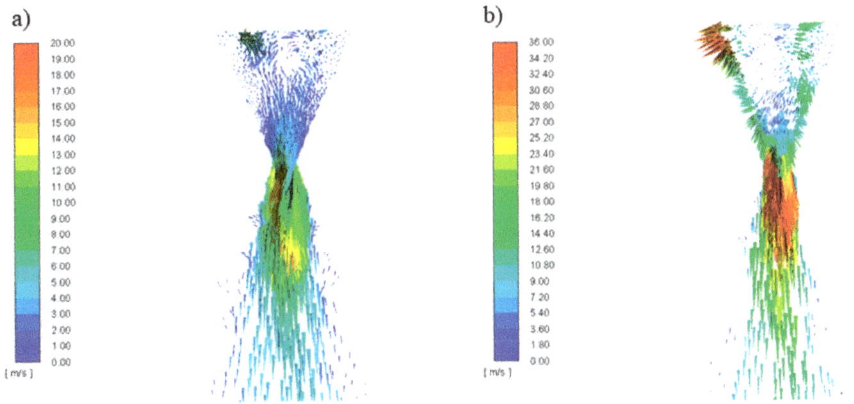

Figure 10. Vector velocity field for SAN-1 atomizer: (**a**) $Re_L = 5640$, $Re_G = 2850$, (**b**) $Re_L = 5640$, $Re_G = 9900$.

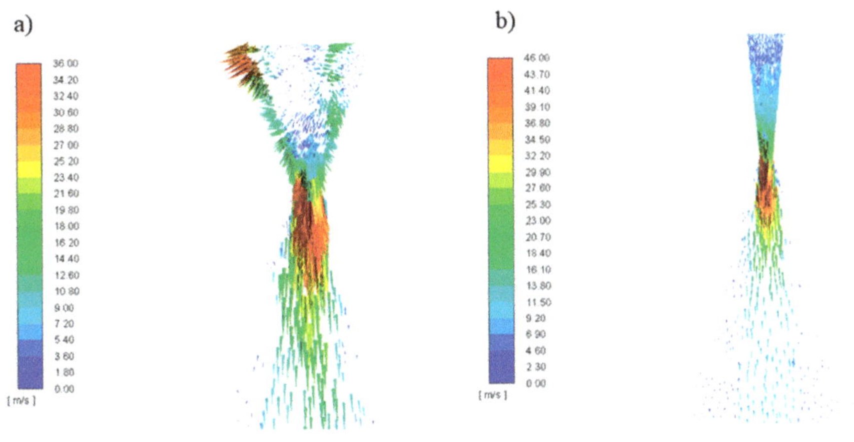

Figure 11. Vector velocity field ($Re_L = 5640$, $Re_G = 9450$): (**a**) SAN-1 atomizer, (**b**) SAN-2 atomizer.

On the basis of the obtained images, using the AutoCAD software, the spray angle was measured, defined as the angle formed between two straight lines along the stream flowing out of the atomizer. It can be noticed that for the SAN-2 atomizer the spray angle was smaller than for the SAN-1 atomizer. It is related to the increase in the height of the atomizer swirl chamber while maintaining the same diameter (increase in the H_S/D_S ratio). The specific elongation of the swirl chamber reduces the angle of the transition cone in the atomizer (transition section of vortex chamber), which clearly translates into the spray effect obtained. This confirms the earlier observations suggesting that, when the value of the geometric invariant $H_S/D_S \geq 3$, lower spray angle values are observed. The increase in the height of the swirl chamber causes the generation of additional frictional resistances, which results in a decrease in the momentum of the quantity of motion, which translates into a reduction of the spray angle obtained. In addition, in the case of the SAN-2 design, the effect of obtaining three velocity ranges in the spray is particularly visible. The effect of changing the geometry on the atomization quality is directly related to the nature of the flow inside the atomizer. The disintegration of the liquid stream is the key variable influencing the resulting spray [40]. Figures 12 and 13 show the volume fraction of the liquid phase and its velocity in the mixing chamber for the SAN-1 and SAN-2 atomizers, respectively, at the same flow conditions Re_L = 1410, Re_G = 9500.

a) b)

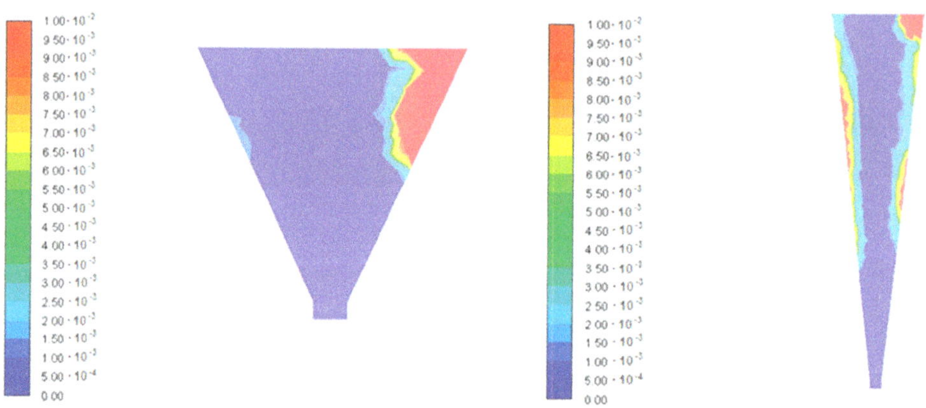

Figure 12. Volume fraction of the liquid phase in the mixing chamber: (**a**) SAN-1 atomizer, (**b**) SAN-2 atomizer.

In the case of the SAN-1 atomizer, the breakdown of the stream inside the swirl chamber is much greater than for the SAN-2 atomizer. For the SAN-1 atomizer, only a narrow field at the top of the mixing chamber is visible, where the concentration of the liquid is significant, further along the entire chamber, we have a negligible fraction of liquid, which proves its significant dispersion. For the SAN-2 atomizer, the fraction of the liquid phase is visible on a considerable length of the swirl chamber, near its wall. Thus, it can be concluded that the liquid stream decomposition in this structure takes place to a lesser extent.

Figure 9 summarizes the velocity of the liquid inside the atomizer. For the SAN-1 system in the outlet orifice, it can be noticed that higher velocities are at its walls, while for the SAN-2, the liquid flows out the same through the entire cross-section of the outlet. This suggests that in the SAN-1 atomizer, the air core is in the central part of the outlet orifice, and in the case of the SAN-2 the liquid flows out through the entire cross-section. The differences in the internal flow for the analyzed atomizers are also related to the air core formed in a different way. In the SAN-1 atomizer, the core is formed with a much larger diameter than in the SAN-2 atomizer. Moreover, in the first case, the core is observed to be kept practically along the entire length of the swirl chamber, while in the second case, the core vanishes relatively quickly.

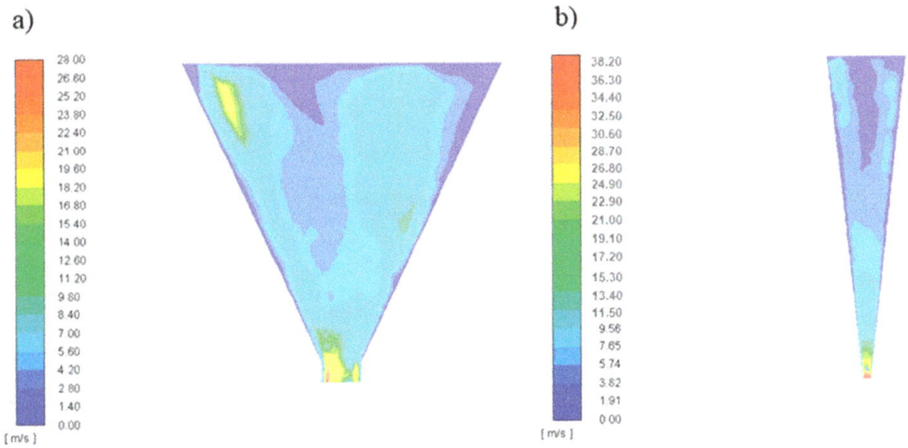

Figure 13. Velocity of the liquid phase in the mixing chamber: (**a**) SAN-1 atomizer, (**b**) SAN-2 atomizer.

Figure 14 shows the effect of changing the operating parameters on the liquid dispersion inside the swirl chamber of the SAN-1 atomizer. The volume fraction of liquid was measured for three projections in the XY plane at different heights of the swirl chamber and in the XZ plane. The images in the left column are for the flow at $Re_L = 1410$, $Re_G = 2850$, while in the right: $Re_L = 5650$, $Re_G = 2850$.

It can be noticed that an increase in the liquid flow rate (liquid Reynolds number) for the same gas flow results in poorer and later liquid dispersion. At high liquid flow rates, we obtain a thicker liquid film (with a thickness of s) on the inner walls of the swirl chamber, surrounding the air core. The air that was outside and was sucked in due to the pressure gradient, as well as the air that is supplied to the atomizer as a second phase, are in the swirl chamber and both contribute to the formation of the air core and affect the resulting spray. The diameter of the air core is smaller than the diameter of the atomizer outlet orifice. The air core primarily affects the actual outlet orifice cross-section occupied by the liquid, which is less than:

$$A_0 < \frac{\pi d_0^2}{4} \tag{9}$$

The more liquid is fed to the atomizer, the more difficult it is to disperse it. A solid stream of liquid is harder to tear. The formation of the air core in connection with the spray angle, droplet diameter and the flow velocity of the liquid in the spray was described, among others, by Durdina and his colleagues in their works [41].

Figure 15 shows that the turbulent kinetic energy is high in the core region due to high shear rates and interphase coupling, since the dispersion is high in this region. The high intensity of turbulence at the orifice is associated with the high flow velocity of the fluid. The turbulent kinetic energy is highest (103 m^2/s^2) at nozzle orifice. The build-up of turbulent kinetic energy is observed in the axis of symmetry of the atomizer outlet, and in the result in the axis of the spray. A better understanding of the turbulence generated by the spray system can be beneficial for the evaluation of several important phenomena such as explosion enhancement.

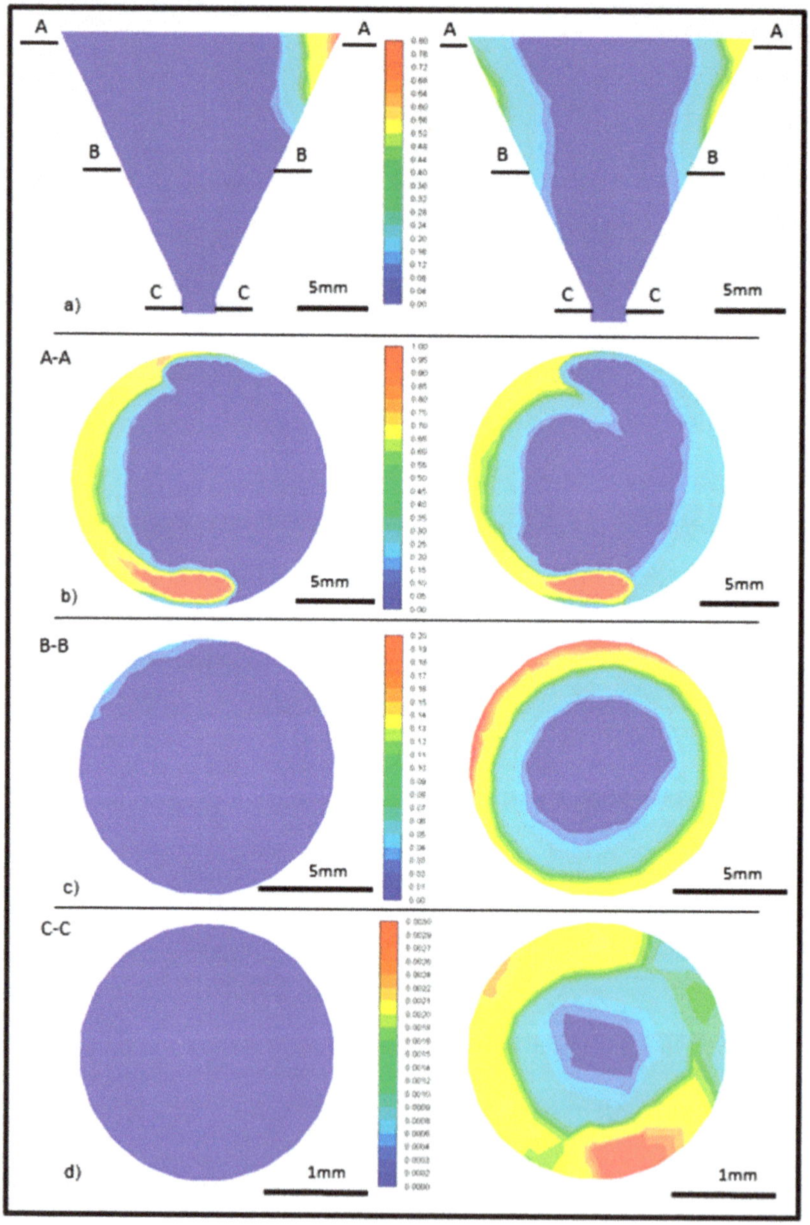

Figure 14. Volume fraction of the liquid phase in the mixing chamber SAN-1 atomizer: (**a**) in the XZ plane (**b**) in the XY plane at the height of liquid and gas inlet, (**c**) in the XY plane at the height of the center of the mixing chamber, (**d**) in the XY plane at the height of the outlet orifice.

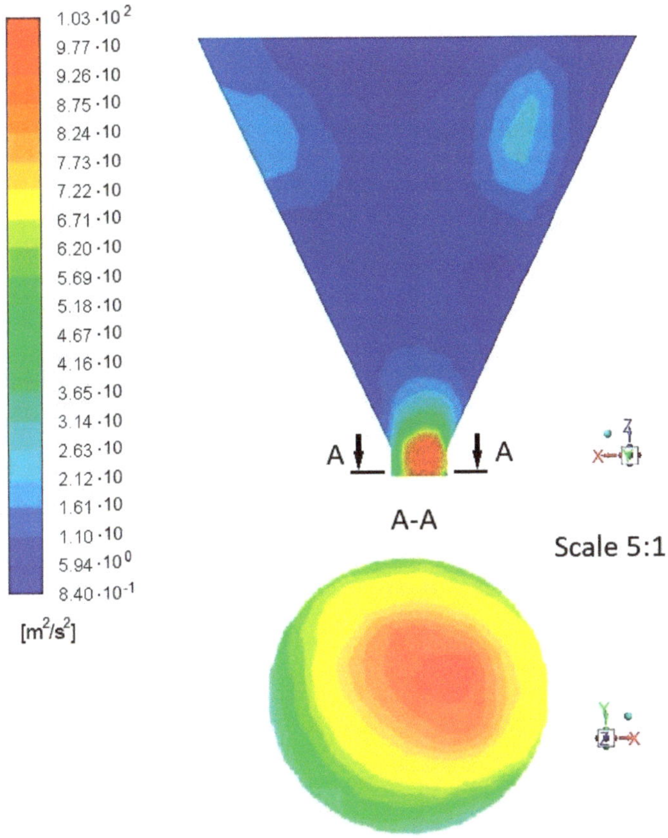

Figure 15. Turbulent kinetic energy in atomizer.

The tested atomizers designs are a very attractive proposition, because they can have outlet openings with very large diameters, and by appropriately changing the geometry of the atomizer, the thickness of the liquid film can be made thicker or thinner. It has been shown that at low flow velocities, the liquid flows out of the atomizer in a dense stream, from which individual drops can detach. As the gas stream increases, the liquid stream is atomized into droplets and the spray angle increases. When the gas stream is relatively large, the spray angle may decrease due to the flowing around the stream of the atomized liquid. Increasing the stream of liquid may increase the volume of the stream, which makes it more difficult to disintegrate.

4. Conclusions

Two-phase flows are characterized by high complexity and difficulty in defining the conditions for the formation of appropriate structures. There are still no theoretical models to describe the phenomenon of two-phase atomization. The conducted research and calculations are aimed at increasing the knowledge about atomization obtained thanks to the use of two-phase atomizers with swirl flow. Numerical calculations were performed with the use of computational fluid dynamics CFD using the ANSYS Fluent R18.1 software. The analysis of the obtained data showed that with the increase of the Reynolds number for the gas, the spray angle value also increases. In the spray axis, the highest values of liquid flow velocity and the greatest swirls (turbulences) were observed. The impact of geometry change on the atomization quality is directly related to the nature of the internal flow and the air core formed. In the atomizer with $H_S/D_S < 3$, the air core is formed with a much larger diameter than in the atomizer $H_S/D_S \geq 3$. Moreover, in the first case the air core is kept practically along the entire length of the swirl chamber, while in the second the air core decays relatively quickly. As the height of the swirl chamber increases, while maintaining its constant diameter, a decrease in the value of the transition section cone angle is observed. This results in less contraction of the stream, which translates into a reduction in the angular momentum. As the value of the H_S/D_S ratio increases, the spray angle value decreases. The increase in the height of the swirl chamber results in additional frictional resistance, which causes a decrease in angular momentum.

The correctness of the performed numerical simulations was verified by measuring the obtained spray angles and comparing its values with the data obtained experimentally. CFD analysis allowed for a more detailed analysis of the atomization mechanism providing information about the droplet velocity. Moreover, it made it possible to characterize the two-phase flow along the atomizer by determining the proportion of phases in the two-phase mixture. It is also a basis for further analysis of mass and energy exchange along the atomizer. The outcomes from this numerical calculation allow the optimization of the atomizer construction for specific applications and extend the classical analysis without time-consuming and costly experimental tests. Numerical simulations in modelling the phenomenon of spray formation and spreading can reduce costly tests and experimental studies. They can be helpful in the search for an atomizer technical solution that will obtain a specific spray (spraying intentionally) or be used by designers in the search for an optimal solution.

Author Contributions: Conceptualization, M.O. and A.K.; methodology, M.O., A.K., S.W. and D.J.; software, D.J.; validation, A.K., T.W. and S.W.; formal analysis, A.K. and R.O.; investigation, A.K., M.O., R.O., T.W. and S.W.; data curation, A.K.; writing—original draft preparation, A.K. and M.O.; writing—review and editing, M.O.; visualization, A.K., S.W. and D.J.; supervision, M.O. All authors have read and agreed to the published version of the manuscript.

Funding: This research was funded by the Ministry of Education and Science of Poland.

Institutional Review Board Statement: Not applicable.

Informed Consent Statement: Not applicable.

Data Availability Statement: The data presented in this study are available on request from the corresponding author.

Conflicts of Interest: The authors declare no conflict of interest.

References

1. Lefebvre, A.; McDonell, V. *Atomization and Sprays*; CRC Press Taylor & Francis Group: Boca Raton, FL, USA, 2017.
2. Orzechowski, Z.; Prywer, J. *Production and Use of Spray Liquid*; WNT: Warsaw, Poland, 2008. (In Polish)
3. Ochowiak, M. The Analysis of Liquid Atomization in Effervescent and Effervescent-Swirl Atomizers, Habilitation. Ph.D. Thesis, Wydawnictwo Politechniki Poznańskiej, Poznań, Poland, 2014. (In Polish).
4. Sirignano, W.A. *Fluid Dynamics and Transport of Droplets and Spray*; Cambridge University Press: Irvine, CA, USA, 1999.
5. Liu, H. *Science and Engineering of Droplets–Fundamentals and Applications*; William Andrew Publishing: New York, NY, USA, 2000.
6. Edgar, P.; Herrero, E.M.; Del Valle, M.; Galán, M.A. Instability study of a swirling annular liquid sheet of polymer produced by air-blast atomization. *Chem. Eng. J.* **2007**, *133*, 69–77. [CrossRef]
7. Shao, C.X.; Luo, K.; Yang, Y.; Fan, J.R. Detailed numerical simulation of swirling primary atomization using a mass conservative level set method. *Int. J. Multiph. Flow* **2017**, *89*, 57–68. [CrossRef]
8. Shervani-Tabar, M.T.; Parsa, S.; Ghorbani, M. Numerical study on the effect of the cavitation phenomenon on the characteristics of fuel spray. *Math. Comput. Modell.* **2012**, *56*, 105–117. [CrossRef]
9. Jarrahbashi, D.; William, S.; Popov, P.; Hussain, F. Early spray development at high gas density: Hole, ligament and bridge formations. *J. Fluid Mech.* **2016**, *792*, 186–231. [CrossRef]
10. Rajamanickam, K.; Basu, S. Insights into the dynamics of spray–swirl interactions. *J. Fluid Mech.* **2017**, *810*, 82–126. [CrossRef]
11. Zhou, W.; Zhao, T.; Wu, T.; Yu, Z. Application of fractal geometry to atomization process. *Chem. Eng. J.* **2000**, *78*, 193–197. [CrossRef]

12. Schick, R.J. *Spray Technology Reference Guide: Understanding Drop Size*; Spraying Systems Co.: Wheaton, IL, USA, 2008.
13. Dziubiński, M.; Prywer, J. *Mechanics of Two-Phase Fluids*; WNT: Warsaw, Poland, 2010. (In Polish)
14. Ochowiak, M.; Krupińska, A.; Włodarczak, S.; Matuszak, M.; Markowska, M.; Janczarek, M.; Szulc, T. The two-phase conical swirl atomizers: Spray characteristics. *Energies* **2020**, *13*, 3416. [CrossRef]
15. Ochowiak, M.; Lytvynenko, O.; Włodarczak, S.; Matuszak, M.; Krupińska, A. Design and study of conical pressure-swirl atomizers. In *Lecture Notes in Mechanical Engineering, Proceedings of the International Conference on Design, Simulation, Manufacturing: The Innovation Exchange, DSMIE-2018, Sumy, Ukraine, 12–15 June 2018*; Springer: Berlin/Heidelberg, Germany, 2018; pp. 472–480.
16. Ochowiak, M. The experimental study on the viscosity effect on the discharge coefficient for effervescent atomizers. *Exp. Therm. Fluid Sci.* **2013**, *50*, 187–192. [CrossRef]
17. Nishida, K.; Tian, J.; Sumoto, Y.; Long, W.; Sato, K.; Yamakawa, M. An experimental and numerical study on sprays injected from two-hole nozzles for DISI engines. *Fuel* **2009**, *88*, 1634–1642. [CrossRef]
18. Nouri-Borujerdi, A.; Kebriaee, A. Numerical Simulation of laminar and turbulent two-phase flow in pressure-swirl atomizers. *AIAA J.* **2012**, *50*, 2091–2101. [CrossRef]
19. Amini, G. Liquid flow in a simplex swirl nozzle. *Int. J. Multiph. Flow* **2016**, *79*, 225–235. [CrossRef]
20. Qin, C.; Loth, E. Numerical description of a pressure-swirl nozzle spray. *Chem. Eng. Process.* **2016**, *107*, 68–79. [CrossRef]
21. Li, H.; Chen, X.; Shu, C.M.; Wang, Q.; Zhang, Y. Experimental and numerical investigation of the influence of laterally sprayed water mist on a methane-air jet flame. *Chem. Eng. J.* **2019**, *356*, 554–569. [CrossRef]
22. Amedorme, S.K.; Burluka, A.A. Numerical prediction of Sauter mean diameter from pressure swirl atomizer using Eulerian model. *Int. J. Eng. Technol.* **2017**, *7*, 484–494.
23. Belhadef, A.; Vallet, A.; Amielh, M.; Anselmet, F. Pressure-swirl atomization: Modelling and experimental approaches. *Int. J. Multiph. Flow* **2012**, *39*, 13–20. [CrossRef]
24. Zhua, S.; Pozarlik, A.; Roekaerts, D.; Rodrigues, H.C.; der Meera, T. Numerical investigation towards HiTAC conditions in laboratory-scale ethanol spray combustion. *Fuel* **2018**, *211*, 375–389. [CrossRef]
25. Lin, A.; Zhou, J.; Fawzy, H.; Zhang, H.; Zheng, Q. Evaluation of mass injection cooling on flow and heat transfer characteristics for high-temperature inlet air in a MIPCC engine. *Int. J. Heat Mass Transf.* **2019**, *135*, 620–630. [CrossRef]
26. Lin, A.; Zheng, Q.; Jiang, Y.; Lin, X.; Zhang, H. Sensitivity of air/mist non-equilibrium phase transition cooling to transient characteristics in a compressor of gas turbine. *Int. J. Heat Mass Transf.* **2019**, *137*, 882–894. [CrossRef]
27. Lin, A.; Zheng, Q.; Fawzy, Y.H.; Luo, M.; Zhou, J.; Zhang, H. Effect of water injection cooling on flow field characteristics in the cooling section of precooled turbine-based combined cycle engine. *Int. J. Heat Mass Transf.* **2019**, *141*, 615–626. [CrossRef]
28. ANSYS. *Fluent User's Guide, Version 18.1*; ANSYS: Canonsburg, PA, USA, 2017.
29. Vashahi, F.; Dafsari, R.A.; Rezaei, S.; Lee, J. CFD Simulation of Two-Phase Flow Behaviour of a Hybrid Pressure Swirl Atomizer. In Proceedings of the KSME Fluid Engineering Division Spring Conference, BEXCO, Busan, Korea, 25–26 May 2017.
30. Tonini, S.; Conti, P.; Cossali, G.E. Numerical modelling of internal flow in water mist injectors: Effect of nozzle geometry and operating conditions. *Fire Technol.* **2019**, *55*, 2395–2417. [CrossRef]
31. Rashad, M.; Yong, H.; Zekun, Z. Effect of geometric parameters on spray characteristics of pressure swirl atomizers. *Int. J. Hydrog. Energy* **2016**, *41*, 15790–15799. [CrossRef]
32. Dafsari, R.A.; Vashahi, F.; Lee, J. Effect of swirl chamber length on the atomization characteristics of a pressure-swirl nozzle. *At. Sprays* **2017**, *27*, 859–874. [CrossRef]
33. Bae, J.; Lee, H.J.; Choi, H.; Park, D.C. Hydraulic and internal flow characteristics of swirling superheated hydrocarbon liquid jets. *Int. J. Heat Mass Transf.* **2019**, *137*, 1014–1026. [CrossRef]
34. Xue, R.; Ruan, Y.; Liu, X.; Cao, F.; Hou, Y. The influence of cavitation on the flow characteristics of liquid nitrogen through spray nozzles: A CFD study. *Cryogenics* **2017**, *86*, 42–56. [CrossRef]
35. Yu, L.; Yuan, S.; Pang, Z.; Wang, Y. Analysis of CFD for design method of swirl atomizer. *Appl. Mech. Mater.* **2012**, *226–228*, 812–815. [CrossRef]
36. Halder, M.R.; Dash, S.K.; Som, S.K. A numerical and experimental investigation on the coefficients of discharge and the spray cone angle of a solid cone swirl nozzle. *Exp. Therm. Fluid Sci.* **2004**, *28*, 297–305. [CrossRef]
37. Sun, Y.; Alkhedhair, A.M.; Guan, Z.; Hooman, K. Numerical and experimental study on the spray characteristics of full-cone pressure swirl atomizers. *Energy* **2018**, *160*, 678–692. [CrossRef]
38. Datta, A.; Som, S.K. Numerical prediction of air core diameter, coefficient of discharge and spray cone angle of a swirl spray pressure nozzle. *Int. J. Heat Fluid Flow* **2000**, *21*, 412–419. [CrossRef]
39. Shi, H.; Kleinstreuer, C. Simulation and analysis of high-speed droplet spray dynamics. *J. Fluids Eng.* **2007**, *129*, 621–633. [CrossRef]
40. Faazil, A.B.; Eapen, A.M.; Sudhakaran, S.K. Disintegration of liquid sheets and quantification of its dynamics. In Proceedings of the 2020 Advances in Science and Engineering Technology International Conferences (ASET), Dubai, United Arab Emirates, 4–6 February 2020; pp. 1–6.
41. Durdina, L.; Jedelsky, J.; Jicha, M. Spray structure of a pressure-swirl atomizer for combustion applications. *EPJ Web Conf. Exp. Fluid Mech.* **2012**, *25*, 1–10. [CrossRef]

energies

MDPI

Article

Investigation of Thermal-Flow Characteristics of Pipes with Helical Micro-Fins of Variable Height

Piotr Bogusław Jasiński, Michał Jan Kowalczyk, Artur Romaniak, Bartosz Warwas *, Damian Obidowski and Artur Gutkowski

Institute of Turbomachinery, Lodz University of Technology, 90-924 Lodz, Poland; piotr.jasinski@p.lodz.pl (P.B.J.); michal.kowalczyk.1@dokt.p.lodz.pl (M.J.K.); artur.romaniak@dokt.p.lodz.pl (A.R.); damian.obidowski@p.lodz.pl (D.O.); artur.gutkowski@p.lodz.pl (A.G.)

* Correspondence: bartosz.warwas@dokt.p.lodz.pl; Tel.: +48-42-631-23-71

Abstract: The results of numerical investigations of heat transfer and pressure drops in a channel with 30° helical micro-fins are presented. The main aim of the analysis is to examine the influence of the height of the micro-fins on the heat-flow characteristics of the channel. For the tested pipe with a diameter of 12 mm, the micro-fin height varies within the range of 0.05–0.40 mm (with 0.05 mm steps), which is equal to 0.4–3.3% of its diameter. The analysis was performed for a turbulent flow, within the range of Reynolds numbers 10,000–100,000. The working fluid is water with an average temperature of 298 K. For each tested geometry, the characteristics of the friction factor $f(Re)$ and the Nusselt number $Nu(Re)$ are shown in the graphs. The highest values of Nusselt numbers and friction factors were obtained for pipes with the micro-fins H = 0.30 mm and H = 0.35 mm. A large discrepancy is observed in the friction factors $f(Re)$ calculated from the theoretical relationships (for the irregular relative roughness values shown in the Moody diagram) and those obtained from the simulations (for pipes with regular roughness formed by micro-fins). The *PEC* (Performance Evaluation Criteria) heat transfer efficiency analysis of the geometries under study is also presented, taking into account the criterion of the same pumping power. The highest *PEC* values, reaching 1.25, are obtained for micro-fins with a height of 0.30 mm and 0.35 mm and with Reynolds numbers above 40,000. In general, for all tested geometries and for large Reynolds numbers (above 20,000), the *PEC* coefficient reaches values greater than 1, while for lower Reynolds numbers (less than 20,000), its values are less than 1.

Keywords: heat transfer coefficient; micro-fins; friction factor; numerical methods; CFD

Citation: Jasiński, P.B.; Kowalczyk, M.J.; Romaniak, A.; Warwas, B.; Obidowski, D.; Gutkowski, A. Investigation of Thermal-Flow Characteristics of Pipes with Helical Micro-Fins of Variable Height. *Energies* **2021**, *14*, 2048. https://doi.org/10.3390/en14082048

Academic Editor: Pouyan Talebizadeh Sardari

Received: 12 March 2021
Accepted: 2 April 2021
Published: 7 April 2021

1. Introduction

The pipe, as a flow channel, is one of the simplest and most commonly used in power engineering and thermal devices. It is widely applied in heat exchangers and to transport the medium in installations. The phenomena occurring in it are related to both fluid mechanics and thermal processes; therefore, the key issue is to design the geometry in such a way as to maximize heat transfer while limiting a negative effect of flow resistance.

There are many methods to intensify the heat transfer process in the pipe. Among others, intensifying inserts of various shapes and other flow turbulence devices are widely used. Wijayanta et al. [1] conducted a numerical study to evaluate the thermal hydraulic performance of a turbulent flow inside a tube equipped with a square-cut twisted tape and a classical twisted tape insert. In the range of Reynolds number from 8000 to 18,000 under investigation, the tube with a square-cut twisted tape had the highest values of heat transfer rate, friction factor, and thermal performance factor for the twisted ratio y/W = 2.7. In [2], Wijayanta et al. investigated the thermal performance of a tube heat exchanger with punched delta-winglet vortex generators. The numerical studies for Re = 9100–17,400 and the attack angles of 30°, 50°, and 70° allowed one to determine values of the Nusselt number, friction factor, and thermal performance factor, which increased with an increasing value of attack angle. Jasiński [3–5] conducted experimental and numerical investigations of a flow in the circular tube with ball turbulators. The investigations for different diameters of the balls, different distances between them, and *Re* = 10,000–300,000 were carried out. The results showed that the highest increase in the Nusselt number and the friction factor was observed for the insert ball with largest diameters; however, the highest values of thermal performance factor were observed for the smallest balls. Arjmandi et al. [6] conducted a numerical investigation of applying twisted tape turbulator and Al_2O_3/water nanofluid in double pipe heat exchanger. The studies of heat transfer coefficient and pressure drop were carried out for different pitches ratios 0.09–0.18, angles 0–30°, and Reynolds numbers in the range 5000–20,000. According to obtained results, the optimization accomplished by the response surface methodology was performed, thanks to which the optimal vortex generator geometry was created.

Another method to intensify heat transfer is to apply nano-liquids as a modification of the working fluid. Patil et al. [7] made a review and presented investigations on synthesis, thermo-physical properties, and a heat transfer mechanism of nanofluids. Shajahan et al. [8] carried out experimental studies with a combination of nanofluids and inserts under the conditions of laminar flow. The results allowed for the determination of highest values of the Nusselt number, friction factor, and thermal performance factor according to different twist ratios of inserts and various types of nanofluids. Kristiawan et al. [9] investigated numerically an influence of micro-fins and TiO_2/water nanofluids on thermo-hydraulic performance. The highest *PEC* (Performance Evaluation Criteria) achieved the squared mini-channel with micro-fins and the nanoparticle concentration of 0.01 vol.%. Asirvatham et al. [10] presented the results of experimental studies of convective heat transfer with a low volume fraction of the CuO/water nanofluid. According to the gained experimental data, a correlation for the Nusselt number was evolved.

The third method of heat transfer intensification is surface finning, which is considered in this article. The problem of similar geometry has been already addressed widely in the scientific literature. In most cases, researchers perform experimental investigations, but an increasing number of publications based on combined numerical and experimental tests is observed. For example, Mann and Eckels [11], in order to improve heat transfer and reduce a pressure drop, optimized the following geometrical pipe parameters: the number of fins, their height and helical angle, for a Reynolds number ranging from 30,000 to 60,000. They used the ANSYS Fluent software and the NSGA II algorithm in the study. After comparing the numerical results with the experimental results, the researchers concluded that for helical angles above 45°, the relationship between heat transfer and geometry was chaotic rather than ordered. Wen-Tao et al. [12] conducted an experiment for a developed turbulent flow in the Reynolds number range from 10,000 to 100,000, for 16 different pipes with internal grooves. They compared the experimental results with the analytical method using the Gnielinski equation, which was then extended by the Nusselt number. The proposed extended equation yielded a good agreement of the results, where the deviation for 93% of the data was within ±20%.

Li et al. [13] used PIV (Particle Image Velocimetry) technology in the experimental investigations. They determined a relationship between flow characteristics and heat transfer for the geometry of a micro-finned rectangular duct. They observed an influence of the formation of coherent structures in the fluid on an increase in the heat transfer coefficient. Additionally, for low Reynolds numbers, characteristic of the laminar and transitional flow, they observed a lack of vortices filling the spaces between the fins, which was equivalent to the deterioration of heat transfer. Guo-Dong et al. [14] compared the experimental results for a plain pipe with an internally finned pipe using the (HMiM) BF_4 medium (1-hexyl-3-methyl-imidazolium-tetrafluoroborate). Based on the obtained results, they determined empirical coefficients to calculate a friction coefficient and a heat transfer coefficient for a laminar flow (*Re* = 60–600). Brognaux et al. [15] developed the characteristics of a single-phase flow through the pipe with two types of internal fins. They analyzed an influence of geometry on the Prandtl number and determined a dependence of the Reynolds number on the friction coefficient. On the basis of experimental results, they observed that the Prandtl number exponent agreed well with other correlations for the geometry with two-dimensional roughness.

Jasiński [16] used numerical simulations to model a flow in the internally finned pipe for various helical angles of the micro-fins in the range of 0–90°. Using the EGM (Entropy Generation Minimization) method to assess the flow, he showed that for each geometry, the minimum entropy was generated for the range Re = 60,000–90,000. Tang and Li [17] developed a correlation of the friction coefficient for data from various experiments. The data include old and new refrigerants that were used in internally finned tube heat exchangers. Jensen and Vlakancic [18] compared the experimental results for different geometries of internally fined pipes and compared them with theoretical relationships. They developed an empirical correlation to calculate the friction coefficient and the Nusselt number. The developed criteria differed for high fins and micro-fins, which allowed for the appropriate use of the resulting correlations depending on the geometry of the pipe. Dastmalchi et al. [19] optimized the geometry of micro-finned tubes using numerical methods to increase heat transfer and minimize flow resistance. The research concerned double pipe heat exchangers during the turbulent flow. In the study, they changed the geometric parameters of the pipe, i.e., its internal diameter, number of fins, fin height and their helix angle, for various Reynolds numbers from 3000 to 100,000. Among other findings, the authors noted that the optimal micro-fin height increases with an increasing Reynolds number for all internal pipe diameters considered. In addition, Dastmalchi et al. [20] conducted oil flow tests with low Reynolds numbers, ranging from 100 to 1000, in a tube with internal micro-fins. As before, they investigated an influence of the tube geometry on heat transfer efficiency and pressure drop during flow. The sample results for Re = 1000 show the maximal heat transfer increase of 44%, but also a 69% increase in the friction coefficient, for the flow through micro-finned tubes compared to smooth tube flow.

Filho and Jabardo [21] investigated an influence of the geometry of three selected types of pipes: smooth, micro-finned, and herringbone on the thermo-hydraulic flow characteristics with the use of refrigerants. On the basis of the conducted research, they found that the thermal efficiency was the best for the herringbone pipe; however, it had the highest pressure drop. Raj et al. [22] investigated the properties of various types of internally and externally finned pipes for two fluids: water and 46% glycol solution. They showed that the Prandtl number had a large impact on the intensification of heat transfer, and at the same time, it was dependent on the temperature of fluids.

The present paper attempts to investigate the problem of thermal efficiency of several pipe geometries with micro-fins. By changing the height of inner fins, the researchers searched for the highest values of the PEC coefficient while maintaining the same pumping power. The tests were performed using numerical simulations, whereas for one pipe geometry, experimental tests were carried out as well. The paper presents in detail and discusses an influence of the height of micro-fins on the friction factor, heat transfer, and thermal efficiency of pipes, as well as a correlation of these parameters with mathematical functions. The main goal of this article, which provides the novelty, was the discovery of the best geometry of a pipe in terms of thermal efficiency by determining the highest value of PEC coefficient.

2. Materials and Methods

2.1. Experimental Stand

Before the numerical simulations were carried out, one of the pipe geometries was tested on the experimental stand by Zawadzki et al. [23], and the results obtained from the experiment were used to validate the numerical model. The object of the investigations was an internal finned pipe, produced industrially by KME Germany AG & Co. KG under the working name "TECTUBE fin 12736CV50/65D" and with the dimensions shown in Figure 1, which is used for the production of both single-phase and two-phase heat exchangers, e.g., evaporators in the refrigeration equipment.

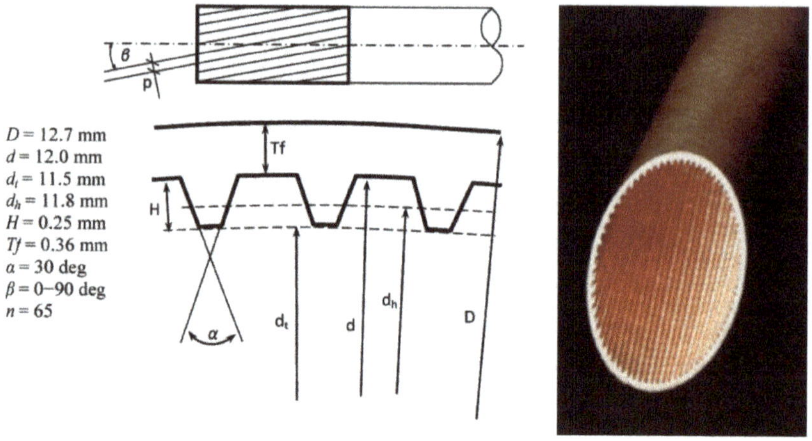

Figure 1. Scheme and dimensions of the micro-finned tube "TECTUBE fin 12736CV50/65D" produced by KME Germany AG & Co. KG company.

A schematic diagram of the experimental stand on which the test was performed is presented in Figure 2. The list of used devices in the experimental stand is introduced in Table 1. It consists of cold and hot water circulations, a data acquisition system, and a test section, which is a double pipe heat exchanger. During the measurements, constant temperatures in the circulation of cold and hot water were maintained at the inlets to the exchanger. In order to obtain thermal-flow characteristics, the water flow rate was gradually changed within the range of Reynolds numbers 10,000–60,000.

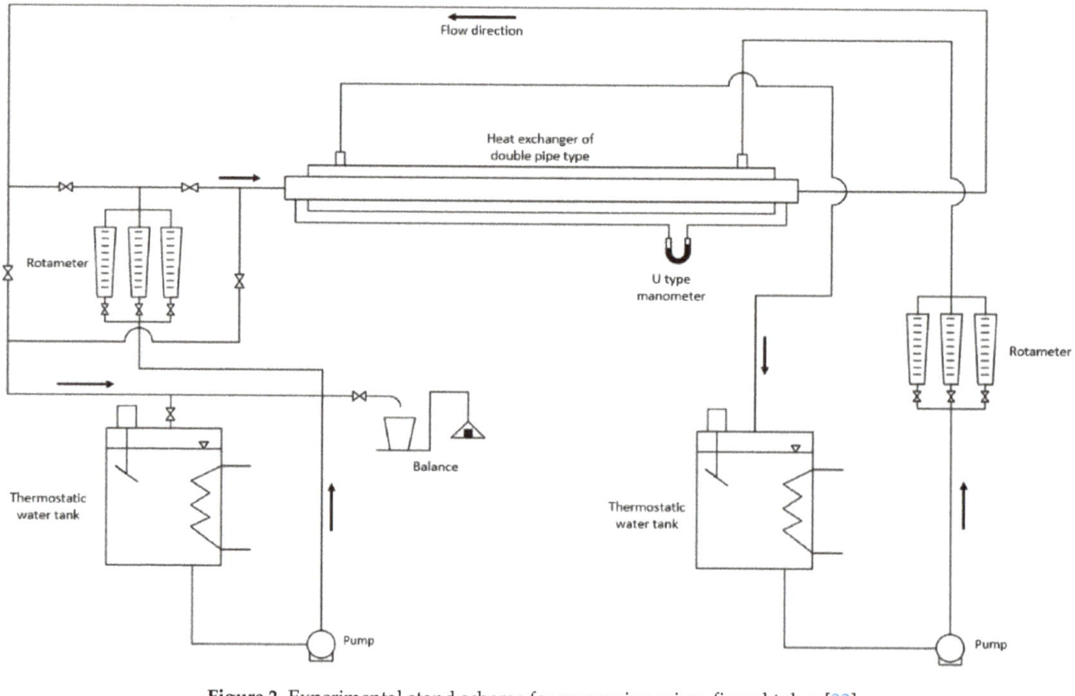

Figure 2. Experimental stand scheme for measuring micro-finned tubes [23].

Table 1. List of content used in experimental stand.

Legend Name	Type	Parameter	Operating Range
Balance	ZAO Gdansk—WT 1002	Weight [g]	10–1000 g
Thermostatic water tank	MLW—U10	Temperature [°C]	20–90 °C
Rotameter	Yokogawa—310142/002	Flow [l/h]	0–25 L/h
Pump	Lowara—2HMS4T/A	Flow rate [l/min]	20–70 L/min
U type manometer	Metalchem—MUR 1200	Pressure [Pa]	782.62–11,739.25 Pa

In Figure 3, the results of the experiment and numerical tests for the micro-finned tube from the previous studies are shown. This procedure also verified the two most popular turbulence models—the classic k-ε and the SST k-ω (Shear Stress Transport). It can be clearly seen that the best fit data and a good agreement with the experimental values for both pressure drop and heat transfer were obtained with the SST k-ω turbulence model. This is the reason why that model was used for further numerical calculations presented in this paper.

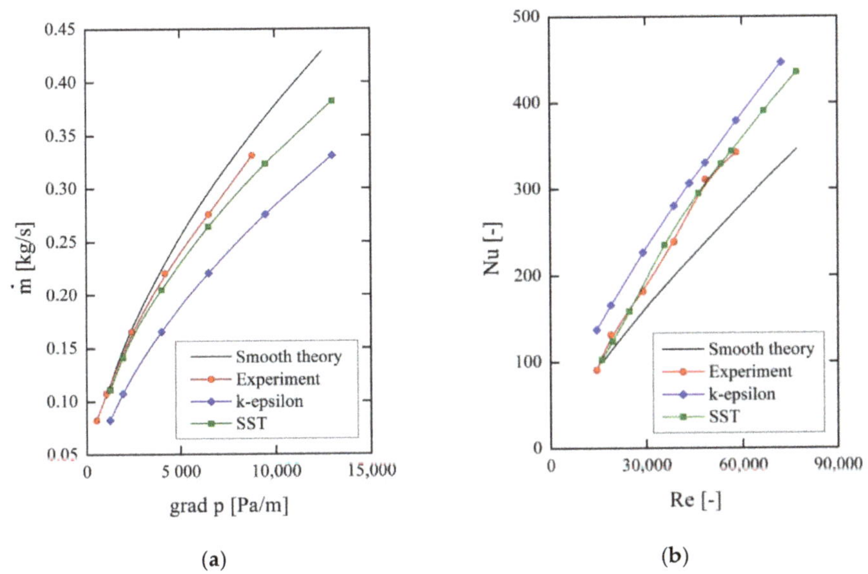

Figure 3. Comparison between results of the simulation and the experiment for the micro-fin tube and the SST and k-ε turbulence model: (**a**) mass flow vs. *grad p*, (**b**) *Nu* vs. *Re* [16,24].

2.2. Geometrical Model

The aim of the numerical calculations was to investigate the thermal-flow characteristics of pipes with internal micro-fins, with a very similar geometry as shown in Figure 1 for the industrial pipe "TECTUBE fin 12736CV50/65D". The only geometric parameter that was changed was the height of the micro-fin H. All other parameters, including the fin helix angle $\alpha = 30°$ and the helical tooth line angle $\beta = 30°$, remained unchanged, as in the mentioned real pipe. Modifications of the tooth height H were made every 0.05 mm in the range of 0.05–0.40 mm, changing its size in relation to the height $H = 0.25$ mm, which the actual pipe has (see Figure 4). The relative roughness of the pipe, which is defined as the ratio of the height of unevenness on the surface of the pipe to its diameter $\varepsilon = H/d$ changes with a change of the micro-fin height. In this case, the "unevenness" are micro-fins of a given height H. Table 2 lists all the fin heights tested, together with the corresponding relative roughness ε. The models of selected cases are shown in Figure 5.

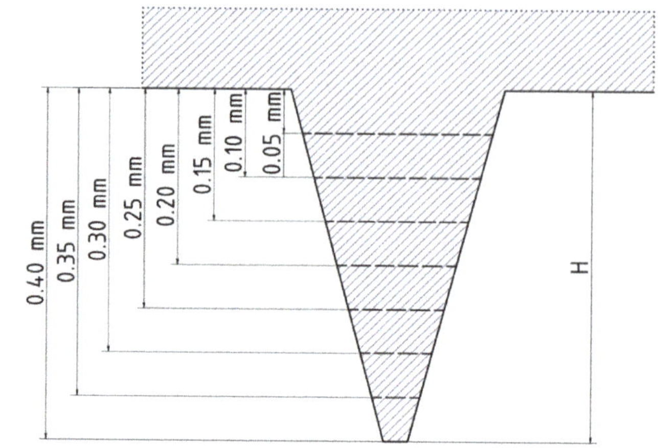

Figure 4. Scheme of changes in the fin height tested in the simulation.

Table 2. Results of the calculations of relative roughness.

H [mm]	**0.05**	**0.10**	**0.15**	**0.20**	**0.25**	**0.30**	**0.35**	**0.40**
ε [-]	0.004	0.008	0.013	0.017	0.021	0.025	0.029	0.033

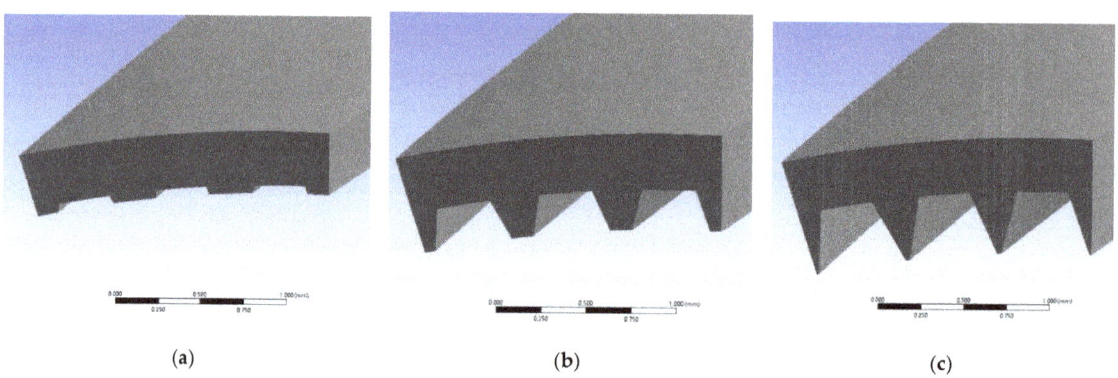

Figure 5. Different height of fins in the tube fragment model. (**a**) $H = 0.05$ mm, (**b**) $H = 0.25$ mm, (**c**) $H = 0.40$ mm.

2.3. Numerical Model

As the computational domain, a part of the spiral extruded pipe geometry was used, with a width of three fins as in Figure 5 and a length corresponding to its single helix pitch through an angle of 360° (Figure 6). The use of the width of three fins was due to the more convenient procedure of generating a computational mesh, which allowed us to maintain its better geometrical quality (cells angles, aspect ratio, etc.) in the central part of the channel near its axis, than for the width of one fin, but also applicable in such geometry. The list of used boundary conditions is introduced in Table 3.

Generally, instead of a full, long pipe, a repeating and periodic fragment of it, representative of the entire channel and reflecting the same heat-flow phenomena, was used for the calculations. This approach to the problem is correct, and it is commonly used in the numerical analysis [25,26], as long as the resulting flow is fully developed, both hydraulically and thermally. It is known that under normal conditions, a fully developed flow can be obtained only on a pipe length equal to approximately 40–50 diameters from its inlet. In the case of using a short repeating domain fragment, appropriate boundary conditions have to be applied to obtain such a flow structure. According to this, translational periodicity was used at the inlet and outlet of the domain, where the fluid flow was forced by a pressure gradient, corresponding to the range of numbers Re = 10,000–100,000. Due to the twist of the micro-fins by an angle of 30°, there is also a component of the rotational velocity during the flow. Therefore, the rotational periodicity was used on the flank surfaces instead of the normal plane of symmetry. The concept of reducing the domain size, and thus also the number of computational mesh nodes, enables a significant reduction in computation time while maintaining high mesh quality and accuracy of results. This method of investigations has also been presented in [24,27,28].

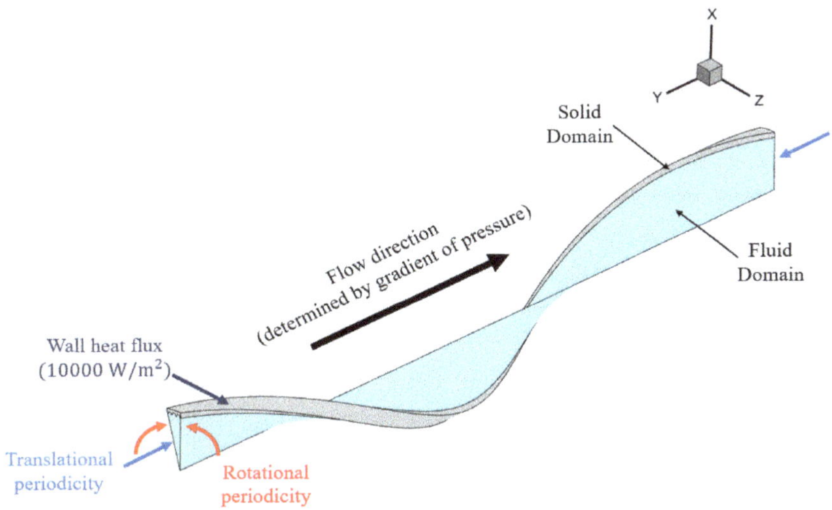

Figure 6. Boundary conditions on the models.

Table 3. Boundary conditions and the parameters values.

Boundary Condition	Description	Parameter	Value/Type
Fluid Domain	Water	Temperature Reference pressure Turbulence	298 K 1 atm SST
Subdomain	Subdomain was set in domain of water. The gradient of pressure determined the flow in Y component.	Gradient of pressure Volumetric heat flux	440–24,779 Pa $-Q_{vol} = \frac{q \cdot A_{wall}}{V_{water}}$ W
Solid Domain	Copper	Temperature	298 K
Wall	Boundary condition set on exterior area of pipe in the form of constant heat flux	Heat flux	10,000 $\frac{W}{m^2}$
Translational periodicity	Translational periodicity set on the inlet and outlet areas of fluid and solid domain.	-	-
Rotational periodicity	Rotational periodicity set on the both sides of fluid and solid domain. As the rotation axis, Global Y was set.	-	-

The pipe segment forming the computational domain was "extended" to the length of a full revolution, as shown along with other boundary conditions in Figure 6. As previously mentioned, the boundary condition forcing the fluid flow was a pressure gradient.

The simulation was conducted in Ansys CFX 2020 R2. The numerical model consisted of two computational domains. In the flow domain, water is defined as the working fluid with an average temperature of 298 K. The solid-state domain was defined as copper, which is the pipe material in the experiment. Additionally, a negative volumetric heat source was set in the fluid domain to obtain a fully thermally developed flow and to calculate the correct temperature field. This solution made it possible to balance the thermal energy supplied to the domain wall as a heat flux of q = 10,000 W/m^2.

The SST k-ω turbulence model is one of the most popular models used in many CFD (Computational Fluid Dynamics) applications. Its main attribute is the ability to solve the viscous sublayer by applying the k-ω model near the wall and the standard k-ε model in the turbulent core area. Switching between the two models is controlled by a special built-in blending function [29,30]. Correct use of the SST model requires several mesh nodes inside the turbulent sublayer to maintain the condition of dimensionless distance $y^+ < 2$ in the entire boundary layer of the computational domain [31]. In the results presented in this work, the maximum value of y^+ did not exceed the mentioned value in any of the geometries.

One of the criteria for the uniqueness of the numerical solution was to achieve the appropriate convergence for residues: momentum, energy, and turbulence. In all simulations, a convergence of 1×10^{-4} or maximal residuals and an order of magnitude lower (1×10^{-5}) for mean RMS (Root-Mean-Square) residuals were obtained. The second criterion for the uniqueness of the solution was the stabilization of flow-thermal parameters, such as velocity, pressure, and temperature, which were monitored both as mean values and in several selected points in the computational domain. The computation process was finished if the above parameters did not change for several consecutive iterations.

Before the actual simulations, checking calculations were performed for different mesh qualities. For further calculations, a structural (in solid) and hexagonal mesh of such density, at which its further densification gives results differing less than 1%, was selected. For all geometries, efforts were made to maintain a computational mesh of approximately the same average cell volume. In the area of the hydraulic boundary layer, the mesh was additionally compacted to obtain the appropriate y^+ value. After the test, it can be concluded that the mesh used and the test results are independent of its density, which is consistent with the recommendations in [32]. The obtained value of the GCI (Grid Convergence Method) parameters in relation to the average temperature is GCI_{fine} = 1.8%, and in case of the wall heat flux, it is GCI_{fine} = 0.001%.

In Figure 7, the mesh used for the numerical calculations for H = 0.25 mm is shown.

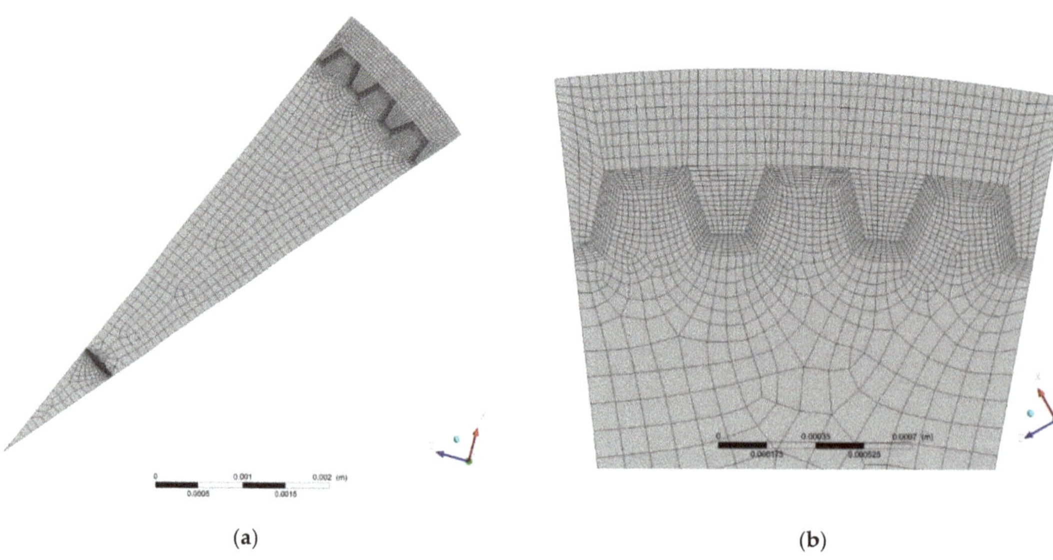

Figure 7. Mesh used in the numerical simulations for H = 0.25 mm: (**a**) full view of mesh; (**b**) mesh region between solid and fluid domain.

3. Results

After performing numerical simulations, the correctness of the obtained results was assessed by analyzing the distribution of selected physical quantities. In Figure 8, results from the numerical simulations are shown. The contours of temperature and vectors of velocity distributions are presented for the cases in Figure 5, respectively.

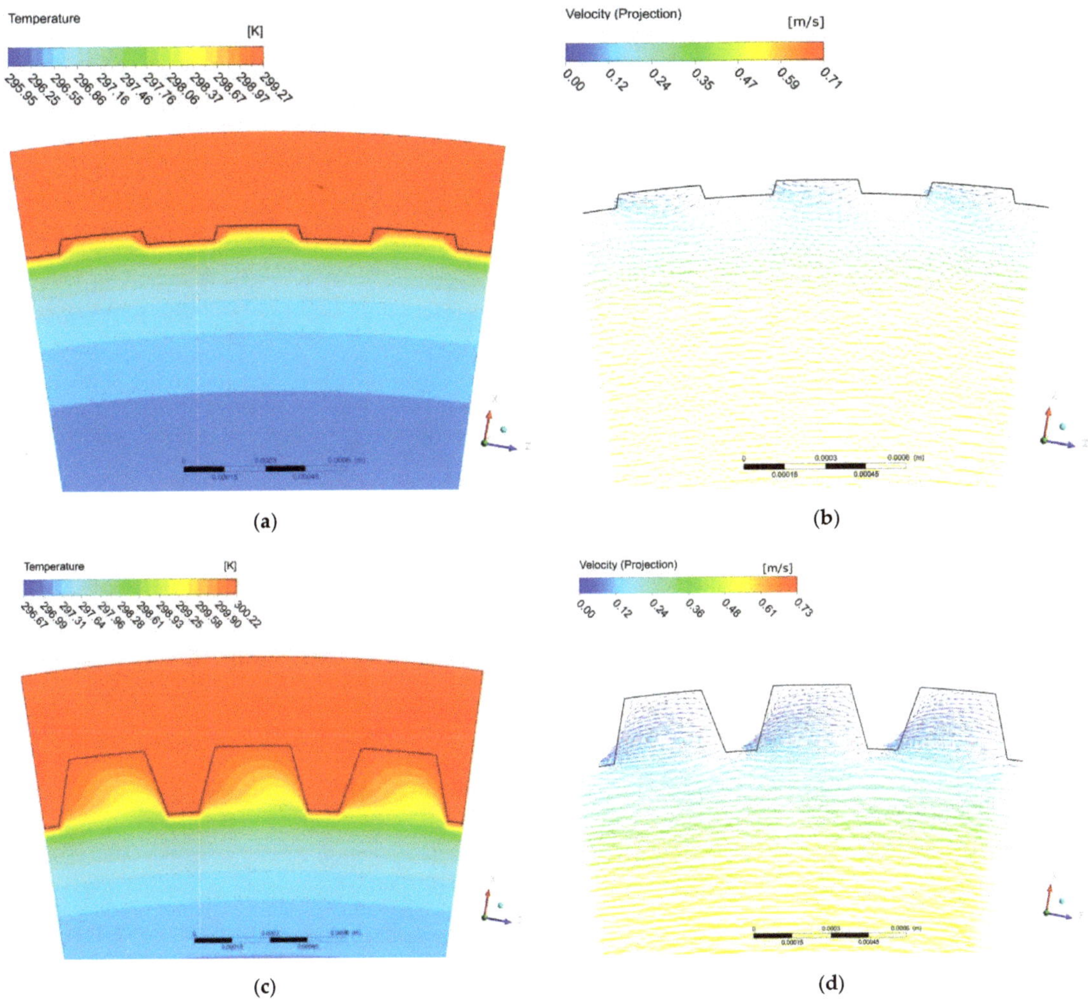

Figure 8. *Cont.*

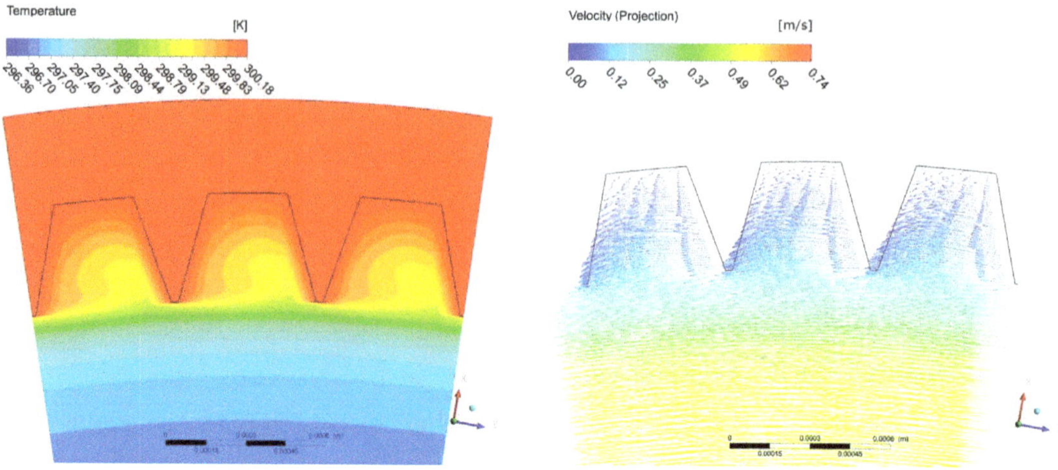

(e) (f)

Figure 8. Results from the numerical simulations: (**a**) contour of temperature for H = 0.05 mm; (**b**) vectors of velocity (tangential projection) for H = 0.05 mm; (**c**) contour of temperature for H = 0.25 mm; (**d**) vectors of velocity (tangential projection) for H = 0.25 mm; (**e**) contour of temperature for H = 0.40 mm; (**f**) vectors of velocity (tangential projection) for H = 0.40 mm.

3.1. Data Processing

Based on the experimental data obtained from the examination of the industrial pipe TECTUBE fin 12736CV50/65D, the numerical model was validated and verified. The most important parameters in terms of the performance of the analyzed pipes are the friction factor and the heat transfer coefficient, as expressed by the Nusselt number. To create such characteristics, it is necessary to define basic flow parameters such as velocity, temperature, and pressure drop for each pipe geometry, which were obtained as a result of the computer simulations.

In numerical calculations, the pressure gradient described by Equation (1) was used to force the flow:

$$grad\ p = \frac{\Delta p}{L} = f \cdot \frac{u_{av}^2 \cdot \rho}{2 \cdot d}. \tag{1}$$

For each tested geometry, the friction factor was calculated using the Darcy–Weisbach Equation (2), which is a modification of Equation (1):

$$f = \frac{2 \cdot \Delta p \cdot d}{\rho \cdot u_{av}^2 \cdot L}. \tag{2}$$

The theoretical value of the friction factor for a plain pipe, as the reference level for the numerical results, was calculated from the Blasius correlation (3):

$$f_{plain} = 0.3164 \cdot Re^{-0.25}. \tag{3}$$

Similarly, for the plain pipe, the Nusselt number was calculated from the well-known Dittus–Boelter [33,34] Equation (4):

$$Nu_{plain} = 0.023 \cdot Re^{0.8} \cdot Pr^{0.4}. \tag{4}$$

For the investigated cases of a finned tube, the formula (5) was used to calculate the Nusselt number:

$$Nu = \frac{h \cdot d}{k}. \tag{5}$$

The heat transfer coefficient h used in Equation (5) was determined from the formula for heat flux (6), from the obtained results of the numerical tests.

$$h = \frac{q}{T_{wall} - T_{bulk}} \tag{6}$$

3.2. Friction Factor

The results of numerical simulations, in the form of the characteristics of the friction factor $f(Re)$, are shown in Figure 9. The graph also features a curve for a smooth pipe, which was calculated from the Blassius correlation (3), to show the reference level.

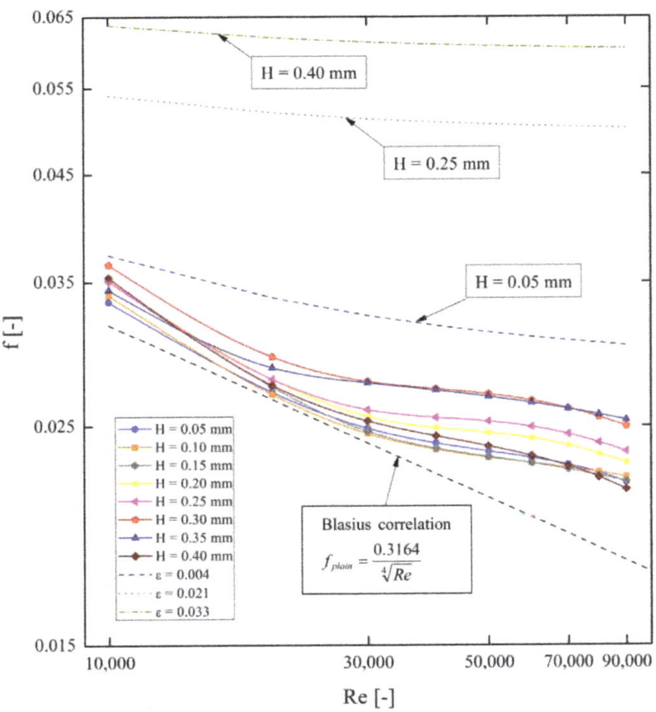

Figure 9. Results from the numerical simulations for various fin heights in the tube—f vs. Re.

It is difficult to find any simple approximation function that would express the variation in the height of micro-fins with a mathematical formula. For this reason, the friction factor was approximated separately for each pipe geometry, using an exponential third-order decay function (7), which makes the best fitting of the research results. The calculated correlation coefficients of the function are given in Table 4.

$$f = y_0 + A_1 \cdot \exp\left(\frac{Re}{t_1}\right) + A_2 \cdot \exp\left(\frac{Re}{t_2}\right) + A_3 \cdot \exp\left(\frac{Re}{t_3}\right) \tag{7}$$

Table 4. Fitting parameters for Equation (7). For all geometries coefficient, $y_0 = 0.0208$.

	$H = 0.05$	$H = 0.10$	$H = 0.15$	$H = 0.20$	$H = 0.25$	$H = 0.30$	$H = 0.35$	$H = 0.40$
A_1	0.02839	−0.2313	2.38×10^{-1}	−0.786	−0.2992	0.04275	0.6984	−0.5991
t_1	−8956	9.99×10^{6}	-1.50×10^{4}	-2.06×10^{4}	-3.03×10^{4}	-7.74×10^{3}	-9.73×10^{4}	3.24×10^{5}
A_2	0.1788	0.03154	−0.2564	0.5124	0.2837	0.3272	−0.6903	0.6042
t_2	1.34×10^{5}	-8.61×10^{3}	-1.76×10^{4}	-1.76×10^{4}	-3.27×10^{4}	-4.41×10^{4}	-9.68×10^{4}	3.33×10^{5}
A_3	−0.1754	0.2348	5.36×10^{-2}	0.3076	0.05362	−0.3259	0.05387	0.04127
t_3	1.30×10^{5}	-9.54×10^{92}	-2.78×10^{4}	-2.55×10^{4}	-1.04×10^{4}	-4.21×10^{4}	-4.53×10^{3}	-6.97×10^{3}

For the micro-fins of height $H = 0.05$ mm, $H = 0.25$ mm, and $H = 0.40$ mm (minimal, medium, and maximal micro-fin height, respectively) the curves were calculated for several ε values from Table 2, using the empirical formula (8) given by Swamee and Jain [35].

In Figure 10, an influence of the pipe geometry on a value of the friction factor $f(H)$ for several Reynolds numbers is shown. The individual curves show the friction factor with respect to micro-fins height H for specific Reynolds numbers, i.e., for the same flow parameters.

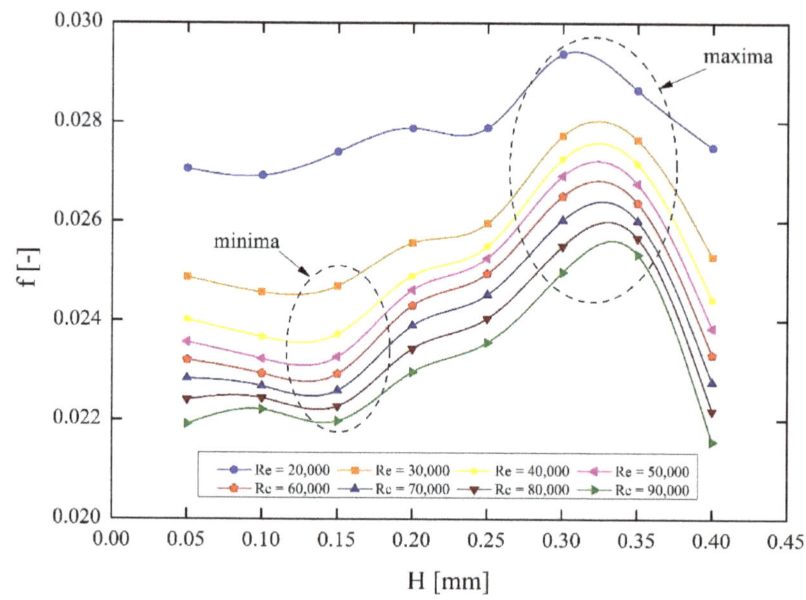

Figure 10. Results from the numerical simulations of the friction factor for various micro-fin heights in the tube—f vs. H (for a specific value of Reynolds Number).

An analysis of the obtained results allowed us to observe a large deviation from the characteristics of rough pipes presented in the Moody diagram [36]. In the case of pipes, the relative roughness is defined as a ratio of the height of unevenness to the diameter, and for the tested geometries, it is closely related to the height of the micro-fins. For the presented channels, it is 0.004–0.033, which was presented in Table 2.

$$f = \frac{0.25}{\left(\log\left(\frac{\varepsilon}{3.7} + \frac{5.74}{Re^{0.9}}\right)\right)^2} \quad (8)$$

3.3. Heat Transfer

In Figure 11, thermal characteristics for all tested geometries are shown, in the form of the $Nu(Re)$ function, and one curve for a plain pipe, calculated from (4) as the reference level.

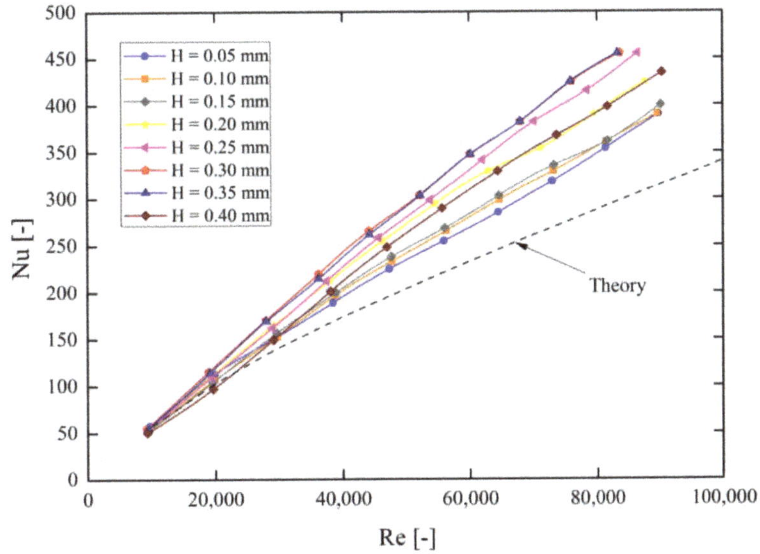

Figure 11. Results from the numerical simulations for various fin heights in the tube—*Nu* vs. *Re*.

Between the functions of Nusselt numbers shown in Figure 11, there is no simple geometric dependence (similar to the friction factor); nevertheless, certain mathematical functions can be adjusted to these data.

For the correlation of the Nusselt number function for the studied cases, the most suitable formula is the exponential function (9):

$$Nu = A \cdot Re^{B} \cdot Pr^{0.4}. \tag{9}$$

In Table 5, values of the A and B coefficients for each fin height are shown.

Table 5. Fitting parameters for Equation (9).

	$H = 0.05$	$H = 0.10$	$H = 0.15$	$H = 0.20$	$H = 0.25$	$H = 0.30$	$H = 0.35$	$H = 0.40$
A	0.014370	0.013610	0.013760	0.013940	0.006390	0.009170	0.007314	0.006306
B	0.8402	0.8470	0.8475	0.8544	0.9301	0.9014	0.9218	0.9242

3.4. PEC (Performance Evaluation Criteria)

The *PEC* (Performance Evaluation Criteria) thermal efficiency rating for the tested pipes was calculated from Equation (10). This parameter combines the Nusselt number and the friction factor obtained from the tests of channels with geometries other than the smooth pipe, and the *Nu* and f numbers calculated for a smooth pipe with the same Reynolds number. The *PEC* is an indicator of how the heat transfer will increase in the tested channel relative to a plain pipe for the same pumping power. *PEC* values above 1 indicate a greater impact of intensification of the heat transfer in the pipe than flow resistance, while values below 1 indicate greater flow resistance in relation to the benefit obtained from intensifying the heat transfer for the tested geometry (Figure 12).

$$PEC = \frac{\frac{Nu}{Nu_{plain}}}{\left(\frac{f}{f_{plain}}\right)^{\frac{1}{3}}} \quad (10)$$

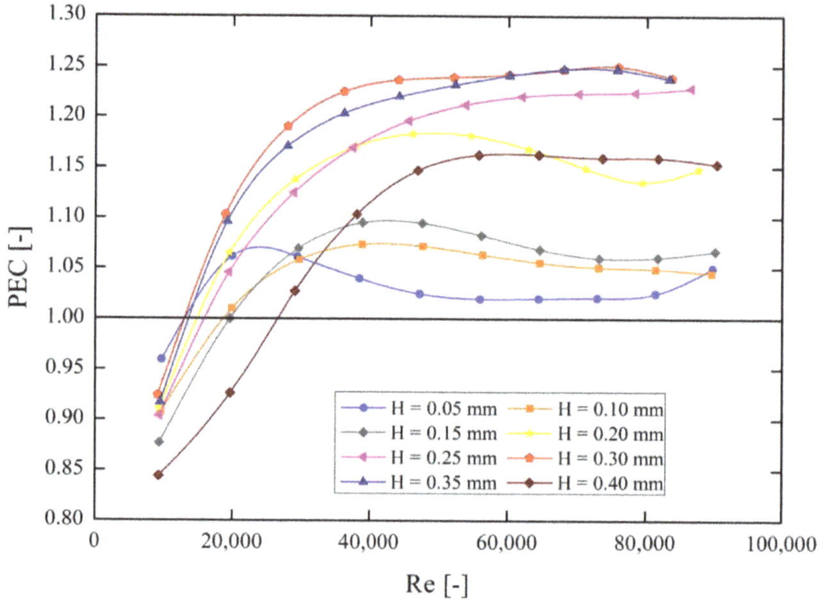

Figure 12. PEC (Performance Evaluation Criteria) coefficient of effectiveness for each tested model—*PEC* vs. *Re*.

4. Discussion

A validation of the numerical model with the experimental data from the tests for a pipe with the fin height of H = 0.25 mm was presented. When analyzing the results, the discrepancies between the experiment and the obtained numerical results found are as follows: for the friction factor—maximum 7%; for a Nusselt number—maximum 12%, as indicated in Figure 3.

Using the Blasius formula determining the friction factor, a comparison of the obtained numerical results for different micro-fin heights with the plain pipe was made (Figure 9). For each case of micro-finned pipes, the obtained values are greater than the values for a smooth pipe, which indicates the physical correctness of the results obtained. For Re = 10,000–25,000, the lowest values of the friction factor are achieved by the pipe with micro-fins of the height of H = 0.05 mm; whereas the highest values for the height of micro-fins are H = 0.30 mm. All curves are rather regular and straight lines on a logarithmic plot. For Reynolds numbers above 25,000, the lowest values of the friction factor are achieved by tubes with micro-fins of the height of H = 0.10 mm and H = 0.15 mm; in turn, the highest values are achieved by two geometries for pipes with micro-fins of the height of H = 0.30 mm and H = 0.35 mm. In this range, functions change their character, and it is difficult to find a regularity in their position.

Considering the dependence of the friction factor in relation to the fin height for different Reynolds numbers, its value decreases with an increase in the Reynolds number for each flow channel. On the basis of Figure 10, apart from minor irregularities in the characteristics, one can notice their quite clear trend. For the micro-fins height H = 0.30–0.35 mm, a clear maximum can be seen for all the characteristics, which means that these pipe geometries give the highest flow resistance. On the other hand, for the height of approximately H = 0.15 mm, one can observe a "slight" minimum of these curves and a decrease in the value for the highest height of micro-fins H = 0.40 mm. The decrease in the value of the friction factor for H = 0.40 mm is probably due to low thickness of the fin compared to other dimensions (Figure 5). At the same time, a small contact area with the main turbulent core, where the highest flow velocities occur, exerts also an influence on a decrease in the friction factor.

Each tested tube had a different relative roughness related to the height of the micro-fins. In Figure 9, a complete discrepancy between the positions of the curves obtained from the numerical simulations and those calculated theoretically on the basis of the well-known formula (8) was shown for the same relative roughness. One can notice that it is not possible to calculate the friction factors from Equation (8) for the tested geometries as the model derives significantly different values than the ones obtained in the tests. Therefore, one of the fundamental conclusions resulting from the numerical tests carried out is a lack of correlation of the friction factor between the theoretically calculated (for irregular roughness) and the one obtained from the tests (for the same roughness but with regular shapes). The same fact was recognized by Wang et al. in their research [37].

When analyzing the obtained results of the heat transfer intensity for the geometries under investigation, several phenomena can be observed. The presented results show an irregular order of the Nusselt number characteristics for various fluid flow rates (Figure 11). For Reynolds numbers above 20,000, pipes with micro-fins having the height of H = 0.20 mm and higher achieve significantly larger values of the Nusselt number than for the plain pipe, compared to the cases with micro-fins below H = 0.20 mm, for which the characteristics are very similar to those of the smooth pipe. In the entire range of Reynolds numbers, the highest values of Nusselt numbers are achieved by pipes with micro-fin heights equal to H = 0.30 mm and H = 0.35 mm, and the same pipes for which the highest friction factor was observed. The irregular position of these characteristics indicates a significant influence of turbulences in the vicinity of the laminar boundary layer and the size of the heat transfer surface related to the height of the micro-fins.

As can be seen in Figure 12, in the range of low Reynolds numbers up to approximately 25,000, the *PEC* value of less than 1 was observed for all geometries. It means that using these pipes in this flow range is less efficient than using the regular plain pipe. For Reynolds numbers higher than 25,000, all characteristics are higher than 1, and it is within this range that the use of such pipes is justified. The highest *PEC* values, up to 1.25, are achieved by tubes with the micro-fin height of 0.30 and 0.35 mm for Reynolds numbers above 50,000. A characteristic feature of these geometries is a virtually constant value of this coefficient in the given Reynolds number range. Therefore, these micro-fins heights can be considered the most optimal for thermal-flow applications among all numerically tested in this work.

5. Conclusions

Based on the numerical investigations presented, the most important conclusions of this work can be drawn as follows:

- A numerical model of the tested pipes was built and verified with the experimental data.
- The mathematical correlations describing the nature of changes in the friction factor and the Nusselt number as a function of the Reynolds number were determined for the examined micro-fins heights.
- Using the PEC (Performance Evaluation Criteria) method of assessing the thermal efficiency of flow channels, the highest values were observed for micro-fins with the height of H = 0.30 mm and H = 0.35 mm.

- The theoretical formulas for the friction factor for rough pipes (Moody's diagram) were not compatible with the obtained numerical results for the same relative roughness, but with a regular shape.
- For Reynolds numbers below 20,000, the use of the investigated type of pipe micro-finning is unjustified in terms of heat transfer efficiency.

Author Contributions: Conceptualization, P.B.J., M.J.K., A.R., B.W., and D.O.; methodology, D.O.; validation, M.J.K., A.R., and B.W.; investigation, M.J.K., A.R., and B.W.; writing—original draft preparation, M.J.K., A.R., and B.W.; writing—review and editing, P.B.J. and A.G.; supervision, P.B.J. and A.G.; All authors have read and agreed to the published version of the manuscript.

Funding: This research received no external funding.

Institutional Review Board Statement: Not applicable.

Informed Consent Statement: Not applicable.

Data Availability Statement: Not applicable.

Acknowledgments: This article has been completed while the second, third, and fourth authors were Doctoral Candidates in the Interdisciplinary Doctoral School at the Lodz University of Technology, Poland. We would like to thank Malgorzata Jozwik for her significant linguistic help during the preparation of the manuscript.

Conflicts of Interest: The authors declare no conflict of interest.

Nomenclature

A_{wall}	exterior area of the pipe [m^2]
d	diameter [mm]
f	friction factor [-]
f_{plain}	friction factor for plain tube [-]
GCI_{fine}	fine-grid convergence index [%]
h	heat transfer coefficient [W/m^{2}K]
H	height of fin [mm]
k	thermal conductivity [W/mK]
L	length of the pipe [m]
Nu	Nusselt number [-]
Nu_{plain}	Nusselt number for plain tube [-]
PEC	performance evaluation criteria [-]
Δp	pressure drop [Pa]
Pr	Prandtl number [-]
Q_{vol}	volumetric heat flux [W/m^3]
q	wall heat flux [W/m^2]
Re	Reynolds number [-]
T_{wall}	average temperature in pipe [K]
T_{bulk}	minimal temperature in water [K]
u_{av}	average velocity [m/s]
V_{water}	volume of water domain [m^3]
α	micro-fin angle [°]
β	helical angle of micro-fin [°]
ρ	density [kg/m^3]
ε	relative roughness [-]

References

1. Wijayanta, A.T.; Pranowo; Mirmanto; Kristiawan, B.; Aziz, M. Internal flow in an enhanced tube having square-cut twisted tape insert. *Energies* **2019**, *12*, 306. [CrossRef]
2. Wijayanta, A.T.; Aziz, M.; Kariya, K.; Miyara, A. Numerical study of heat transfer enhancement of internal flow using double-sided delta-winglet tape insert. *Energies* **2018**, *11*, 3170. [CrossRef]

3. Jasiński, P.B. Numerical study of the thermo-hydraulic characteristics in a circular tube with ball turbulators. Part 1: PIV experiments and a pressure drop. *Int. J. Heat Mass Transf.* **2014**, *74*, 48–59. [CrossRef]
4. Jasiński, P.B. Numerical study of the thermo-hydraulic characteristics in a circular tube with ball turbulators. Part 2: Heat transfer. *Int. J. Heat Mass Transf.* **2014**, *74*, 473–483. [CrossRef]
5. Jasiński, P.B. Numerical study of thermo-hydraulic characteristics in a circular tube with ball turbulators. Part 3: Thermal performance analysis. *Int. J. Heat Mass Transf.* **2017**, *107*, 1138–1147. [CrossRef]
6. Arjmandi, H.; Amiri, P.; Pour, M.S. Geometric optimization of a double pipe heat exchanger with combined vortex generator and twisted tape: A CFD and response surface methodology (RSM) study. *Therm. Sci. Eng. Prog.* **2020**, *18*. [CrossRef]
7. Patil, M.S.; Seo, J.-H.; Kang, S.-J.; Lee, M.-Y. Review on synthesis, thermo-physical property, and heat transfer mechanism of nanofluids. *Energies* **2016**, *9*, 840. [CrossRef]
8. Shajahan, M.I.; Michael, J.J.; Arulprakasajothi, M.; Suresh, S.; Nasr, E.A.; Hussein, H.M.A. Effect of Conical Strip Inserts and ZrO 2 / DI-Water Nanofluid on Heat Transfer Augmentation: An experimental study. *Energies* **2020**, *13*, 4554. [CrossRef]
9. Kristiawan, B.; Wijayanta, A.T.; Enoki, K.; Miyazaki, T.; Aziz, M. Heat transfer enhancement of TiO2/water nanofluids flowing inside a square minichannel with a microfin structure: A numerical investigation. *Energies* **2019**, *12*, 3041. [CrossRef]
10. Asirvatham, L.G.; Vishal, N.; Gangatharan, S.K.; Lal, D.M. Experimental Study on Forced Convective Heat Transfer with Low Volume Fraction of CuO/Water Nanofluid. *Energies* **2009**, *2*, 97–119. [CrossRef]
11. Mann, G.W.; Eckels, S. Multi-objective heat transfer optimization of 2D helical micro-fins using NSGA-II. *Int. J. Heat Mass Transf.* **2019**, *132*, 1250–1261. [CrossRef]
12. Ji, W.-T.; Zhang, D.-C.; He, Y.-L.; Tao, W.-Q. Prediction of fully developed turbulent heat transfer of internal helically ribbed tubes? An extension of Gnielinski equation. *Int. J. Heat Mass Transf.* **2012**, *55*, 1375–1384. [CrossRef]
13. Li, P.; Campbell, M.; Zhang, N.; Eckels, S.J. Relationship between turbulent structures and heat transfer in microfin enhanced surfaces using large eddy simulations and particle image velocimetry. *Int. J. Heat Mass Transf.* **2019**, *136*, 1282–1298. [CrossRef]
14. He, G.-D.; Fang, X.-M.; Xu, T.; Zhang, Z.-G.; Gao, X.-N. Forced convective heat transfer and flow characteristics of ionic liquid as a new heat transfer fluid inside smooth and microfin tubes. *Int. J. Heat Mass Transf.* **2015**, *91*, 170–177. [CrossRef]
15. Brognaux, L.; Webb, R.L.; Chamra, L.M.; Chung, B.Y. Single-phase heat transfer in micro-fin tubes. *Int. J. Heat Mass Transf.* **1997**, *40*, 4345–4357. [CrossRef]
16. Jasiński, P. Numerical optimization of flow-heat ducts with helical micro-fins, using entropy generation minimization (EGM) method. In *Recently Advances in Fluid Mechanics and Heat & Mass Transfer, Proceedings of the 9th IASME/WSEAS International Conference on Fluid Mechanics and Aerodynamic Engineering FMA'11, Proceedings of the 9th IASME/WSEAS International Conference on HTE'11, Florence, Italy, 23–25 August 2011*; WSEAS: Athens, Greece, 2011; pp. 47–54.
17. Tang, W.; Li, W. Frictional pressure drop during flow boiling in micro-fin tubes: A new general correlation. *Int. J. Heat Mass Transf.* **2020**, *159*, 120049. [CrossRef]
18. Jensen, M.K.; Vlakancic, A. Technical Note Experimental investigation of turbulent heat transfer and fluid flow in internally finned tubes. *Int. J. Heat Mass Transf.* **1999**, *42*, 1343–1351. [CrossRef]
19. Dastmalchi, M.; Sheikhzadeh, G.A.; Arefmanesh, A. Optimization of micro-finned tubes in double pipe heat exchangers using particle swarm algorithm. *Appl. Therm. Eng.* **2017**, *119*, 1–9. [CrossRef]
20. Dastmalchi, M.; Arefmanesh, A.; Sheikhzadeh, G. Numerical investigation of heat transfer and pressure drop of heat transfer oil in smooth and micro-finned tubes. *Int. J. Therm. Sci.* **2017**, *121*, 294–304. [CrossRef]
21. Filho, E.P.B.; Jabardo, J.M.S. Experimental study of the thermal hydraulic performance of sub-cooled refrigerants flowing in smooth, micro-fin and herringbone tubes. *Appl. Therm. Eng.* **2014**, *62*, 461–469. [CrossRef]
22. Raj, R.; Lakshman, N.S.; Mukkamala, Y. Single phase flow heat transfer and pressure drop measurements in doubly enhanced tubes. *Int. J. Therm. Sci.* **2015**, *88*, 215–227. [CrossRef]
23. Zawadzki, A.; Plocek, M.; Kapusta, T.; Kasieczka, W. Heat transfer and friction factor characteristics of single-phase flow through a circular, internally micro-finned, horizontal tube fitted with twisted tape inserts—experimental investigations. In Proceedings of the XL Refrigeration Days, Poznań, Poland, 15–17 October 2008; pp. 115–124. (In Polish)
24. Jasiński, P. Numerical Study of Friction Factor and Heat Transfer Characteristics for Single-Phase Turbulent Flow in Tubes with Helical Micro-Fins. *Arch. Mech. Eng.* **2012**, *59*, 469–485. [CrossRef]
25. Sobczak, K.; Obidowski, D.; Reorowicz, P.; Marchewka, E. Numerical investigations of the savonius turbine with deformable-blades. *Energies* **2020**, *13*, 3717. [CrossRef]
26. Obidowski, D.; Stajuda, M.; Sobczak, K. Efficient Multi-Objective CFD-Based Optimization Method for a Scroll Distributor. *Energies* **2021**, *14*, 377. [CrossRef]
27. Fodemski, T.; Górecki, G.; Jasiński, P. Corrugated channels heat transfer efficiency Analysis based on velocity fields resulting from computer simulation and PIV Measurements. In Proceedings of the 8th International Conference on Heat Transfer, Fluid Mechanics and Thermodynamics, HEFAT, Pointe Aux Piments, Mauritius, 11–13 July 2011.
28. Li, X.-W.; Meng, J.-A.; Guo, Z.-Y. Turbulent flow and heat transfer in discrete double inclined ribs tube. *Int. J. Heat Mass Transf.* **2009**, *52*, 962–970. [CrossRef]
29. Di Piazza, I.; Ciofalo, M. Numerical prediction of turbulent flow and heat transfer in helically coiled pipes. *Int. J. Therm. Sci.* **2010**, *49*, 653–663. [CrossRef]

30. Eiamsa-Ard, S.; Wongcharee, K.; Sripattanapipat, S. 3-D Numerical simulation of swirling flow and convective heat transfer in a circular tube induced by means of loose-fit twisted tapes. *Int. Commun. Heat Mass Transf.* **2009**, *36*, 947–955. [CrossRef]
31. Manual ANSYS-CFX, Release 2020 R2. Available online: http://www.ansys.com (accessed on 15 July 2020).
32. Celik, I.B.; Ghia, U.; Roache, P.J.; Freitas, C.J.; Coleman, H.; Raad, P.E. Procedure for estimation and reporting of uncertainty due to discretization in CFD applications. *J. Fluids Eng. Trans. ASME* **2008**, *130*, 78001–78004. [CrossRef]
33. Bejan, A. *Convection Heat Transfer*, 4th ed.; John Wiley & Sons, Inc.: Hoboken, NJ, USA, 2013. [CrossRef]
34. Holman, J.P. *Heat Transfer*, 10th ed.; McGraw-Hill, Inc.: New York, NY, USA, 2010.
35. Swamee, P.K.; Jain, A.K. Explicit Equations for Pipe-Flow Problems. *J. Hydraul. Div.* **1976**, *102*, 657–664. [CrossRef]
36. Moody, L.F. Friction Factors for Pipe Flow. *Trans. Am. Soc. Mech. Eng.* **1944**, *66*, 671–681.
37. Wang, C.C.; Chiou, C.B.; Lu, D.C. Single-phase heat transfer and flow friction correlations for microfin tubes. *Int. J. Heat Fluid Flow* **1996**, *17*, 500–506. [CrossRef]

energies

MDPI

Article

State-of-the-Art Review of Effervescent-Swirl Atomizers

Krystian Czernek [1,*], Michał Hyrycz [2], Andżelika Krupińska [2], Magdalena Matuszak [2], Marek Ochowiak [2], Stanisław Witczak [1] and Sylwia Włodarczak [2]

1 Department of Process and Environmental Engineering, Faculty of Mechanical Engineering, Opole University of Technology, ul. Prószkowska 76, 45-758 Opole, Poland; s.witczak@po.edu.pl

2 Department of Chemical Engineering and Equipment, Poznan University of Technology, 60-965 Poznan, Poland; hyryczmichal@gmail.com (M.H.); andzelika.krupinska@put.poznan.pl (A.K.); magdalena.matuszak@put.poznan.pl (M.M.); marek.ochowiak@put.poznan.pl (M.O.); sylwia.wlodarczak@put.poznan.pl (S.W.)

* Correspondence: k.czernek@po.edu.pl; Tel.: +48-77-449-8778

Abstract: This paper presents issues in the field of theory, construction, calculations, as well as the design of effervescent-swirl atomizers. The results of experimental studies of spraying liquids with different physico-chemical properties for this type of atomizers are discussed. Effervescent-swirl atomization is a complex process and its mechanism is not fully understood. Therefore, the purpose of the manuscript is the complexity of the atomization process and its mechanism as well as the influence of individual parameters on its efficiency were thoroughly analyzed. The analyzed parameters include: atomizer design, outlet shape, gas and liquid flow rate, injection pressure, physicochemical properties of the atomized liquid, pressure drop, outflow coefficient, spray angle, quantitative droplet distributions, and average droplet diameter. Moreover, in the work, on the basis of the literature review, the results of the research related to, inter alia, the phenomenon of air core formation and the influence of a number of parameters on the efficiency of the atomization process are analyzed. The literature review included in the work makes it possible to better understand the atomization process carried out in effervescent-swirl atomizers, and also provides better design criteria and analysis of the efficiency of the tested devices. The article presents correlation equations covering the basic features of the atomization process, which relate a large number of parameters influencing the efficiency of this process and the character of the sprayed liquid, which may be useful in design practice.

Keywords: atomization; micro- and macro-parameters of the atomized liquid; mechanism of effervescent-swirl atomization; efficiency of atomization process; effervescent-swirl atomizer

Citation: Czernek, K.; Hyrycz, M.; Krupińska, A.; Matuszak, M.; Ochowiak, M.; Witczak, S.; Włodarczak, S. State-of-the-Art Review of Effervescent-Swirl Atomizers. *Energies* **2021**, *14*, 2876. https://doi.org/10.3390/en14102876

Academic Editors: Felix Barreras, Goodarz Ahmadi, Kiao Inthavong and Pouyan Talebizadeh Sardari

Received: 12 April 2021
Accepted: 13 May 2021
Published: 16 May 2021

1. Introduction

Devices called swirl flow atomizers are widely used in many industries (for example, in the processes of combustion, painting, fire suppression, and air conditioning). When designing atomizers of this type, it is necessary to analyze the influence of its geometry and flow on the atomization process (among others, on the thickness of the liquid film, the flow rate, droplet diameter, spray angle). It is assumed that the internal flow in the atomizer is treated as a two-phase countercurrent flow, which in turn makes detailed analysis of the atomization process quite complex. This work describes both the design of effervescent and effervescent-swirl atomizers, as well as the results of experimental research together with numerical modeling taking into account the most important quantities describing the atomization process. The paper also presents the influence of the spraying device design and the physicochemical quantities of the atomized liquid and the environment on the atomization process, analyzing the resulting air core and the micro- and macro-parameters of the atomized liquid.

The current state of knowledge on liquids atomization using swirl motion, makes it necessary to plan and conduct scientific research, most often of an experimental nature.

Carrying out research taking into account a wide range of changes in the properties of the liquid, the conditions of the conducted process or the design of atomizers, allows to determine the factors that have a clear impact on the form of the atomized liquid stream [1–4].

It is worth noting that the design of atomizers relies heavily on experimental data, and the design process is based on several stages. In the first stage, the preliminary technical documentation is made, then the atomizer is constructed, and finally, experimental tests should be carried out and the obtained test results have to be analyzed. The analysis of the so far published works devoted to the problem of liquid atomization using the phenomenon of swirl motion proves that the basic calculations performed for single-stage swirl atomizers with simple structure are quite well known. Therefore, when designing these devices, it is allowed to use theoretical formulas. However, it should be remembered that these formulas do not take into account all design features and properties of the sprayed liquid [5,6]. Missing data should be estimated on the basis of the analysis of the results obtained with experimental methods. This is due to the fact that the flow in two-phase atomizers is ambiguous and complex, which results, among others, from interactions between the gas and liquid phases. Despite the fact that the number of scientific publications on this topic has increased in recent years, there is still no work that would allow to summarize the achievements in this field. The analysis of the literature shows that most of the works are illustrative, fragmentary, and do not cover the comprehensively discussed problem (because they concern only specific design solutions of spraying systems) [7].

In the case of swirl flow atomizers, at least one of the factors (gas or liquid) undergoes swirling. These atomizers are characterized by a very good quality of atomization, which is achieved as a result of supplying a single thin film or more thin streams of liquid to the swirled stream of gas. In general, air flow atomizers are classified as external mixing atomizers, where the liquid takes the form of a stream or film before contact with the gas flowing through it. On the other hand, in the case of atomizers with internal mixing, the contact between the sprayed liquid and the flowing gas takes place inside the device [6,8]. Another classification of the atomizers concerns the form created by a liquid when in contact with gas, as a result of which the atomizers are divided into the jet stream and film type atomizers [6].

Figure 1 presents the factors showing the effect on the parameters of the sprayed stream as a result of the atomization process using swirl motion. However, these parameters are independent of each other and it is thanks to them that it is possible to optimize the spraying process. The most frequently analyzed parameters enabling the generation of an aerosol with the desired characteristics include: mass flow rate of gas and liquid and their ratio, as well as the pressure of individual factors (operating parameters). The above-mentioned parameters can be modified while the device is running. The independent parameters concerning the sprayed liquid are, first of all, the Newtonian or non-Newtonian nature of the sprayed liquid, the physico-chemical properties of the liquid, as well as the single-component nature or the degree of complexity of the sprayed liquid. Figure 1 also includes the quantities that have a clear impact on the parameters of the sprayed liquid (describing the internal geometry of the atomizer) [7,8].

Swirl motion atomizers have a number of advantages, which leads to their widespread use in the energy, machinery, food, pharmaceutical, agricultural, and forestry industries, as well as in environmental protection [5,6,8–16]. The process of spraying liquids with different properties (Newtonian and non-Newtonian) is often used in various types of agrotechnical treatments (for example in orchard sprayers), in spray drying, in industrial painting, as well as in the production of many different pharmaceutical preparations [17–21]. Swirl type atomizers are often used in oil burners, where single-stage and circulation atomizers with needle closure of the outlet opening are used [6]. On the other hand, effervescent-swirl atomizers are used in gasoline and diesel engines, gas turbines, combustion processes (for example kerosene and heavy fuel oils) [1,8,13,22–39]. These types of atomizers are equipped with additional structural elements such as inserts enabling the control of the size of the generated droplets [40], inserts that induce turbulence in the flow [41] and tangential inlet nozzles [7,18,19]. A previous paper [32] described the possibility of using effervescent atomizers as an Automatic Hand Sanitizer (AHS). This is illustrated by the wide field of application of this type of construction: in schools, workplaces, and health care facilities.

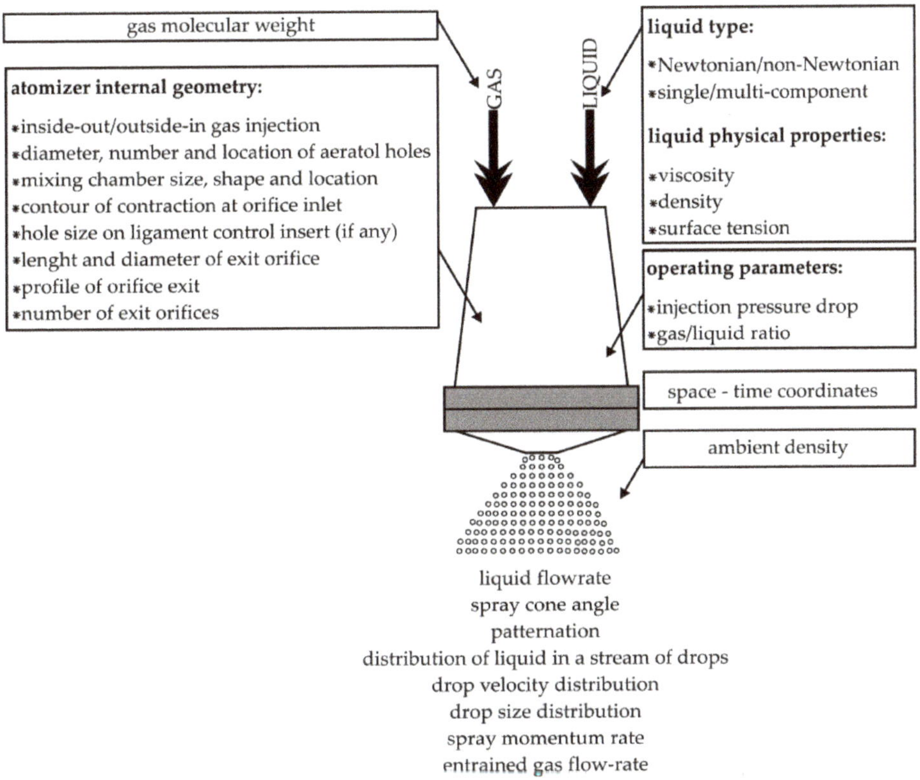

Figure 1. Graphical representation of the liquid atomization process, taking into account the influence of individual parameters [7,8].

The process of the effervescent-swirl atomization is a complex process when its mechanism is not fully analyzed and understood. In view of the above, the aim of this article is to explore the complexity of the atomization process and its mechanism, as well as the influence of certain parameters on its efficiency, which were thoroughly analyzed. The manuscript includes an introduction and conclusion, along with chapters on design and characterization of effervescent-swirl atomizers including building, construction and design, liquid flow structures, discharge coefficient, spray angle, droplet diameter.

2. Design and Characteristics of Effervescent-Swirl Atomizers

2.1. Building, Construction, and Design

The concept of the effervescent-swirl atomizer construction together with the results of the conducted experimental tests are presented in the work [1]. The analysis of the results proves that the application of the new design of the effervescent-swirl atomizer (where the movement of the liquid and the gas-liquid mixture is swirling) leads to an extension of the time of direct contact of both phases, which in turn contributes to the improvement of the spray quality. In addition, the results of the experimental studies included in [7,20] showed that the swirl flow for the gas-liquid system leads to the formation of a more homogeneous gas-liquid system in the mixing chamber, and thus also to a smaller mean diameter of the droplets than in the case of axial flow. As there are relatively few materials in the literature on the subject of effervescent atomization using swirl motion, this chapter also mentions materials related to effervescent atomization, from which effervescent-swirl atomizers are derived.

In the literature on the subject, the issue of effervescent atomizers comes in two design variants: inside-out and outside-in. Both types differ in the way gas is dosed to the system [8,42,43]. An example of the construction of both atomizers is shown in Figure 2 [20,44]. In inside-out atomizers, the gas is introduced in the form of bubbles into the liquid volume through openings in the middle of the mixing chamber [6,8]. On the other hand, in outside-in devices, gas is supplied to the mixing chamber from the surrounding annular space using small holes in the pipe (a perforated pipe is used for this purpose) [6–8,17,18,43]. The effervescent-swirl atomizers can be subdivided in an analogous manner, taking into account the fact that the swirling motion can be induced for the gas phase, the liquid phase, or both at the same time. Figure 3 shows the modified designs of the tested atomizers using the phenomenon of swirl motion [1,7]. The papers [8,20,45] describe atomizers with typical dimensions and atomizers with minimized dimensions (about 5 times shorter).

It is worth emphasizing that an effervescent atomizer is subject to constant modifications in order to improve the atomization process. Hammad et al. [46] showed the new design of the injector that connects the flow characteristics inside a newly designed outside-in-liquid (OIL) atomizer. This study used technique of the flow visualization and digital image processing. The experimental data proves that the new construction of OIL atomizer (whose internal geometry has been properly optimized) can overcome the spray unsteadiness problem in relation to an effervescent atomizer.

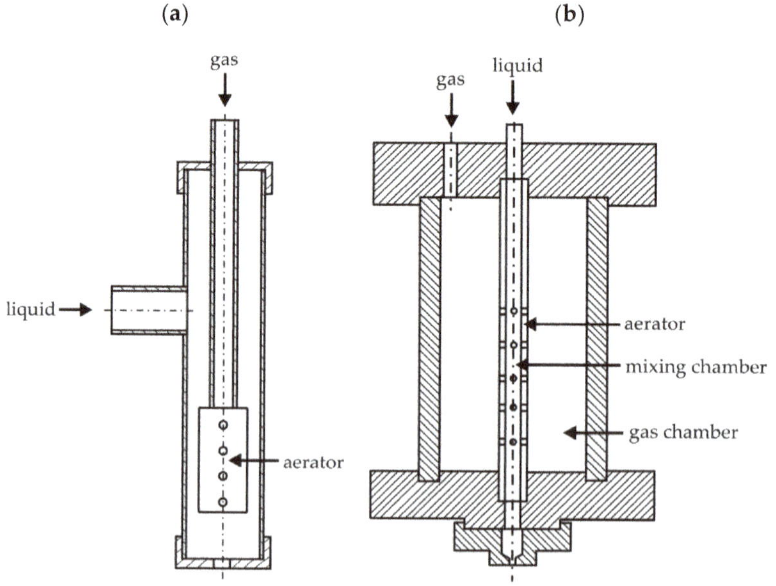

Figure 2. Exemplary schemes of effervescent atomizers of the type [20,44]: (**a**) inside-out, (**b**) outside-in.

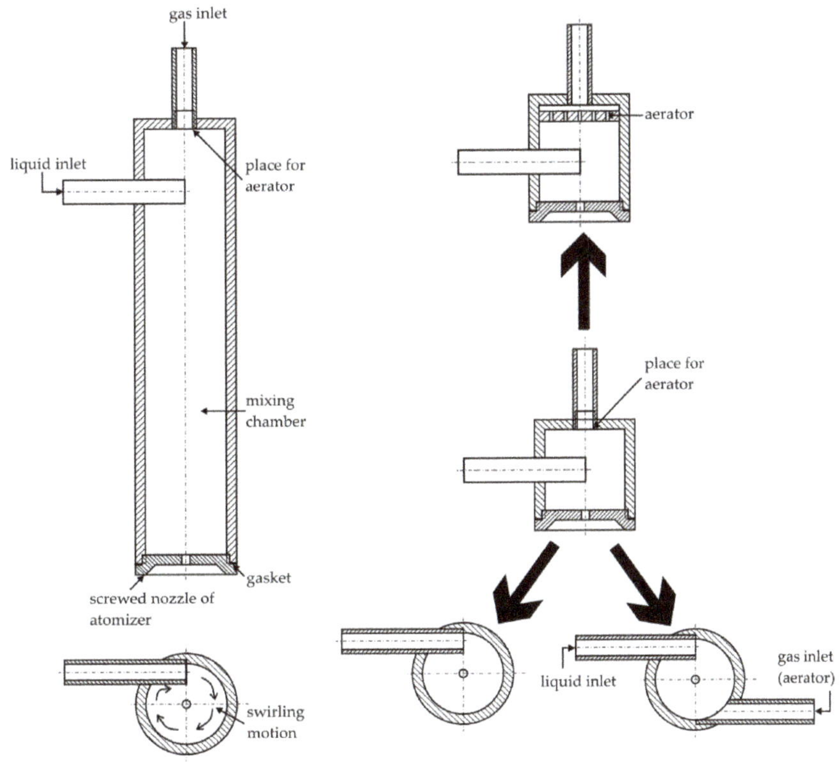

Figure 3. Modifications of effervescent-swirl atomizers proposed in the work [7].

In the work of Kourmatzis et al. [47] is described a hybrid atomizer that uses an effervescent and airblast atomization mechanism. In the effervescent atomizer was added a coaxial shear flow to a central two-phase bubbly flow. This method is a simple extension of effervescent atomization. In the article are used LDA/PDA measurements, high speed microscopic imaging of the atomization zone, and advanced image processing techniques in order to measure and analyze character of produced spray. The research carried out showed that the morphology of a spray obtained by the used hybrid atomizer is different to that of a conventional effervescent atomizer showing a superposition of sinusoidal instabilities onto a bubbly two phase core.

Another paper [48] conducted an experimental study of the atomization process at multi-hole effervescent atomizers. This study used phase Doppler anemometry in order to define the spray quality. The experimental results showed that the *SMD* profiles of these atomizers were inversely bell-shaped, where the minimum was to the nozzle axis. It is worth noting that the radial distance had an effect on droplet diameter, where for larger radial distance we can observe the occurrence of large droplets and the distribution character changes from uni- to bi-modal. Jedelský and Jícha [48] proved that the internal geometry (including inserts, mixing chamber size, and aeration arrangement) has a significant effect on the internal two-phase flow; however, the effect on the droplet diameter is not significant. Furthermore, the experimental data shows that the spray characteristics depends on the internal geometry of the multi-hole effervescent atomizer.

In paper [49], it described the new twin-fluid nozzle. The obtained results showed the droplet characteristics and size distribution depend significantly on the coupling between the gas, liquid, and structure of the twin-fluid nozzle. The efficiency of the spraying process was determined using the method of a phase Doppler particle analyzer. The results proved that the spray atomization characteristics and droplet size distribution depend on the performance of primary and secondary atomization. Accordingly, an important role in the atomization process is played by the parameter, such as atomizing core structure and the gas to liquid mass flow rate ratio. It is worth emphasizing that better atomization efficiency was achieved for the new twin-fluid nozzle than the standard twin-fluid nozzle. The use of an atomizer with a new design of twin-fluid nozzle leads to the decrease in the droplet diameter, and also the increase in the droplet number concentration, axial velocity, and the spray cone angle.

The papers [50–52] described an innovative design of an effervescent-swirl atomizer equipped with various types of swirl inserts (Figure 3). Three insert channels with different inclination angles were tested (Figure 4), for which the said angle was 30, 45, and 60 °C, respectively. The authors of the above works also analyzed the dependence of the spray angle on the diameter of the outlet opening of the atomizer, the value of which ranged from 1.5 to 2.5 mm.

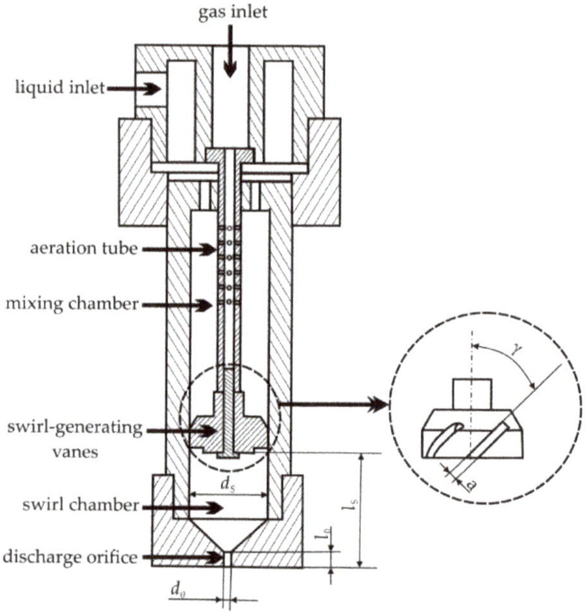

Figure 4. Design of a two-phase atomizer with a swirl insert [52].

The analysis of the obtained results proved that inside-out type atomizers perform better at low values of liquid flow rate, while outside-in type atomizers perform better at higher values of liquid flow rate [53–59]. The advantages of both gas-powered atomizers have not been thoroughly analyzed and clearly defined so far [8]. A standard effervescent atomizer has a diameter of about 50 mm and a length of about 100 mm. In turn, the diameter of the mixing chamber ranges from about 5 to 25 mm, and the diameters of the outlet openings range from 0.1 to 6 mm [8].

It is worth emphasizing that despite the fact that effervescent and effervescent-swirl atomizers are very popular, it has not been possible to develop a detailed calculation method yet. This is explained by the fact that so far it has not been possible to define in an unambiguous manner how the gas and liquid phases interact during the flow. Based on the experimental studies conducted so far, which were carried out in a wide range of parameter variability, it is possible to propose correlation equations. These equations concern the relationship between: the design of the atomizer, the properties of the sprayed liquid, the flow rate of the media and the diameter of the formed aerosol droplets. These dependencies are especially useful in the design aspect of atomizers. Examples of the equations are described in a later part of the article.

A very important issue in the atomization process is its optimization in order to reduce the costs of liquids and chemicals, unplanned production stoppages, energy consumption, and the negative impact on the environment, as well as to extend the operating time of the atomizer and improve the quality of spraying [17,20,60,61]. In the process of designing an effervescent-swirl atomizer that generates droplets of the desired size, the diameter of the atomizer outlet opening, which significantly affects the quality of atomization should be taken into account. The research carried out and published in the work [45] showed that as the diameter of the outlet opening increases, the quality of atomization deteriorates clearly. However, it should be remembered that the large diameter of the outlet opening is its main advantage, as it allows to minimize the problem of its clogging during spraying contaminated liquids or suspensions [62]. Therefore, it is assumed that the selection of the minimum diameter of the outlet opening should be based on the maximum value of the mass flow rate, taking into account also the possibility of clogging the opening or its possible erosion [17,56].

Another important aspect in the atomization process is the shape and diameter of the mixing chamber [8,48,62–64]. Based on the research described in [65–67], one main conclusion can be reached that if the mixing chamber has a small diameter, it does not affect the size of the generated droplets. On the other hand, when the mixing chamber diameter is greater than 5 mm, a significant effect on the mean diameter of the droplets is observed [65–67].

Other studies describe the dependence of the atomization process on the type of aerator used. The research described in the work [68] proves that the design of the aerator itself has practically no effect on the size of the formed droplets. However, the authors emphasize that the use of an atomizer with a multi-hole aerator makes it possible to obtain an aerosol of a slightly more monodisperse character than in the case of using an atomizer with a single-hole aerator. In both cases, the atomization process was carried out for the same total cross-sectional area of the gas inlet openings. In addition, the study also proved that an atomizer with a single-hole aerator is more effective when applying low injection pressure. A clear effect on the size of the generated aerosol droplets was noted for the ratio of the area of the outlet opening to the total area of the aerator's openings [42,45]. This ratio describes the gas velocity coefficient, the value of which depends on the diameter and number of holes in the aerator. It is assumed that its optimal value is expressed by the formula [67]:

$$\left(\frac{A_0}{A_{A,h}}\right)_{opt} = 6.3GLR \tag{1}$$

Accordingly, it can be concluded that the atomization process using effervescent atomizers can be carried out with gas supply aerators with a large opening diameter.

2.2. Liquid Flow Structures

A review of the literature shows that in an effervescent atomization process, a flow structure similar to the one occurring in the case of swirl atomizing may appear at the outlet. This situation also applies to the formation of an air core in the spray axis [7,8]. The introduction of swirl motion to effervescent atomizers intensifies the presence of the air core. This process includes complex two-phase phenomena that are difficult to describe and model [8]. The flow occurring in effervescent atomizers takes a more complicated form than the other one-phase and two-phase atomizers. The atomizing mechanism is based on the mixing of both phases (gas and sprayed liquid) inside the device. The development of the two-phase mixture occurs as it flows through the atomizer until it is sprayed at the outlet opening. An example of the atomization process in this type of devices is shown in Figure 5 [7–9,69–74].

In effervescent atomizers, both gas and liquid are supplied by means of inlet stubs. The gas is dosed into the system at a pressure slightly higher than that of the liquid using the perforated aerator pipe for the liquid-phase, so that the gas supplied takes on a form of bubbles. The two-phase mixture thus formed flows into the nozzle outlet of the atomizer. The method of spraying, the development, and structure of the mixture are mainly dependent on the geometry of the gas injection, the outlet opening of the atomizer, the shape and size of the mixing chamber, the injection pressure, the gas/liquid flow ratio, and the physical properties of the liquid. In the mixing chamber, it is possible to observe the phenomenon of bubble formation, their transport, and the changes taking place in the form of joining or breaking, as well as the movement of the gas and liquid phases. In turn, in the vicinity of the outlet opening, the mixture is compressed. At the outlet of the atomizer, a rapid expansion of the gas phase takes place, which leads to the breakup of the liquid stream, the formation of films and jets of liquid, and the jet breakup (primary and secondary breakup) [8,29,75,76]. The two-phase mixture that exits the atomizer outlet can take various forms (it can be a foam, a stream, or a ring), where the swirling of the liquid causes the formation of an annular structure. The authors of [10–12] analyzed the mechanism of liquid jet breakdown and noticed that larger drops are formed from larger jets. Bar-Kohany and Levy [76] described experiments with flash-boiling atomization, which proved that the nature of bubble growth should be addressed and analyzed, because bubbles lead to more efficient atomization process.

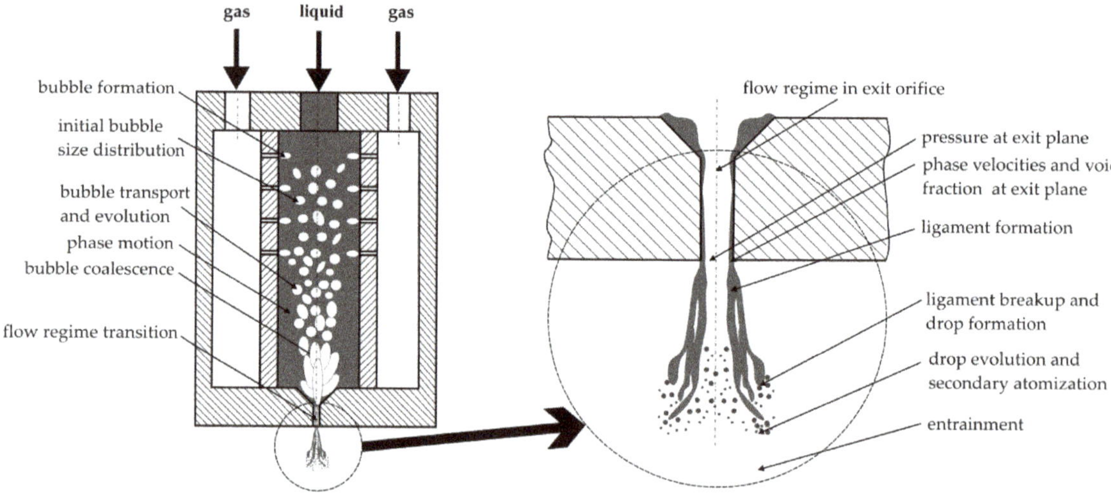

Figure 5. Schematic diagram of the two-phase atomization process in an effervescent atomizer [7,8].

Different flow structures are distinguished in the outlet opening: effervescent, ring-dispersion, annular, cork, and foam, as shown in Figure 6 [75–82]. The effervescent structure is characterized by a small number of tiny bubbles that are evenly distributed throughout the volume. However, in the cork structure there is an object called a cork. In the cork structure, the relative interphase surface decreases with respect to the bubble structure. On the other hand, in the foam structure, the occurrence of a large irregular object can be observed. The number of objects in the tested volume decreases, while the size of the relative interfacial surface remains practically unchanged in relation to the cork structure. The ring structure, also known as a lamellar, ring-bubble, and ring-dispersion structure are characterized by a fully developed regular air core. Small droplets of liquid may be present inside the air core. These structures are characterized by the largest relative volume fraction of the gas phase [78,80,83].

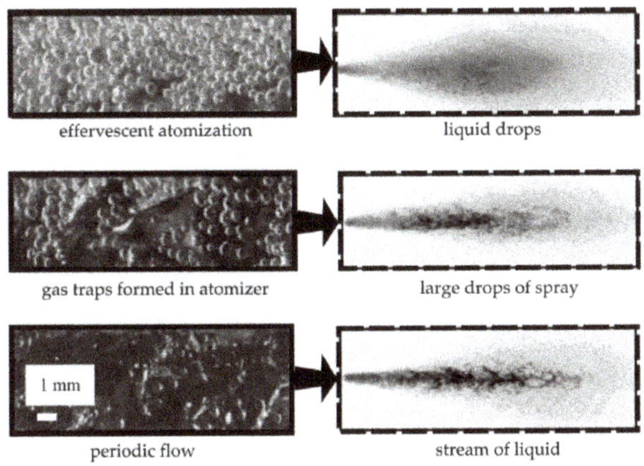

Figure 6. The quality of the generated spray depends on the structure of the internal flow [84].

Kourmatzis et al. [47] analyzed the hybrid atomization (effervescent and air-blast atomization) and showed that the air-blast mode has a significant impact on the sizes of generated ligament at the exit orifice for *GLR* equal to <2.7%. Experimental data shows that the coaxial flow makes it possible to entrainment of air in the central core, thereby influencing the degree of atomization and dispersion in relation to the pure effervescent mode.

The flow regime inside in the new atomizer (outside-in-liquid) is controlled using the perforated chamber geometry and also the exit orifice diameter [46]. Studies showed that the operating conditions do not affect the stability of the annular flow, which generated while a perforated chamber with many small injection holes and a small exit orifice were used. However, in the case where the atomization is carried out with a perforated chamber with a few large injection holes and/or a larger exit orifice, we can observe the existence of four flow regimes inside the mixing chamber (such as inhomogeneous bubbly, slug, wavy-annular, and stable-annular flows). It was also observed in [46] that the transition between defined regimes are dependent on the operating conditions. The obtained results shows the relationship between the gas-to-liquid ratio (*GLR*), the relative mixing pressure ($R_{\Delta p}$) and transition between flow regimes. It was observed that the increase in *GLR* provides faster to the transition to the annular flow; however the increase in $R_{\Delta p}$ causes the delay of transition. Based on the obtained results for the design of OIL atomizer, a new correlation equation is created:

$$GLR^{-1.628} R_{do}^{3.4} R_{\Delta p}^{0.8} \left(\frac{\rho_G}{\rho_L} \frac{d_G^2}{d_L^2} R_{ih} \sin\theta \right)^{0.314} = \frac{11}{300} \quad (2)$$

This correlation is suggested for the transition criterion between the favorable annular flow and the unfavorable intermittent flow, where R_{do} is relative exit orifice diameter, R_{ih} is relative area of a single injection hole, θ is angle, d_G diameter of gas phase, and d_L diameter of liquid phase.

In the work [85], the relationship between the gas flow rate and the angle of inclination of the insert channels and the air core formed during spraying was demonstrated. The conducted tests prove that the diameter of the air core increases with the increase of the gas flow rate and the angle of the swirler channels inclination. Selected results of the research described in the work [85] are presented graphically in Figure 7. The clear influence of the used insert on the diameter of the air core is visible, first of all, at higher values of gas flow rates. Additionally, it has been shown that an increase in the diameter of the atomizer outlet opening contributes to an increase in the diameter of the air core.

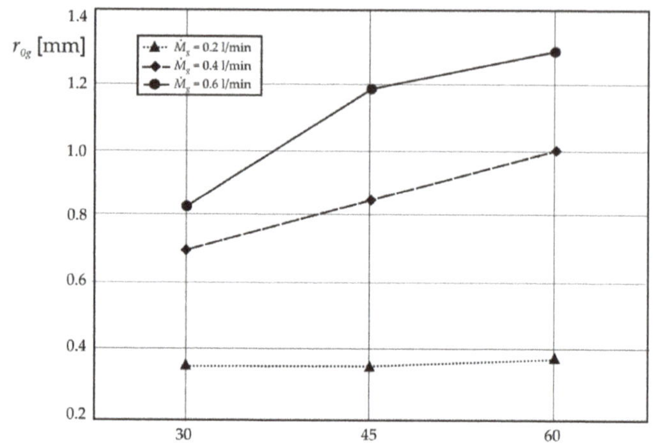

Figure 7. The relationship between the flow swirl insert and the diameter of the air core [85].

The authors of the work [45] conducted an in-depth analysis of the liquid flow structure at the outlet opening of the atomizer. The atomization process was carried out for a single and two-phase system. Selected photographic images of the sprayed liquid for the air-water system are shown in Figure 8, and for the system of air-water glycerin solution with polymer addition in Figure 9.

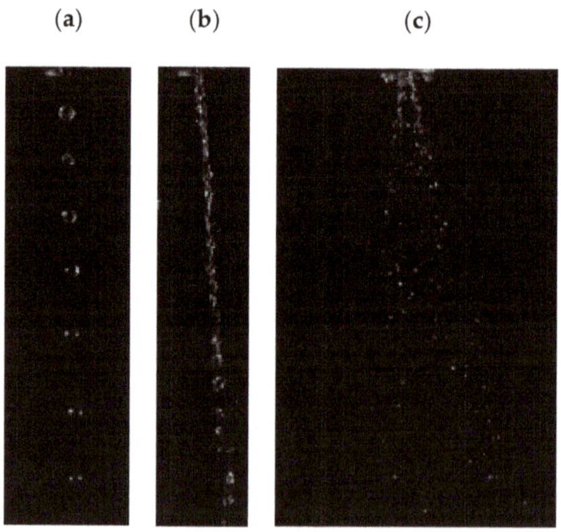

Figure 8. Selected structures of the sprayed liquid observed as a result of spraying water with an effervescent-swirl atomizer [7]: (**a**) $\dot{M}_c = 0.0014$ kg/s, C5, (**b**) $\dot{M}_c = 0.021$ kg/s, C5, (**c**) $\dot{M}_c = 0.008$ kg/s.

During the flow of 59% aqueous glycerin solution with the addition of 0.5% rocrysol WF1 in the effervescent-swirl atomizer, turbulences were observed on the surface of the stream due to swirling of the liquid stream (Figure 9d) [7]. The presented photos show the characteristic beads-on-a-string structure (BOAS) of a stream, i.e., thin threads of the sprayed liquid interconnecting the droplets (droplet-fibrous form of a stream) (Figure 9b). When spraying liquids with significant viscosities, the spray angle is reduced until a stream with large fibers and droplets is formed (Figure 9e). However, the appearance of turbulent flow, as a result of the increase in Reynold's number, contributes to the destruction of the liquid structure and the reduction of the liquid viscosity, which results from the properties characteristic of non-Newtonian liquid properties [7].

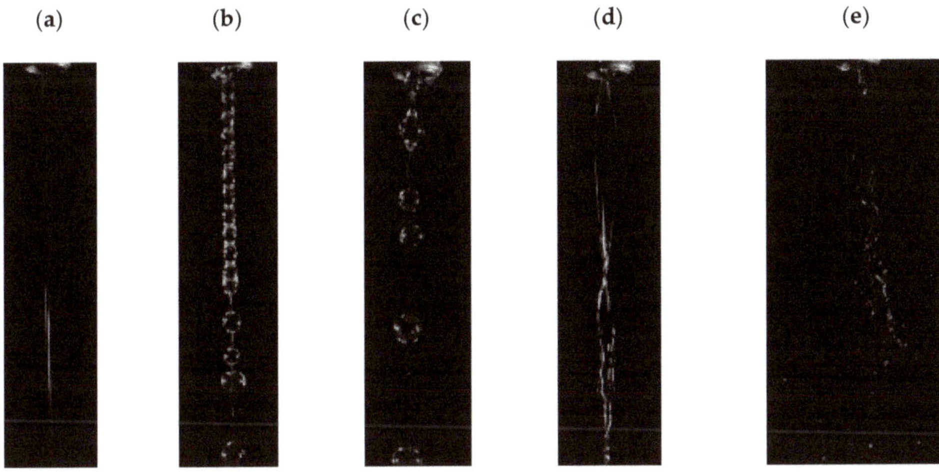

Figure 9. Selected structures of the sprayed liquid observed during one- and two-phase spraying of a 59% water solution of glycerin with the addition of 0.5% rocrysol WF1 with the use of an effervescent-swirl atomizer [7]: (**a**) $\dot{M}_c = 0.0056$ kg/s, C5, (**b**) $\dot{M}_c = 0.011$ kg/s, $\dot{M}_g = 0.000028$ kg/s, P, (**c**) $\dot{M}_c = 0.0056$ kg/s, $\dot{M}_g = 0.000014$ kg/s, LC, (**d**) $\dot{M}_c = 0.028$ kg/s, LC, (**e**) $\dot{M}_c = 0.021$ kg/s, LC.

On the basis of the conducted research, it has been shown that as a result of the atomization process, it is possible to observe, inter alia, the breakup of the jet into droplets, the phase of a bent pencil, and proper spraying [5,7,45]. Moreover, the appearance of a jet with droplets, a jet of sprayed liquid with gas bubbles, a jet of individual bubbles and bubbles connected with liquid fibers was also observed. It is worth emphasizing that the obtained images are characteristic of liquids of significant viscosity (especially longitudinal viscosity). These conclusions confirm previous literature reports [45].

Most likely, the addition of polymer results in an increase in the longitudinal viscosity, which leads to the formation of a coarse-fiber structure and delay of the stream breakup [86,87]. This phenomenon is beneficial in many applications [5,88]. As a result of spraying aqueous polymer solutions, a droplet-fibrous structure is formed. It is worth noting that during the atomization process, the occurrence of the so-called secondary destabilization of the fibers connecting the droplets, which in turn causes the appearance of very small droplets, was observed. Despite the fact that the use of a polymer additive causes an increase in the thickness of the fibers (Figure 9d), as well as the suppression of disturbances causing delayed breakup of the streams, the liquid stream becomes thinner with increasing distance from the nozzle outlet, which in turn leads to secondary growth oscillations on the surface of the liquid and the breakup of the threads into drops of very small diameters [7].

2.3. Discharge Coefficient

In order to fully assess the design of the atomizer, it is also necessary to analyze the pressure drops occurring at its tip and in the body (i.e., drops occurring on the stub-tubes, in the aerator and in the mixing chamber). The studies carried out so far prove that the greatest pressure drops occur at the outlet end, while the smallest pressure drops occur in the pipe, which constitutes the mixing chamber. With the increase of pressure drops in the atomizer, the value of the discharge coefficient determined for the whole atomizer is observed to decrease. In a situation when the value of the mass gas stream is close to or equal to zero, it is assumed that the calculation of the spraying parameters for effervescent-swirl atomizers (including the flow coefficient) is simplified to the calculations used in the case of single-phase atomizers characterized by the same geometry [5,6,89–91].

The relationship between the ratio of length to the diameter of the outlet opening l_0/d_0 the discharge coefficient is very complex. In the range of values $l_0/d_0 < 2$, the liquid stream is contracted. On the other hand, when $l_0/d_0 \geq 2$, the stream expands within the opening, which means that its outflow is the same as in the case of the so-called adapters/attachments. In the narrowest cross-section of the liquid stream there is a negative pressure leading to an increase in the liquid flow. It happens that when appropriate conditions appear, the flow velocity increases the phenomenon of liquid stream detachment from the opening occurring. Due to the above, there is a possibility of the stream narrowing and a marked decrease in the liquid flow [5,14]. In the works [6,7,92], it was proved that the value of the discharge coefficient depends to a large extent on the ratio l_0/d_0, but also on the shape of the inlet section of the atomizer's tip. The values of the liquid discharge coefficient are quite varied and depend on the type of the discharged orifice used, otherwise the so-called diffuser [14,93]. Moreover, experimental studies have shown that the size of the l_0/d_0 ratio plays a key role in a situation where the Reynolds number values are small.

The authors of the work [7] estimated the correlation equation that allows to determine the value of the flow coefficient for single-phase flow, which takes the form:

$$C_D = \left\{ \left[Re_c^{-0.88} 64 \left(\frac{l_0}{d_0} \right)^{1.15} \left(\frac{d_0}{D_s} \right)^{1.15} \left(\frac{\eta_c}{\eta_{water}} \right)^{1.15} + 53.88 \right]^3 + \left[\frac{1 - \left(\frac{d_0}{D_s} \right)}{C_{D,tur}^2} \right]^3 \right\}^{-0.165} \quad (3)$$

It is worth noting that this equation is valid for a Reynolds number in the range from 20 to 40,000, and C_D, tur is the value of the fluid discharge coefficient during a turbulent flow.

It should be emphasized that in the case of effervescent-swirl atomizers, it is more important to determine the value of the flow coefficient for a two-phase flow. It is also worth noting that in the case of effervescent-swirl atomizers, a smaller influence of the tips on the value of the liquid discharge coefficient during a turbulent flow was noted than in the case of standard effervescent atomizers. Due to the above, it is possible to obtain some savings by using tips with lower accuracy and lower quality in the production and use, avoiding the risk of deteriorating the quality of the spray obtained [7].

Figure 10 shows the effect of the l_0/d_0 ratio on C_D during a two-phase flow for the two selected *GLR* values of 0.0038 and 0.046. The results obtained for the process carried out with the use of an effervescent-swirl atomizer showed that the value of the discharge coefficient was increased until reaching the maximum value equal to 2.04. In turn, a further increase in the l_0/d_0 value contributes to a decrease in the C_D value. This phenomenon is explained by the occurrence of an increase in friction forces, which occurs as a result of an increase in the value of l_0 (extension of the outlet opening length). Additionally, the conducted research proves that the geometry of the outlet opening does not affect the discharge coefficient or affects it only slightly in the situation when the *GLR* values are large [7]. Moreover, it was noticed that the values of C_D, determined for all tested atomizers, decrease with the increase in the *GLR* value. The authors of the work [7] developed a correlation equation that allows to estimate the liquid discharge coefficient in a two-phase flow for the outlet opening (atomizer tip) with the following form:

$$C_D = 0.0822 \left(\frac{\eta_c}{\eta_{water}} \right)^B \left(\frac{\sigma_c}{\sigma_{water}} \right)^{0.02} \frac{C_{D,tur}}{GLR^{0.43}} \quad (4)$$

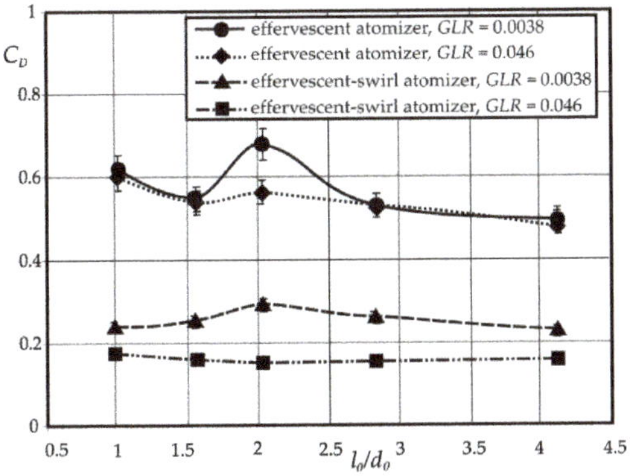

Figure 10. Graph of the dependence of the discharge coefficient on the value of l_0/d_0 using effervescent and effervescent-swirl atomizers [7].

It is worth emphasizing that in the analyzed atomization process, the influence of viscosity is not obvious and it is visible only when *GLR* values are small [14,94,95]. Most likely, this is due to the fact that while the *GLR* takes low values, the nature of the two-phase flow is similar to that of a single-phase liquid flow (where the gas addition is minimal), and therefore the effect of viscosity on the flow can be observed. On the other hand, when the *GLR* values are higher, the viscosity of the gas phase plays a significant role. Similar conclusions can be drawn after analyzing the results of research on classic swirl atomizers [14]. These studies prove the occurrence of a certain liquid viscosity value in the atomization process, which is assigned the maximum value of the discharge coefficient [14]. The maximum value of the mass stream of liquid occurs when the liquid turbulence disappears (with a certain liquid viscosity), which leads to the liquidation of the air core and the outflow of the liquid itself out of the atomizer outlet opening. On the other hand, a continuous increase in the viscosity of the liquid leads to a decrease in the stream mass caused by an increase in frictional resistance [14].

A review of the literature revealed several papers devoted to the analysis of flow resistances [7]. The data contained in the studies make it possible to estimate the value of the discharge coefficient on the basis of semi-empirical [96] or empirical [44,97] correlation equations. By contrast, [98] describes the combination of two analytical models in order to determine the value of the discharge coefficient. For this purpose, the homogeneous flow model (HMF) and the separated flow model (SFM) were used [98].

In the work [96], the authors, on the basis of the conducted research, proposed a model equation for the discharge coefficient of atomizers, taking into account the external gas supply:

$$C_D = c\left(1 - \frac{\dot{V}_g}{\dot{V}_g + \dot{V}_c}\right)^{0.3}\left(1 + \frac{1}{GLR}\right)^{0.15} \tag{5}$$

In this equation, the value of C_D depends on the properties of the liquid (viscosity and density) and takes various values, which are listed in Table 1 [96]. This equation is limited to the value $GLR < 0.12$.

Table 1. Summary of the constant c z values estimated on the basis of Equation (4) [96].

η_c [Pa · s]	ρ_c [kg/m^3]	c
RSP1	0.020	0.015
RSP2	0.020	0.020
RSP3	0.020	0.025
RSP4	0.020	0.020
RSK1	0.020	0.015

According to the authors of the work [99], the liquid flowing through the outlet opening does not fill it in the entire cross-section, and the formed air core is surrounded by a liquid ring. This situation is analogous to the phenomenon that occurs when spraying with swirl atomizers. Based on the conducted research and analysis of the obtained results, a correlation equation was proposed in the form (valid for *GLR* ranging from 0.02 to 0.46):

$$C_D = 0.0088\left(GLR\frac{d_0}{D_s}\right)^{-0.75} \pm 14\% \tag{6}$$

In turn, Jedelsky and Jicha developed another form of the model equation to determine the value of the discharge coefficient [98]. This equation is based on two flow models: homogeneous flow and stratified two-phase flow. This relationship takes the following form:

$$C_D = 0.62\left(\frac{\eta_c}{\eta_{water}}\right)^{0.04}\left(\frac{\sigma_c}{\sigma_{water}}\right)^{0.02}\left(\frac{l_0}{d_0}\sin(2\varphi)^{0.5}\right)^{-0.11}\frac{\dot{M}_c}{A_0(2\rho_c\Delta P)^{0.5}}\frac{1}{(1+GLR)} \tag{7}$$

where φ is the inclination angle of the mixing chamber wall. This equation is consistent with the data described in the work [4]. Equation (7) allows to determine the C_D value with an accuracy of ± 10%, thanks to which it is currently the most universal and takes into account the largest number of variables.

Figure 11 summarizes the dependencies of C_D on *GLR*, drawn on the basis of a literature review [1,7,44,96,98]. Due to the fact that the atomization process was carried out with the use of atomizers of a different design, no direct comparative analysis can be performed. The analysis of the data presented in the graph shows one common tendency, where the value of the discharge coefficient decreases with the increase of the *GLR* value [100]. An exception to this rule is the research results presented by Ramamurthi et al. [97], where a practically constant value of C_D can be observed.

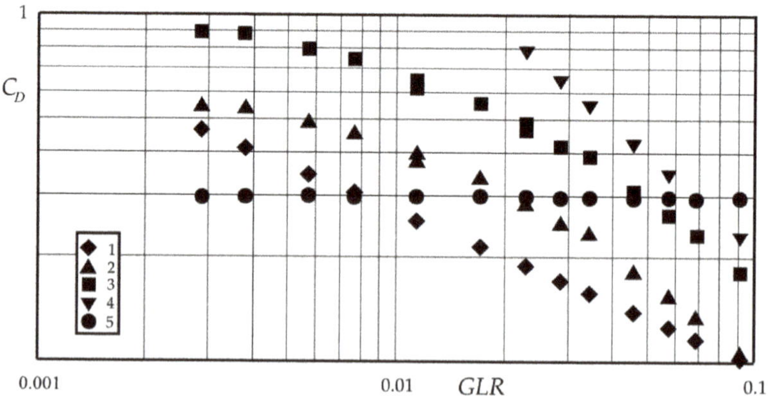

Figure 11. Summary of literature values of the discharge coefficient (d_0 = 0.002 m; l_0/d_0 = 1; β = 0°) [7]: 1-Chen and Lefebvre [96], 2-Jedelsky and Jicha [98], 3-Ochowiak [1], 4-Ochowiak et al. [44], 5-Ramamurthi et al. [97].

The dependence of the discharge coefficient on the liquid viscosity is theoretically included in the equations containing the Reynolds number and in Equation (6). On the other hand, the analysis of literature data related to the effervescent and effervescent-swirl atomization process [4,75] proves that the influence of liquid viscosity (for Newtonian and non-Newtonian liquids) on C_D can, as a rule, be neglected. Studies have shown that the influence of liquid viscosity is greater when *GLR* takes lower values. At the same time, at higher GLR values (>0.07), we do not observe the viscosity influence. The obtained results are consistent with the studies published in the works [94,101,102].

2.4. Spray Angle

Chen and Lefebvre [53] conducted research on the spray angle of liquids of different viscosity and surface tension. The obtained results prove that the size of the estimated spray angle is much greater than the values obtained for pressure atomizers of similar design. In turn, in the work [36] it was shown that the value of the spray angle rapidly decreases in the vicinity of the outlet opening, and then it assumes an approximately constant value. The spray angle increases as the gas pressure increases and the viscosity and surface tension of the liquid decrease. The work [53] says that the spray angle takes values lower than 23°. This is confirmed by the studies described in the work [35,36,58,60,103], which show that the values of the liquid spray angle with the use of effervescent atomizers with a single-hole aerator take values not greater than 22–23. The literature review reports that the spray angle values obtained for effervescent atomizers are lower values compared to other types of atomizers (except pressure ones) [5,6,53,103].

In the literature on the subject, works devoted to the analysis of the spray angle are very rarely found. However, the published works determine congruently that the spray angle size increases monotonically as a result of the injection pressure increase [35,53,72,101,104]. In the work [35], the following correlation equation for the spray angle was proposed:

$$\frac{\alpha}{2} = (15GLR) + (0.039P_{in}) + \left(0.0451P_{ot}^4 - 0.6211P_{ot}^3 + 2.7551P_{ot}^2 - 3.62P_{ot}\right) + 7.0 \quad (8)$$

where the pressure values are included in the unit of megapascal.

In several studies, a dependence of the increase in the size of the spray angle with the increase in the GLR ratio was noted [35,53,105,106]. Moreover, Chen and Lefebvre [53] and Jedelsky et al. [106] observed a decrease in the spray angle value after reaching a certain maximum value, which most probably results from a gradual change in the internal flow structure (transition from bubble to annular structure). Additionally, it was noted in the work [106] that reducing the spray angle may cause an increase in the ratio of the outlet opening length to its diameter from 0.08 to 1.2.

Loebker and Empie [107,108] published their studies of the process of spraying highly viscous liquids and obtained relatively large spray angle values of about 60°, where GLR was greater than 0.003. Moreover, the analysis of the results proved that not only the viscosity of the liquid, but also its flow rate has a significant impact on the spray angle value [7,107,109,110]. The highest spray angle values published in the works [109,110] were about 80°, and in the work [111] the angle value reached even 150° (at $GLR \approx 0.07000$), where Newtonian liquids were sprayed using an atomizer with an internal gas flow. The maximum angle value was reached with $GLR \approx 0.07000$, after exceeding this value, a sharp decrease in the spray angle value was observed.

The research on the atomization process carried out in effervescent-swirl atomizers with the use of different gas and liquid flow velocities presented in the work [4] showed that the increase in gas flow velocity has a beneficial effect on the spray angle value (especially in a situation where the value of the mass flow rate of liquid is small). This is explained by the fact that the use of low flow velocities causes the liquid to flow as a compact jet, from which individual drops are detached. With a further increase in the value of the gas flow velocity, the stream of liquid breaks down into droplets, and thus the spray angle increases. At the moment when the stream of gas is significant, we observe the phenomenon of flowing around the stream of the sprayed liquid, so the spray angle is reduced. It should be remembered that the use of a too large stream of liquid may lead to an increase in the volume of the stream and difficulty in its spraying. Accordingly, we can observe a decrease or increase in the size of the spray angle. These observations were confirmed in the work [75], where the possibility of the maximum spray angle occurrence and its non-linear dependence on the *GLR* value was demonstrated. The described process can be explained by the flow transition from the bubble to the cork (or annular) range inside the atomizer, caused by the presence of low *GLR* values and operating pressure. Another explanation for the phenomenon is its occurrence as a result of a rapid increase in the volume of gas bubbles right at the outlet of the atomizer orifice, which may contribute to increasing the spray angle.

It should be remembered that the spray angle is also influenced by the internal geometry and dimensions of the atomizer [5,96,112], which include, e.g., the diameter of the atomizer outlet opening [45,111,113], and the diameter of the swirl chamber [5,6,114]. One cannot forget about the viscosity of the sprayed liquid either [45,111,113,115,116].

In the work [4], another correlation equation taking into account the spray angle was proposed:

$$\alpha = AC_{D,tur}^{0.9} D^{0.39} d_k^{-0.78} \eta_c^{-0.31} d_0^{1.02} \dot{M}_c^{0.25} GLR^{0.21} \tag{9}$$

where: A is the constant for the design of the atomizer, and d_k is the diameter of the liquid inlet stub-tube [5,6]. For the analyzed atomizers, the value of the constant A which is equal to 901, was determined experimentally [7]. It is worth noting that the above equation is correct for the entire examined range of the considered variables. Figure 12 shows graphically examples of the relationship between the spray angle and the viscosity of the liquid [7].

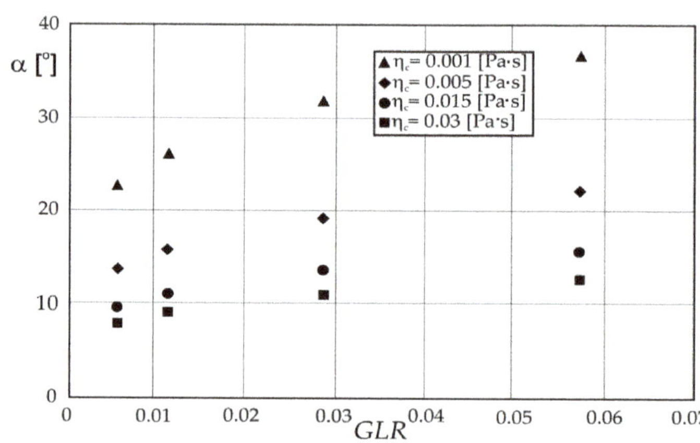

Figure 12. Dependence of the liquid spray angle on *GLR* for liquids of different viscosity, sprayed with an effervescent-swirl atomizer [7].

As mentioned before, the spray angle depends, apart from the process conditions, also on the design of the atomizer itself; however, all variables cannot be taken into account in the correlation equation. This is confirmed by the research conducted by Jedelski and Jichy [117]. The study showed that the α value for effervescent atomizers increases due to the increase in *GLR* value, and then reaches its maximum value at the *GLR* value of about 0.1, after which this value decreases. Moreover, it has been proven that the atomizers additionally equipped with swirl inserts have larger spray angle sizes than the atomizers with cylindrical outlet openings [117].

In summary, it can be stated that in order to obtain relatively large spray angles, the process should run within a *GLR* range of about 0.02 to 0.1.

In the work [54] it was observed that the spray angle significantly depends on the diameter of the atomizer's outlet opening. It has been shown that with the increase in the opening diameter, the spray angle value increases (Figure 13). The authors of the work [7] reached the same conclusions.

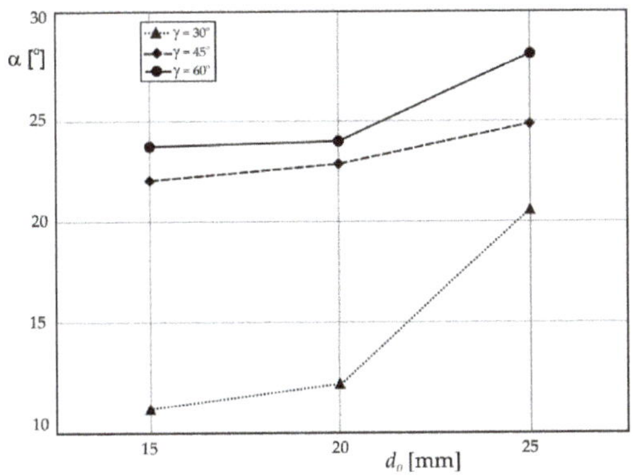

Figure 13. Influence of the outlet opening diameter of the atomizer on the spray angle for inserts of various designs [54].

The research carried out by Jedelsky et al. [118] showed the influence of the atomizer design and *GLR* values on the size of the spray angle obtained as the result of carrying out the atomization process both in a standard effervescent atomizer and in an effervescent-swirl atomizer with a swirl insert (Figure 14). The subject of the atomization were suspensions. The dimensions of the individual components of the device are summarized in Table 2.

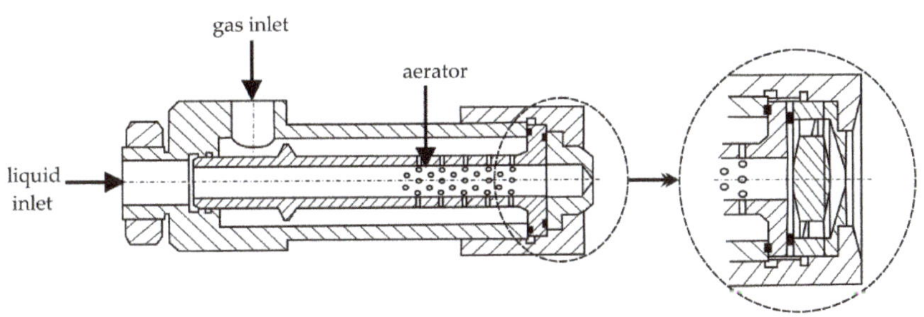

Figure 14. Diagram of the effervescent-swirl atomizer from the study [118,119].

Table 2. Summary of dimensions of individual components of atomizers [118].

Ç$_c$ [Pa · s]	ρ_c [kg/m^3]	c
RSP1	0.020	0.015
RSP2	0.020	0.020
RSP3	0.020	0.025
RSP4	0.020	0.020
RSK1	0.020	0.015

Figure 15 presents the relationship between the spray angle and *GLR* for both analyzed atomizers [118]. Based on the analysis of the diagram, it can be noticed that in the case of $GLR \approx 0$, large values of the spray angle are obtained for atomizers equipped with a swirling element (atomizers I and II). When analyzing the dependence curve obtained for atomizer I, an increasing tendency of the spray angle with the increase of the *GLR* value, to the maximum value of about 0.006 can be noticed. Then, a further increase in the *GLR* value leads to a decrease in the value of the angle under analysis.

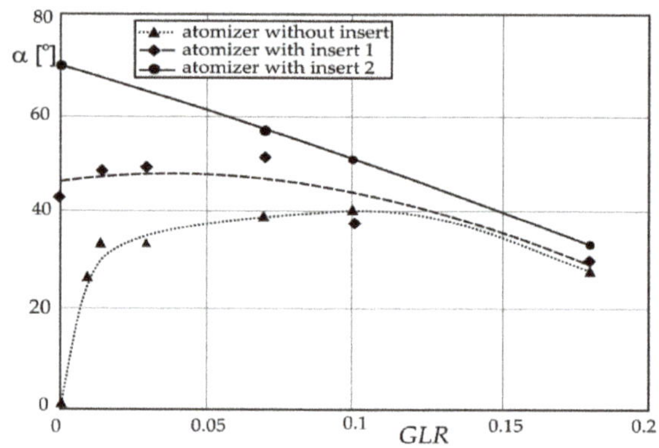

Figure 15. Diagram of the dependence of spray angle on *GLR* determined with a classical bubble atomizer and its modifications [118].

The situation is different in the case of atomizer II, where a downward trend in the size of the spray angle with the increase in *GLR* value can be observed. The downward nature of the dependence curve α on *GLR* was also observed in the case of an effervescent atomizer without a swirl insert, where at $GLR \approx 0$ the atomizer produced an angle of approximately 0°. In turn, for $GLR = 0.1$, the maximum spray angle value was obtained, which was 40°. It is also worth emphasizing that the research showed that the sprayed liquid stream took the form of a hollow cone.

2.5. Droplet Diameter

The paper [66] describes a model that enables the estimation of the average diameter of sprayed liquid droplets at specific physico-chemical properties of the sprayed liquid and set *GLR* values. The analysis of the performed experimental tests showed that the change in volume causes the expansion of the cone of liquid flowing from the device outlet opening. During the analysis of the results, it was assumed that the total energy introduced into the volume is the same as the total energy exiting the system. It is worth adding that the total energy introduced into the volume was assumed to be the sum of the gas energy, the kinetic energy of the liquid and the surface energy of the gas bubbles embedded in the liquid. Finally, the following correlation equation was proposed to enable the determination of the *SMD* values:

$$SMD = \frac{12\sigma_c}{\rho_c\left\{w_c^2 + \kappa GLR w_g^2 - \frac{\left[\left(w_c+\kappa_1 GLR w_g\right)^2\right]}{(1+\kappa_1 GLR)}\right\}} \tag{10}$$

where wc is the liquid velocity, and wg is the gas velocity, which were measured across the atomizer outlet opening cross-section. The value of the coefficient κ_1 was determined based on the experimental data. It should be emphasized that Equation (10) is correct for certain conditions, where $\dot{M}_g$, $\dot{M}_c > 1.5$ g/s, $P < 336$ kPa, and $GLR < 0.02$. Equation (10) proposed by Buckner and Sojek takes into account the deviations of the values determined from the model from the experimental values of ≤25% [66].

A different correlation equation allowing to determine the value of the mean droplet diameter was proposed in paper [55]. This equation includes the knowledge of the gas and liquid mass flow rate, the physico-chemical properties of the liquid, and the geometry of the atomizer outlet opening. This equation is compatible with the research carried out by Santangelo and Sojki [120,121], where the analysis of structure stability was used to estimate the size of the sprayed liquid droplets formed. For this model it was assumed that the two-phase gas and liquid mixture leaves the atomizer in the form of an air core surrounded by an annular layer of liquid. This layer breaks down into streams and films of liquid, which then break down forming droplets. The authors of the work [44,55] used the correlations published in the subject literature in order to determine the part of the gas-filled cross-section and the gas-liquid inter-phase slip in the annular flow. Lund et al. [55] initially estimated the thickness of the annular layer of liquid, and then concluded that it is divided into several cylindrical streams called ligaments, of the same diameters as the thickness of the liquid ring. In order to determine the size of streams and droplets, the stability analysis described by Weber [122] was used, which concerns the breakup of liquid ligaments [123]. The size of the formed droplets was determined on the basis of the assumption that each object stabilizes and forms a single droplet. The aspect of secondary atomization has been omitted. Therefore, the authors of the paper [55] obtained the following expression:

$$SMD = \left[\frac{3}{2}\sqrt{2}\pi d_l^3\left(1+\frac{3\eta_c}{\sqrt{\rho_c\sigma_c d_l}}\right)^{1/2}\right]^{1/3} \tag{11}$$

where d_l is the diameter of the liquid ligament, the value of which is compared to the thickness of the liquid film, thanks to which the following relationship [124] is obtained:

$$d_l = 0.18\left(GLR\dot{V}_c\frac{\rho_c}{\rho_g}\right)^{-0.62} \tag{12}$$

It should be noted that the equation proposed by Lund et al. (11) is particularly interesting due to the fact that it is entirely based on the fundamental principles of conservation of energy and momentum and does not contain any empirical constants. The accuracy of the proposed expression increases with increasing *GLR* value. Droplet size is estimated with an accuracy of 25% (with $GLR < 0.02$) and with an accuracy of about 5% (with $GLR > 0.04$). In this model, the aerodynamic effects of the gas surrounding the ligaments of liquid were not taken into account. During the two-phase atomization process, there is always a relative velocity between the atomizing gas and the liquid, called the inter-phase slip, which may ultimately have a significant impact on the liquid stream breakup [55].

Sutherland et al. [40] improved the model proposed by Lund et al. [55] by taking into account the relative velocity between the dosed gas and the liquid, as well as replacing the Weber stability analysis [122] with the Sterling and Sleicher stability analysis [125]. In the work [40], measurements of the atomized liquid momentum were performed in order to determine the velocity of the atomized gas and liquid at the device outlet. The model described by the authors of the work [40] also includes changes in the mean diameter of droplets as a result of changes in the viscosity of the liquid, surface tension, injection pressure, the ratio of the mass flow rate of gas to liquid, as well as the pore size of a given atomizer insert.

Sovani et al. [126,127] also made an attempt to improve the model proposed by Lund et al. [55] to estimate the probability of the distribution of the size of the generated spray droplets at the outlet of the atomizer. It is known that each atomizer generates droplets of different sizes, which depends on many factors that we can control as well as on various random phenomena that take place during atomization and we have no influence on them. The authors of the work [126,127] took into account just these random phenomena in the model of Lund et al. [55] as changes in the relative velocity of gas and liquid and physical properties of the liquid. The random changes included in the equation took the form of the probability distribution function (PDF for short) and the size distribution of the droplets of the generated spray. The results obtained on the basis of the proposed new model prove that the width of the droplet size distribution is a non-linear function of the variables and the relative velocity of the gas and liquid. Additionally, it was shown that changes in the physical properties of liquids did not have a significant effect on the droplet size distribution.

Liu et al. [128] described that the effervescent atomization was unstable at a small *GLR* while the atomization process proved gradually by increasing the *GLR* values. On the basis of the conducted research, it was determined the optimal atomization region, which was at a *GLR* equals 0.1. In [128], it described that the design parameters of atomizer have significant influence on *SMD* and atomization cone angle. The method of phase Doppler analyzer points out that the velocity and Sauter mean diameter distributions of the droplets are symmetrical on the discharge orifice center for analyzed swirl atomizers.

Kourmatzis et al. [129] described the near-field characteristics of effervescent sprays are examined by the used a method of the advanced image processing. The researches provide new quantitative insights into the character of the regime transitions that occur as a function of the *GLR*. The obtained research results prove that character of the distribution is variable and depends on the gas-to-flow ratio, where the distribution of the outer diameter of the ejected air laden liquid jet shows a bimodal character at low *GLR* equal to <0.5%, and mono-modal character at higher *GLRs* values.

Panchagnula and Sojka [57] developed their own model allowing to estimate the velocity profile of droplets in a sprayed liquid. This model concerned atomization of a turbulent stream of liquid. It was based on the velocity profile equation taken from the work [130].

In the work [131], research on atomizing a liquid was carried out using an unusual design of an atomizer with colliding inside it streams of gas, which then floats the liquid. The obtained test results showed different relationships than in the previous studies, because it was shown that with the increase in the diameter of the atomizer outlet opening and the inclination angle of the inlet wall of the atomizer outlet, the value of the mean surface-volume diameter decreases. The authors of the work [131] proposed the following correlation equation:

$$SMD = (4.4 - 0.4d_0)(\tan\varphi)^{-0.4}\left(\frac{P_g + 0.3}{P_c + 2.3}\right)^{(-6.3+2.6d_0-0.4d_0^2)} \tag{13}$$

The work [62] also included experimental analysis of the process of liquid spraying in terms of the impact of the atomizer design and the gas and liquid flow rate on the size of the spray droplets generated. The results of the research proved that *SMD* decreases with the increase of the diameter of the atomizer outlet opening and the liquid flow rate [45,62]. In the analysis of the obtained results, the equation proposed in the work [132] was used:

$$SMD = B(We_g GLR)^A \tag{14}$$

where the power exponent A was –0.12, and the constant B referred to the internal geometry of the atomizer, the value of which was 2.465×10^{-3} m for $d_0 = 3$ mm and 2.249×10^{-3} m for $d_0 = 4$ mm.

Mulhelm et al. [133] also used the formula (14) proposed by Harari and Shera [132] to determine *SMD*. However, in their case, the A value was -0.4, and the B value was dependent on the properties of the liquid and the diameter of the atomizer. In order to estimate the relationship between the constant B and the physical properties of the liquid, the Ohnesorge number and the atomizer geometry (or more precisely: the diameter of the atomizer outlet opening) were used. In the case of *SMD*, the relationship between the B/d_0 ratio and the Ohnesorge number takes the following form:

$$\frac{B}{d_0} = 0.21 Oh^{0.0622} \tag{15}$$

After carrying out appropriate transformations, a correlation equation was obtained in the form:

$$SMD = 0.21 Oh^{0.0622}(We_g GLR)^{-0.4} \tag{16}$$

A very good match between the experimental data and the correlation data was obtained from the dependence [62]:

$$SMD = 1.35 \times 10^{-3}\left[GLR\left(\frac{d_0}{D}\right)\right]^{-0.26} \tag{17}$$

It is worth paying attention to two issues, Equation (17) takes into account the diameter of the mixing chamber amounting to 0.02 m and the fact that it is valid only for the analyzed atomizer designs.

Hammad et al. [46] proved that the internal flow significantly controls the spray character and proposed a correlation equation taking into account the relationship between the Sauter mean diameter and the internal flow transition parameter ($We_{LS}/{\Phi_{ih}}^{-0.314}$):

$$\frac{SMD}{d_o} = \frac{1.42}{1 + e^{(1.08 - (\frac{We_{LS}}{\varnothing_{ih}^{-0.314}}))/0.8355}} - 0.252 \tag{18}$$

where Φ_{ih} is liquid/gas momentum ratio per liquid injection hole, We_{LS} is Weber number of liquid superficial, and a correlation coefficient R^2 equals to 0.75.

Yet another form of the correlation equation was proposed by Ramamurthi et al. [97], where the Sauter diameter also depends on the mass flow rate of gas and liquid and the Reynolds number:

$$SMD = 2.61 \times 10^{-3}[GLR(Re_c)\ ]^{-0.66} \tag{19}$$

Equation (19) is valid for annular flow and Rec < 10000. For the bubbly flow, the equation takes a modified form [97]:

$$SMD = 4 \times 10^{-12}\left[Re_c(GLR)^{0.5}\right]^{-1.14} \tag{20}$$

true for GLR values in the range of 0.005 to 0.04 and for Re_c values from 10,000 to 15,000.

In the work [111], a correlation equation was proposed for the determination of the SMD value of a spray created as a result of spraying aqueous silica solutions (Aerosil 300):

$$SMD = 9.79 \times 10^{-4}(We_g GLR)^{-0.11} \tag{21}$$

This equation is correct when the following assumptions are met: Aerosil 300 concentration in the solution is <4%, d_0 = 0.0017 m, Weg value ranges from 80 to 1700, and GLR ranges from 0.014 to 0.46.

Broniarz-Press et al. [45] conducted research related to the process of spraying aqueous solutions of poly (ethylene oxide) with molar mass from 1×10^6 to 8×10^6 kg/kmol. The tested aqueous polymer solutions showed the characteristics of non-Newtonian liquids. Based on the conducted research, the authors proposed a correlation equation in the form:

$$SMD = 1.2 \times 10^{-2} d_0 M_w^{0.25} GLR^{-0.23} \tag{22}$$

where M_W is the average molar mass of the polymer. This equation is valid when $d_0 \in (0.003; 0.006)$ m and $GLR \in (0.028; 0.92)$.

On the other hand, in the work [110], the study of the atomization of liquids with different rheological properties was carried out, which showed the same dynamic viscosity (at shearing), and were different in longitudinal viscosity (at stretching). The analysis of the obtained tests showed a clear influence of longitudinal viscosity both on the change of the internal and external flow structure and on the increase in the diameter of the formed droplets [110,134,135]. The literature on the subject describes that the longitudinal viscosity also significantly affects the spray angle. The mathematical analysis of the obtained results allowed to estimate the relationship:

$$SMD = f\left(\eta_e^{0.28}\right) \tag{23}$$

Figure 16 shows a diagram of the effervescent atomization process and the structure of the effervescent atomizer used in the research by Qian et al. [72,136]. On the basis of the figure illustrating the course of the atomization process, three main stages can be distinguished: atomization with internal mixing (a two-phase system is formed: gas bubbles in the liquid), primary breakup (breakup of the liquid into small drops), and secondary breakup (the resulting drops are subjected to a series of events, i.e., collisions, breakup, coalescence).

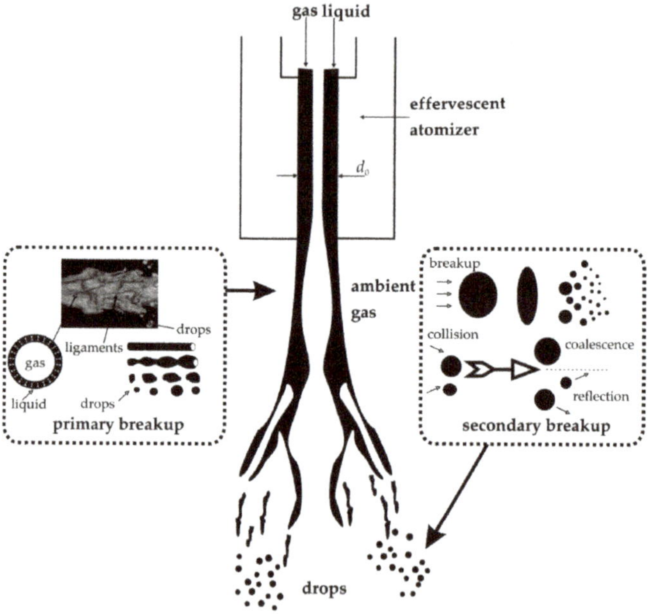

Figure 16. Schematic diagram of the atomization process with the use of an effervescent atomizer [7,72].

A comprehensive three-dimensional model including primary and secondary breakup is described in the works [72,136]. This model was used to estimate the diameter of droplets with the value smaller than the diameter of the atomizer outlet opening. In order to describe the gas phase, the mean values of *Re* from the Navier-Stokes equation (*k*-ε model) were used, while the phase of the dispersed droplets was presented as Lagrange elements in relation to physical phenomena [72,137].

Accordingly, Lin et al. [137] and Qian et al. [72] obtained the equation taking into account the secondary breakup of droplets of the form:

$$SMD = 0.038\left(\frac{GLR}{0.12}\right)^{-0.4787}\left(\frac{P_{in}}{4\times10^6}\right)^{-0.1639}\left(\frac{d_0}{0.2}\right)^{0.7039} + 10^{-3}(ky + SMD_w)\ [mm] \tag{24}$$

where y is the axial distance, the k-factor defines the operating parameters of the atomizer. In Equation (24), the first segment is related to the primary atomization and the second segment is related to the *SMD* change along with the axial distance from the outlet opening. In turn, the third segment of the equation (SMD_w value) determines the change in *SMD* value during secondary atomization. However, it should be remembered that the proposed equation includes variables and constants limiting its application.

In order to determine a convenient mathematical description, the effervescent atomization process underwent modelling based on the appropriate Navier-Stokes equations combined with the particle monitoring method. Furthermore, the external gas flow was assumed to be a turbulent flow. This model also takes into account primary and secondary breakup, and the average diameter of spray droplets was estimated with the use of various operating conditions and properties of the sprayed liquid [72]. Qian et al. [72] finally proposed three correlation equations to estimate the value of the mean surface-volume diameter:

- immediately at the atomizer outlet opening:

$$SMD_{(y\to 0)} = 0.00505\left(\frac{GLR}{0.12}\right)^{-0.4686}\left(\frac{P_{in}}{5\times 10^{6}}\right)^{-0.1805}\left(\frac{d_0}{0.2}\right)^{0.6675}\left(\frac{\eta_c}{0.2}\right)^{0.1714}\left(\frac{\sigma}{46}\right)^{0.1382} \quad (25)$$

- in a situation where the distance from the atomizer outlet is small and does not exceed 10 mm:

$$\begin{gathered} SMD_{(0<y<10)} = 10^{-4}y\left[1.103\left(\tfrac{GLR}{0.12}\right)^{-0.218} + 14.72\left(\tfrac{GLR}{0.12}\right)^{-0.3952}\left(\tfrac{\eta_c}{0.2}\right)^{0.1571}\left(\tfrac{\sigma}{46}\right)^{0.8199}\right] + \\ +0.00505(1-y)\left(\tfrac{GLR}{0.12}\right)^{-0.4686}\left(\tfrac{P_{in}}{5\times 10^{6}}\right)^{-0.1805}\left(\tfrac{d_0}{0.2}\right)^{0.6675}\left(\tfrac{\eta_c}{0.2}\right)^{0.1714}\left(\tfrac{\sigma}{46}\right)^{0.1382} \end{gathered} \quad (26)$$

- when the distance from the atomizer outlet is greater than 10 but less than 200 mm:

$$SMD_{(10<y<200)} = 10^{-4}\left[1.103y\left(\frac{GLR}{0.12}\right)^{-0.218} + 14.72\left(\frac{GLR}{0.12}\right)^{-0.3952}\left(\frac{\eta_c}{0.2}\right)^{0.1571}\left(\frac{\sigma}{46}\right)^{0.8199}\right] \quad (27)$$

As can be seen, in Equations (25)–(27) an important role is played by the value y, which defines the distance from the atomizer outlet opening. In order to clarify the notation of the equations, it is worth adding that in the above equations the values of y and d_0 are defined in centimeters, P_{in} in grams per centimeter and square of a second, η_c in grams per centimeter and second, and σ in grams per square of a second.

In the work [138], a three-dimensional model of liquid droplets and gas bubbles in a two-phase flow was described. The authors of the work conducted research on the development of the atomized liquid structure along the atomizer outlet opening. It was evidenced in the work that *GLR* is one of the most important parameters enabling the control of the atomization process, and the increase in the value of this ratio contributes to a gradual reduction in the size of the analyzed droplets. Moreover, it has been proven that the use of a smaller atomizer outlet opening promotes primary breakup and that high injection pressure has a greater effect on the secondary breakup of the drops.

The internal geometry of the atomizer also plays an important role in the atomization process. It determines the nature of the internal flow of gas and liquid and can significantly affect the efficiency of the entire process. The conducted studies of the influence of the size and number of aerator holes on the average droplet size proved that the use of an aerator with many holes leads to a slightly narrower diameter distribution in relation to the aerator with one hole and the same effective area of gas inlet openings. However, this phenomenon is not fully understood and explained [68].

On the basis of the mean value of the droplet velocity and the distribution of their velocity, it is possible to determine the effectiveness of the process [8,56]. Based on the conducted research, Panchagnula and Sojka [57], Sankar et al. [139], and Jedelsky et al. [106] concluded that the droplet velocity increases as a result of the increase in the set injection pressure. A steady increase in the droplet velocity with the increase in *GLR* value was also observed. It has also been found that the use of higher mass gas flow rates leads to droplets acceleration [106,137]. The literature review indicates that the value of the droplet velocity reaches its maximum in the aerosol axis (y = 0), and it quickly decreases with the increase of the axial distance from the atomizer and the increase of the radial distance from the aerosol axis [40,57,106,137,140–142]. The highest mass flow rate is observed in the axis of the generated spray [115]. Lund et al. [116] analyzed the effect of gas molecular weight on droplet velocity and proved that droplet velocity decreases due to an increase in gas molecular weight.

Numerous studies published in the papers have shown that the use of a polymer additive to the sprayed liquid contributes to the reduction in the number of average droplets, and thus to the increase in the number of droplets with large diameters [7,45,143,144]. In turn, the number of small droplets decreases as a result of an increase in the concentration of the polymer and its molar mass [144]. This phenomenon is important from the practical point of view, because, for example, in agrotechnical spraying processes, droplets with very small diameters are undesirable due to the fact that they can be easily moved by the wind [144], as well as in the process of reducing spontaneous combustion of fuels carried out during emergency landings in aviation [14]. Zhu et al. [145] proved that the mean droplet diameter increases with the increase in the dynamic viscosity of the liquid. Mun et al. [144] also observed some discrepancies in conducted research. They showed that, depending on the type of atomizer used for spraying a 50% aqueous solution of glycerin, a lower, the same, or higher value of the average droplet diameter was obtained with regard to water. This proves that, based on the literature data, it is difficult to unambiguously determine the effect of the change in liquid viscosity on the mean droplet diameter.

It is worth emphasizing, however, that on the basis of all correlation equations proposed in the literature, it appears that the value of the Sauter's mean diameter decreases as a result of the increase in *GLR* value. However, it should be remembered that the influence of the atomizer outlet opening diameter on this value was not taken into account in all cases. It has been shown, in turn, that the longitudinal viscosity of the sprayed liquid is a very important factor influencing the value of the mean surface-volume diameter [7,109]. The increase in the longitudinal viscosity of the sprayed liquid contributes to the increase in the size of the droplets of the spray produced. This is confirmed by the research by Zhu et al. [145], carried out with the use of hydraulic atomizers for spraying. In the work [145] it was evidenced that the droplet diameter increases in a non-linear manner with the increase in the longitudinal viscosity (according to the power function with an exponent of 0.15). This was also confirmed by the studies published in the work [146]. The authors of this work subjected aqueous solutions containing polyacrylamide to the atomization process. In the work [7], a correlation equation was proposed that allows to determine the *SMD* value based on the formula:

$$SMD = Cd_0^{1.12}\dot{M}_c^{0.64}GLR^{-0.80}We_g^{0.21}Oh^{0.06}\left(\frac{\eta_c}{\eta_{water}}\right)^{0.16}Tr^{0.28}\ [m] \tag{28}$$

where C is a parameter (with a value of 6.98×10^{-4}) depending on the internal geometry of the atomizer and other parameters affecting the mean diameter of the droplets (however, it should be remembered that their influence has not been analyzed). Equation (28) is the first one that covers the longitudinal viscosity of the atomized liquid.

The purpose of this article is to summarize the results obtained from scientific researches of effervescent-swirl atomizers performance embracing wide variations in atomizer design, liquid properties, and operating conditions. The results of research on atomizers of this type may contribute to the spread of their use and suggest possible areas for future practical applications, including completely new areas.

3. Conclusions

The paper presents the theoretical basis of the atomization process, the use of atomizers, and an analysis of the collected literature data. Effervescent-swirl atomizers were characterized. The authors of the work, on the basis of the literature review, analyzed the results of research related to, inter alia, the phenomenon of air core formation and the influence of a number of parameters on the efficiency of the atomization process. This work includes the analysis of the results of the research on the influence of gas and liquid flow rates; injection pressure, or the dimensions and design of the atomizer, the shape of the outlet opening, and the physicochemical properties of the atomized liquid on the pressure drop, discharge coefficient, spray angle, droplet size, quantitative distributions, and mean volume-surface diameter. The paper presents correlation equations covering the basic features of the atomization process, which bind a large number of parameters affecting the efficiency of the atomization process and the nature of the spray produced, which may be useful in design practice.

An ability to predict atomizer performance is desired since it would significantly reduce time and cost in the design process. To the areas that we feel warrant further investigations belongs:

- Because, a comprehensive model of the atomization process is not yet available, there is considerable scope for efforts to model the subprocesses involved in effervescent-swirl atomization and to integrate them into a comprehensive model;
- Due to different and different physico-chemical properties, the atomization of non-Newtonian liquids, suspensions and liquid metals should be tested;
- In the future, it would be necessary to investigate and analyze the impact of various independent parameters on the unsteadiness of effervescent-swirl sprays, fluid mechanics of effervescent-swirl atomization, and perform modeling (CFD) of the process.

Further work is necessary to gain a deeper understanding of the fundamental mechanisms and two-phase flow phenomena involved in effervescent-swirl atomization.

Author Contributions: Conceptualization, M.O., M.M., K.C., and S.W. (Stanisław Witczak); formal analysis, S.W. (Stanisław Witczak), K.C., M.M., S.W. (Sylwia Włodarczak), A.K., and M.H.; writing—original draft, M.M., M.O., K.C., and S.W. (Stanisław Witczak). All authors have read and agreed to the published version of the manuscript.

Funding: This research was funded by the Ministry of Education and Science of Poland.

Institutional Review Board Statement: Not applicable.

Informed Consent Statement: Not applicable.

Data Availability Statement: The data presented in this study are available in the article.

Acknowledgments: The study was supported by the Ministry of Education and Science of Poland.

Conflicts of Interest: The authors declare no conflict of interest.

References

1. Ochowiak, M. Koncepcja atomizera pęcherzykowo-wirowego z analizą oporów przepływu. *Inżynieria Apar. Chem.* **2012**, *51*, 360–361.
2. Lefebvre, A.; McDonell, V. *Atomization and Sprays*; CRC Press Taylor & Francis Group: Boca Raton, FL, USA, 2017.
3. Tonini, S.; Conti, P.; Cossali, G.E. Numerical modelling of internal flow in water mist injectors: Effect of nozzle geometry and operating conditions. *Fire Technol.* **2019**, *55*, 2395–2417. [CrossRef]
4. Ochowiak, M.; Krupińska, A.; Włodarczak, S.; Matuszak, M.; Markowska, M.; Janczarek, M.; Szulc, T. The two-phase conical swirl atomizers: Spray characteristics. *Energies* **2020**, *13*, 3416. [CrossRef]
5. Lefebvere, A.H. *Atomization and Sprays*; Hemisphere Publishing Corporation: New York, NY, USA, 1989.
6. Orzechowski, Z.; Prywer, J. *Wytwarzanie i Zastosowanie Rozpylonej Cieczy*; WNT: Warszawa, Poland, 2008. (In Polish)
7. Ochowiak, M. *Analiza Procesu Rozpylania Cieczy w Rozpylaczach Pęcherzykowych i Pęcherzykowo-Wirowych*; Habilitation Thesis; WPP; Poznan University of Technology: Poznań, Poland, 2014; p. 519. (In Polish)
8. Sovani, S.D.; Sojka, P.E.; Lefebvere, A.H. Effervescent atomization. *Prog. Energy Combust. Sci.* **2001**, *27*, 483–521. [CrossRef]
9. Dhivyaraja, K.; Gaddes, D.; Freeman, E.; Tadigadapa, S.; Panchagnula, M.V. Dynamical similarity and universality of drop size and velocity spectra in sprays. *J. Fluid Mech.* **2019**, *860*, 510–543. [CrossRef]
10. Jarrahbashi, D.; Sirignano, W.A.; Popov, P.P.; Hussain, F. Early spray development at high gas density: Hole, ligament and bridge formations. *J. Fluid Mech.* **2016**, *792*, 186–231. [CrossRef]
11. Chaussonnet, G.; Geppertha, S.; Holza, S.; Kocha, R.; Bauera, H.J. Influence of the ambient pressure on the liquid accumulation and on the primary spray in prefilming airblast atomization. *Int. J. Multiph. Flow* **2020**, *125*, 1–24. [CrossRef]

12. Shao, C.; Luo, K.; Chal, M.; Fan, J. Sheet, ligament and droplet formation in swirling primary atomization. *AIP Adv.* **2018**, *8*, 045211. [CrossRef]
13. En'urga Inc. High Flow Rate Effervescent Atomizer. Available online: www.enurga.com (accessed on 12 October 2010).
14. Orzechowski, Z.; Prywer, J. *Rozpylanie Cieczy w UrząDzeniach Energetycznych*; WNT: Warszawa, Poland, 1994. (In Polish)
15. Patil, D.P.; Laskowski, J.S. Development of zero conditioning procedure for coal reverse flotation. *Miner. Eng.* **2008**, *21*, 373–379. [CrossRef]
16. Tricou, C.; Knasiak, K.F. Development of a High Transfer Efficiency Painting Technology Using Effervescent Atomization. In Proceedings of the ILASS–Americas, 18th Annual Conference on Liquid Atomization and Spray Systems, Irvine, CA, USA, 22–25 May 2005.
17. Petersen, F.J.; Worts, O.; Schaefer, T.; Sojka, P.E. Design and atomization properties for and inside-out type effervescent atomizer. *Drug Dev. Ind. Pharm.* **2004**, *30*, 319–326. [CrossRef]
18. Petersen, F.J.; Wørts, O.; Schaefer, T.; Sojka, P.E. Effervescent atomization of aqueous polymer solutions and dispersions. *Pharm. Dev. Technol.* **2001**, *6*, 201–210. [CrossRef] [PubMed]
19. Piątkowski, M.; Zbiciński, I. Płomieniowe suszenie rozpyłowe. *Inżynieria I Apar. Chem.* **2010**, *49*, 87–88. (In Polish)
20. Petersen, F.J. A New Approach for Pharmaceutical Sprays. Effervescent Atomization. Atomizer Design and Spray Characterization. Ph.D. Thesis, Department of Pharmaceutics, The Danish University of Pharmaceutical Sciences, Copenhagen, Danmark, 2004.
21. O'Sullivan, J.J.; Norwood, E.A.; O'Mahony, J.A.; Kelly, A.L. Atomisation technologies used in spray drying in the dairy industry: A review. *J. Food Eng.* **2019**, *243*, 57–69. [CrossRef]
22. Jayapal, S.; Kadiresh, R.M. Spray characteristics of plug type swirl injector. In Proceedings of the Advances in Science and Engineering Technology International Conferences (ASET), Dubai/Sharjah/Abu Dhabi, United Arab Emirates, 6 February–5 April 2018.
23. Kim, K.; Lim, O. Investigation of the spray development process of gasoline-biodiesel blended fuel sprays in a constant volume chamber. *Energies* **2020**, *13*, 4819. [CrossRef]
24. Panchasara, H.; Ashwath, N. Effects of pyrolysis bio-oils on guel atomization—A review. *Energies* **2021**, *14*, 794. [CrossRef]
25. Kim, H.Y.; Ge, J.C.; Choi, N.J. Effects of fuel injection pressure on combustion and emission characteristics under low speed conditions in a diesel engine fueled with palm oil biodiesel. *Energies* **2019**, *12*, 3264. [CrossRef]
26. Holz, S.; Braun, S.; Chaussonnet, G.; Koch, R.; Bauer, H.J. Close nozzle spray characteristics of a prefilming airblast atomizer. *Energies* **2019**, *12*, 2835. [CrossRef]
27. Shinjo, J. Recent advances in computational modeling of primary atomization of liquid fuel sprays. *Energies* **2018**, *11*, 2971. [CrossRef]
28. Payri, R.; García-Oliver, J.M.; Mendoza, V.; Viera, A. Analysis of the influence of diesel spray injection on the ignition and soot formation in multiple injection strategy. *Energies* **2020**, *13*, 3505. [CrossRef]
29. Liu, C.; Liu, F.; Yang, J.; Mu, Y.; Hu, C.; Xu, G. Experimental investigations of spray generated by a pressure swirl atomizer. *J. Energy Inst.* **2019**, *92*, 210–221. [CrossRef]
30. Lee, M.Y.; Lee, G.S.; Kim, C.J.; Seo, J.H.; Kim, K.H. Macroscopic and microscopic spray characteristics of diesel and gasoline in a constant volume chamber. *Energies* **2018**, *11*, 2056. [CrossRef]
31. Fan, X.; Liu, C.; Mu, Y.; Wang, K.; Wang, Y.; Gang Xu, G. Experimental investigations of flow field and atomization field characteristics of pre-filming air-blast atomizers. *Energies* **2019**, *12*, 2800. [CrossRef]
32. Isa, K.M.; Osman, K.; Yahya, A.; Ghaffar, Z.A.; Hamid, A.H.A.; Kasolang, S. Studies on the spray characteristics of pressure-swirl atomizers for automatic hand sanitizer application. *J. Adv. Res. Fluid Mech. Therm. Sci.* **2019**, *55*, 51–64.
33. Golovanevsky, B.; Levy, Y. Fluid Image Velocimetry of the Flow in the Recirculation Zone of a Bluff Body Stabilized and Controlled Burner. In Proceedings of the 10th International Symposium on Applications of Laser Techniques to Fluid Mechanics, Lisbon, Portugal, 10–13 July 2000; pp. 1–10.
34. Golovanevsky, B.; Levy, Y. Suppression of Combustion Instability Using an Aerodynamically Exited Atomizer. In Proceedings of the 11th International Symposium on Applications of Laser Techniques to Fluid Mechanics, Lisbon, Portugal, 8–11 July 2002; pp. 1–14.
35. Sovani, S.D.; Chou, E.; Sojka, P.E.; Gore, J.P.; Eckerle, W.A.; Crofts, J.D. High pressure effervescent atomization: Effect of ambient pressure on spray cone angle. *Fuel* **2001**, *80*, 427–435. [CrossRef]
36. Sovani, S.D.; Crofts, J.D.; Sojka, P.E.; Gore, J.P.; Eckerle, W.A. Structure and steady-state spray performance of an effervescent diesel injector. *Fuel* **2005**, *84*, 1503–1514. [CrossRef]
37. Spicher, U. *Direkteinspritzung im Ottomotor, III: Forschungsergebnisse und Aktueller Entwicklungsstand bei der Benzin-Direkteinspritzung*; Expert Verlag: Renningen, Germany, 2001.
38. Wu, S.R.; Cheng, W.C.; Chiao, J. Low NOx heavy fuel oil combustion with high temperature air. *Fuel* **2007**, *86*, 820–828. [CrossRef]
39. Yu, G.; Li, J.G.; Zhao, J.R.; Yue, L.J.; Chang, X.Y.; Sung, C.J. An experimental study of kerosene combustion in a supersonic model combustor using effervescent atomization. *Proc. Combust. Inst.* **2005**, *30*, 2859–2866. [CrossRef]
40. Sutherland, J.J.; Sojka, P.E.; Plesniak, M.W. Entrainment by ligament-controlled effervescent atomizer-produced sprays. *Int. J. Multiph. Flow* **1997**, *23*, 865–884. [CrossRef]

41. Gong, J.S.; Fu, W.B. The experimental study on the flow characteristics for swirling gas-liquid spray atomizer. *Appl. Therm. Eng.* **2007**, *27*, 2886–2892. [CrossRef]
42. Chin, J.S.; Lefebvre, A.H. A design procedure for effervescent atomizers. *J. Eng. Gas Turbines Power* **1995**, *117*, 226–271. [CrossRef]
43. Sher, E.; Bar-Kohany, T.; Rashkovan, A. Flash-boiling atomization. *Prog. Energy Combust. Sci.* **2008**, *34*, 417–439. [CrossRef]
44. Ochowiak, M.; Broniarz-Press, L.; Różański, J. The discharge coefficient of effervescent atomizers. *Exp. Therm. Fluid Sci.* **2010**, *34*, 1316–1323. [CrossRef]
45. Broniarz-Press, L.; Ochowiak, M.; Woziwodzki, S. Atomization of PEO aqueous solutions in effervescent atomizers. *Int. J. Heat Fluid Flow* **2010**, *31*, 651–658. [CrossRef]
46. Hammad, F.A.; Sun, K.; Che, Z.; Jedelsky, J.; Wang, T. Internal two-phase flow and spray characteristics of outside-in-liquid twin-fluid atomizers. *Appl. Therm. Eng.* **2021**, *187*, 116555. [CrossRef]
47. Kourmatzis, A.; Lowe, A.; Masri, A.R. Combined effervescent and airblast atomization of a liquid jet. *Exp. Therm. Fluid Sci.* **2016**, *75*, 66–76. [CrossRef]
48. Jedelský, J.; Jícha, M. Spray characteristics and liquid distribution of multi-hole effervescent atomisers for industrial burners. *Appl. Therm. Eng.* **2016**, *96*, 286–296. [CrossRef]
49. Chen, B.; Gao, D.; Li, Y.; Chen, C.; Yuan, X.; Wang, Z.; Sun, P. Investigation of the droplet characteristics and size distribution during the collaborative atomization process of a twin-fluid nozzle. *Int. J. Adv. Manuf. Technol.* **2020**, *107*, 1625–1639. [CrossRef]
50. Ghaffar, Z.A.; Kasolang, S.; Hussein, A.; Hamid, A. Characteristics of swirl effervescent atomizer spray angle. *Appl. Mech. Mater.* **2014**, *607*, 108–111. [CrossRef]
51. Ghaffar, Z.A.; Kasolang, S.; Hussein, A.; Hamid, A.; Ahmed, D.I.; Sainan, K.I.; Roselina, N.; Roseley, N. Gas core characteristics of swirl effervescent atomizer. *J. Teknol.* **2015**, *76*, 58–62.
52. Ghaffar, Z.A.; Kasolang, S.; Hussein, A.; Hamid, A.; Chee Sheng, O.; Azlina, M.; Bakar, A. Effect of geometrical parameters interaction on swirl effervescent atomizer spray angle. *J. Teknol.* **2015**, *76*, 63–67.
53. Chen, S.K.; Lefebvre, A.H. Spray cone angles of effervescent atomizers. *At. Sprays* **1994**, *4*, 291–301.
54. Lefebvre, A.H.; Wang, X.F.; Martin, C.A. Spray characteristics of aerated-liquid pressure atomizers. *J. Propuls. Power* **1988**, *4*, 293–298. [CrossRef]
55. Lund, M.T.; Sojka, P.E.; Lefebvre, A.H.; Gosselin, P.G. Effervescent atomization at low mass flow rates. Part 1: The influence of surface tension. *At. Sprays* **1993**, *3*, 77–89.
56. Ochowiak, M.; Broniarz-Press, L. Atomization performance of effervescent atomizers with gas-liquid internal mixing. *Pol. J. Chem. Technol.* **2008**, *10*, 38–41. [CrossRef]
57. Panchagnula, M.V.; Sojka, P.E. Spatial droplet velocity and size profiles in effervescent atomizer-produced sprays. *Fuel* **1999**, *78*, 729–741. [CrossRef]
58. Whitlow, J.D.; Lefebvre, A.H. Effervescent atomizer operation and spray characteristics. *At. Sprays* **1993**, *3*, 137–156. [CrossRef]
59. Whitlow, J.D.; Lefebvre, A.H.; Rollbuhler, J.R. Experimental studies on effervescent atomizers with wide spray angles. *Advis. Group Aerosp. Res. Dev. Conf. Proc.* **1993**, *536*, 1–11.
60. Optimizing Your Spray System. Spray Nozzle Maintenance and Control for Improved Production Efficiency. In Proceedings of the ILASS–Americas, 17th Annual Conference on Liquid Atomization and Spray Systems, Irvine, CA, USA, 13–17 July 2003.
61. Lefebvre, A.H. Twin-Fluid Atomization: Factors Influencing Mean Drop Size. In Proceedings of the 5th International Conference Liquid Atomization & Spray Systems, Gaithersburg, MD, USA, 15–18 July 1991; pp. 49–64.
62. Ochowiak, M. Experimental study on effect of aqueous suspension of silica additives on the atomization process in effervescent nozzles. *Chem. Process Eng.* **2010**, *31*, 35–42.
63. Jicha, M.; Jedelsky, J.; Otahal, J.; Slama, J. Influence of some geometrical parameters on the characteristics of effervescent atomization. In Proceedings of the 18th European Conference on Liquid Atomization and Spray Systems ILASS-Europe, Zaragoza, Spain, 9–11 September 2002; pp. 1–6.
64. Lefebvre, A.H. Some recent developments in twin-fluid atomization. *Part. Part. Syst. Charact.* **1996**, *13*, 205–216. [CrossRef]
65. Buckner, H.E.; Sojka, P.E. Effervescent atomization of high viscosity fluids. Part 1: Newtonian liquids. *At. Sprays* **1991**, *1*, 239–252. [CrossRef]
66. Buckner, H.E.; Sojka, P.E. Effervescent atomization of high viscosity fluids. Part 2: Non-Newtonian liquids. *At. Sprays* **1993**, *3*, 157–170. [CrossRef]
67. Chin, J.S.; Lefebvre, A.H. Flow patterns in internal-mixing, twin-fluid atomizers. *At. Sprays* **1993**, *3*, 463–475. [CrossRef]
68. Wang, X.F.; Chin, J.S.; Lefebvre, A.H. Influence of gas injector geometry on atomization performance of aerated-liquid nozzles. *Int. J. Turbo Jet Engines* **1989**, *6*, 271–279. [CrossRef]
69. Kim, H.G.; Yano, T.; Song, K.K.; Shuichi, T. Microscopic spray characteristics in the effervescent atomizer with two aerator tubes. *Ksme Int. J.* **2004**, *18*, 1661–1667. [CrossRef]
70. Kushari, A. Effect of injector geometry on the performance of an internally mixed liquid atomizer. *Fuel Process. Technol.* **2010**, *91*, 1650–1654. [CrossRef]
71. Li, J.; Lefebvre, A.H.; Rollbuhler, J.R. Effervescent atomizers for small gas turbines. *Am. Soc. Mech. Eng.* **1994**, *78859*, V003T06A048.
72. Qian, L.; Lin, J.; Xiong, H. A fitting formula for predicting droplet mean diameter for various liquid in effervescent atomization spray. *J. Therm. Spray Technol.* **2010**, *19*, 586–601. [CrossRef]

73. Ulbrich, R. *Wykorzystanie Metod Analizy Obrazu do Oceny Nierównomierności Przepływu Gaz–Ciecz w Elementach Płaszczowo-Rurowego Wymiennika Ciepła*; Oficyna Wydawnicza Politechniki Opolskiej: Opole, Poland, 2006. (In Polish)
74. Zając, D.; Ulbrich, R. *Nieinwazyjne Metody Badań Przepływów Dwufazowych Gaz–Ciecz*; Oficyna Wydawnicza Politechniki Opolskiej: Opole, Poland, 2005. (In Polish)
75. Jedelsky, J.; Maly, M.; Del Corral, N.P.; Wigley, G.; Janackova, L.; Jicha, M. Air–liquid interactions in a pressure-swirl spray. *Int. J. Heat Mass Transf.* **2018**, *121*, 788–804. [CrossRef]
76. Bar-Kohany, T.; Levy, M. State of the art review of flash-boiling atomization. *At. Sprays* **2016**, *26*, 1259–1305. [CrossRef]
77. Czernek, K. *Hydrodynamiczne Aspekty Projektowania Aparatów Cienkowarstewkowych dla Cieczy Bardzo Lepkich*; Oficyna Wydawnicza Politechniki Opolskiej: Opole, Poland, 2013. (In Polish)
78. Dziubiński, M.; Prywer, J. *Mechanika Płynów Dwufazowych*; WNT: Warszawa, Poland, 2009. (In Polish)
79. Orzechowski, Z. *Przepływy Dwufazowe Jednowymiarowe Ustalone Adiabatyczne*; PWN: Warszawa, Poland, 1990. (In Polish)
80. Orzechowski, Z.; Prywer, J. *Przepływy Dwufazowe*; Wydawnictwo Politechniki Łódzkiej: Łódź, Poland, 1991. (In Polish)
81. Orzechowski, Z.; Prywer, J. *Rozpylanie Cieczy*; WNT: Warszawa, Poland, 1991. (In Polish)
82. Orzechowski, Z.; Prywer, J.; Zarzycki, R. *Mechanika Płynów w Inżynierii Środowiska*; WNT: Warszawa, Poland, 1997. (In Polish)
83. Bhagwat, S.M. Study of Flow Patterns and Void Fraction in Vertical Downward Two Phase Flow. Master's Thesis, Oklahoma State University, Tulsa, OK, USA, 2011.
84. Rahman, M.A.; Balzan, M.; Heidrick, T.; Fleck, B.A. Effects of the gas phase molecular weight and bubble size on effervescent atomization. *Int. J. Multiph. Flow* **2012**, *38*, 35–52. [CrossRef]
85. Hussein, A.; Hafiz, M.; Rashid, H.; Halim, A.; Wisnoe, W.; Kasolang, S. Characteristics of Hollow Cone Swirl Spray at Various Nozzle Orifice Diameters. *Jurnal Teknologi* **2012**, *58*, 1–4. [CrossRef]
86. Różańska, S. Extensional viscosity of w/o emulsions. *Procedia Eng.* **2012**, *42*, 742–752. [CrossRef]
87. Różańska, S.; Broniarz-Press, L.; Różański, J.; Mitkowski, P.; Ochowiak, M.; Woziwodzki, S. Extensional viscosity and stability of oil-in-water emulsions with addition poly(ethylene oxide). *Procedia Eng.* **2012**, *42*, 733–741. [CrossRef]
88. Sutherland, J.J.; Sojka, P.E.; Plesniak, M.W. Ligament controlled effervescent atomization. *At. Sprays* **1997**, *7*, 383–406. [CrossRef]
89. Asihmin, V.I.; Geller, Z.I.; Skobel'cyn, Y.A. Discharge of a real fluid from cylindrical orifices. *Oil Ind.* **1961**, *9*, 135–172.
90. Broniarz-Press, L.; Ochowiak, M.; Różański, J.; Szaferski, W.; Woziwodzki, S. *Przepływ Trójfazowy Gaz–Ciecz–Ciecz w Dyszach Rozpryskowych*; Wydawnictwo Politechniki Poznańskiej: Poznań, Poland, 2010. (In Polish)
91. Nakayama, Y. Action of the fluid in the air micrometer: First report, characteristics of small diameter nozzle and orifice. *Bull. Jpn. Soc. Mech. Eng.* **1961**, *4*, 516–524. [CrossRef]
92. Koch, R.; Noworyta, A. *Procesy Mechaniczne w Inżynierii Chemicznej*; WNT: Warszawa, Poland, 1998. (In Polish)
93. Marcinkowski, A.; Dziubiński, M. Discharge coefficient for discharge of non-newtonian liquids form vessel. *Chem. Eng. Proc.* **2004**, *25*, 1297–1302. (In Polish)
94. Ochowiak, M. The experimental study on the viscosity effect on the discharge coefficient for effervescent atomizers. *Exp. Therm. Fluid Sci.* **2013**, *50*, 187–192. [CrossRef]
95. Dafsari, R.A.; Lee, H.J.; Han, J.; Park, D.C.; Lee, J. Viscosity effect on the pressure swirl atomization of an alternative aviation fuel. *Fuel* **2019**, *240*, 179–191. [CrossRef]
96. Chen, S.K.; Lefebvre, A.H. Discharge coefficients for plain-orifice effervescent atomizers. *At. Sprays* **1994**, *4*, 275–290.
97. Ramamurthi, K.; Sarkar, U.K.; Raghunandan, B.N. Performance characteristics of effervescent atomizer in different flow regimes. *At. Sprays* **2009**, *19*, 41–56. [CrossRef]
98. Jedelsky, J.; Jichma, M. Prediction of discharge coefficient of internally-mixed twin-fluid atomizers. In Proceedings of the 24th European Conference on Liquid Atomization and Spray Systems ILASS-Europe, Lisbon/Estoril, Portugal, 5–7 September 2011; pp. 1–6.
99. Ochowiak, M.; Matuszak, M.; Włodarczak, S.; Krupińska, A.; Markowska, M.; Gościniak, A.; Szulc, T. The concept design and study of twin-fluid effervescent atomizer with air stone aerator. *Chem. Eng. Proc. Process Intensif.* **2018**, *124*, 24–28. [CrossRef]
100. Jedelsky, J.; Jichma, M.; Slama, J. Discharge coefficient and operational flow characteristics of multihole effervescent atomizer. In Proceedings of the 9th International Conference on Liquid Atomization and Spray Systems ICLASS, Sorrento, Italy, 13–17 July 2003; pp. 1–6.
101. Dziubiński, M.; Przelazły, L. Liquid discharge coefficient for irregularly shaped orifices. *Chem. Eng. Equip.* **2009**, *48*, 31–33.
102. Dikshit, S.; Channiwala, S.; Kulshreshtha, D.; Chaudhari, K. Experimental investigations of performance parameters of pressure swirl atomizer for kerosene type fuel. In Proceedings of the ASME Turbo Expo 2009: Power for Land, Sea, and Air, Orlando, FL, USA, 8–12 June 2009; GT2009-59084. Volume 2, pp. 61–78.
103. Mishra, D.P.; Singh, G. Experimental study of an internally mixed liquid atomizer for an air-breathing engine application. In Proceedings of the 11th International Conference on Liquid Atomization and Spray Systems ICLASS, Vail, CO, USA, 26–30 July 2009; pp. 1–6.
104. Dikshit, S.B.; Kulshreshta, D.B.; Channiwala, S.A. Factors affecting spray cone angle of pressure swirl atomizer for gas turbine combustion chamber: Theoretical and experimental analysis. *Indian J. Sci. Technol.* **2018**, *11*, 1–4. [CrossRef]
105. Wade, R.A.; Weerts, J.M.; Sojka, P.E.; Gore, J.P.; Eckerle, W.A. Effervescent atomization at injection pressures in MPa range. *At. Sprays* **1999**, *9*, 651–667. [CrossRef]

106. Jedelsky, J.; Landsmann, M.; Jichma, M.; Kuritka, I. Effervescent atomizer: Influence of the operation conditions and internal geometry on spray structure; study using PIV-PLIF. In Proceedings of the 22th European Conference on Liquid Atomization and Spray Systems ILASS-Europe, Como Lake, Italy, 8–10 September 2008; pp. 1–8.
107. Loebker, D.W.; Empie, H.J. *Effervescent Spraying of a Medium to High Viscosity Newtonian Fluid at High Mass Flow Rates*; Technical Paper; Institute of Paper Science and Technology: Atlanta, GA, USA, 1998; Volume 62, pp. 1–14.
108. Loebker, D.W.; Empie, H.J. *Independently Controlled Drop Size in Black Liquor Sprays to the Kraft Recovery Boiler Using Effervescent Atomization*; Technical Paper; Institute of Paper Science and Technology: Atlanta, GA, USA, 1999; Volume 823, pp. 1–18.
109. Ochowiak, M.; Broniarz-Press, L.; Różańska, S. The analysis of liquid jet break-up in a one- and two-phase flows. *Chem. Eng. Technol.* **2012**, *35*, 1685–1691. [CrossRef]
110. Ochowiak, M.; Broniarz-Press, L.; Różańska, S.; Różański, J. The effect of exten-sional viscosity on the effervescent atomization of polyacrylamide solutions. *J. Ind. Eng. Chem.* **2012**, *18*, 2028–2035. [CrossRef]
111. Ochowiak, M.; Broniarz-Press, L.; Woziwodzki, S. The analysis of silica suspensions atomization. *Int. J. Heat Fluid Flow* **2011**, *32*, 1208–1215. [CrossRef]
112. Gao, D.; Sun, Y.; Zhang, G. Numerical simulation of atomization characteristics of swirl nozzle. *Icic Express Lett.* **2018**, *12*, 277–285.
113. Ochowiak, M. The effervescent atomization of oil-in-water emulsions. *Chem. Eng. Process. Process Intensif.* **2012**, *52*, 92–101. [CrossRef]
114. Lefebvre, A.H.; McDonell, V.G. *Atomization and Sprays*, 2nd ed.; CRC Press, Taylor and Francis Group: Boca Raton, FL, USA, 2017.
115. Ochowiak, M.; Broniarz-Press, L.; Włodarczak, S.; Markuszewska, M. Effervescent atomization of glicerol aqueous solutions. *Trans. Inst. Fluid-Flow Mach.* **2013**, *125*, 29–38.
116. Gemci, T.; Yakut, K.; Chigier, N.; Ho, T.C. Experimental study of flash atomization of binary hydrocarbon liquids. *Int. J. Multiph. Flow* **2004**, *30*, 395–417. [CrossRef]
117. Jedelsky, J.; Jichma, M. Novel modifications of twin-fluid atomizers: Performance, advantages and drawbacks. In Proceedings of the 23th European Conference on Liquid Atomization and Spray Systems ILASS-Europe, Brno, Czech Republic, 6–8 September 2010; pp. 1–5.
118. Jedelsky, J.; Otahal, J.; Jichma, M. Effervescent atomizer for atomization of suspensions containing large particles. In Proceedings of the 22th European Conference on Liquid Atomization and Spray Systems ILASS-Europe, Como Lake, Italy, 8–10 September 2008; pp. 1–4.
119. Ghaffar, Z.A.; Hamid, A.H.A.; Rashid, M.S.F.M. Spray characteristics of swirl effervescent injector in rocket application: A review. *Appl. Mech. Mater.* **2012**, *225*, 423–428. [CrossRef]
120. Santangelo, P.J.; Sojka, P.E. A holographic investigation of the near nozzle structure of an effervescent atomizer produced spray. *At. Sprays* **1995**, *5*, 137–155. [CrossRef]
121. Santangelo, P.J.; Sojka, P.E. Focused image holography as a dense spray diagnostic. *Appl. Opt.* **1994**, *33*, 4132–4136. [CrossRef] [PubMed]
122. Weber, C. Disintegration of liquid jets. *Z. Für Angew. Math. Und Mech.* **1931**, *11*, 136–159. [CrossRef]
123. Smith, M.W. Utilization of effervescent spray technology to eliminate volatile and toxic diluents. In *Paper of Low- and No-VOC Coating Technologies: 2nd Biennial International Conference*; Research Triangle Institute: Durham, NC, USA, 1995.
124. Esfarjani, S.A.; Dolatabadi, A. A 3D simulation of two phase flow in an effervescent atomizer for suspension plasma spray. *Surf. Coat. Technol.* **2009**, *203*, 2074–2080. [CrossRef]
125. Sterling, A.; Sleicher, C. The instability of capillary jets. *J. Fluid Mech.* **1975**, *68*, 477–495. [CrossRef]
126. Sovani, S.D.; Sojka, P.E.; Sivathanu, Y.R. Prediction of drop size distributions from first principles: The influence of fluctuations in relative velocity and liquid physical properties. *At. Sprays* **1999**, *9*, 133–152. [CrossRef]
127. Sovani, S.D.; Sojka, P.E.; Sivathanu, Y.R. Predictions of drop size distributions from first principles: Joint PDF effect. *At. Sprays* **2000**, *10*, 587–602. [CrossRef]
128. Liu, L.; Pei, N.; Zhao, R.; Tian, L.; Duan, R.; Zhang, Y.; Li, M.; Zhang, X. Effect of the two-phase hybrid mode of effervescent atomizer on the atomization characteristics. *Open Phys.* **2019**, *17*, 960–965. [CrossRef]
129. Kourmatzis, A.; Lowe, A.; Masri, A.R. Conditioned analysis of effervescent atomization. *J. Energy Eng.* **2017**, *143*, 1–33. [CrossRef]
130. White, F.M. *Viscous Fluid Flow*; McGraw-Hill: New York, NY, USA, 1992.
131. Wang, M.R.; Lin, T.C.; Lai, T.S.; Tseng, I.R. Atomization performance of an atomizer with internal impingement. *Jsme Int. J. Ser. B Fluids Therm. Eng.* **2005**, *48*, 858–864. [CrossRef]
132. Harari, R.; Sher, E. Optimization of a plain-jet atomizer. *At. Sprays* **1996**, *6*, 97–113.
133. Mulhem, B.; Schulte, G.; Fritsching, U. Solid-liquid separation in suspension atomization. *Chem. Eng. Sci.* **2006**, *61*, 2582–2589. [CrossRef]
134. Liu, M.; Duan, Y.; Zhang, T. Evaluation of effervescent atomizer internal design on the spray unsteadiness using a phase/Doppler particle analyzer. *Exp. Therm. Fluid Sci.* **2010**, *34*, 657–665. [CrossRef]
135. Liu, M.; Duan, Y.; Zhang, T.; Xu, Y. Evaluation of unsteadiness in effervescent sprays by analysis of droplet arrival statistics—The influence of fluids properties and atomizer internal design. *Exp. Therm. Fluid Sci.* **2011**, *35*, 190–198. [CrossRef]
136. Qian, L.J.; Lin, J.Z. Modeling on effervescent atomization: A review. *Phys. Mech. Astron.* **2011**, *54*, 2109–2129. [CrossRef]
137. Lin, J.Z.; Qian, L.J.; Xiong, H. Relationship between deposition properties and operating parameters for droplet onto surface in the atomization impinging spray. *Powder Technol.* **2009**, *191*, 340–348. [CrossRef]

138. Xiong, H.B.; Lin, J.Z.; Zhu, Z.F. Three-dimensional simulation of effervescent atomization spray. *At. Sprays* **2008**, *19*, 75–90. [CrossRef]
139. Sankar, S.V.; Robart, D.M.; Bachalo, W.D. Swirl effervescent atomizer for spray combustion. *Asme Htd* **1995**, *317*, 175–182.
140. Lund, M.T.; Sojka, P.E. Effervescent atomization at low mass flow-rates. Part 2: The structure of spray. In Proceedings of the 5th Annual Conference on Liquid Atomization and Spray Systems, San Ramon, CA, USA, 18–20 May 1992; pp. 233–237.
141. Jedelsky, J.; Jichma, M. Unsteadiness in effervescent sprays—Measurement and evaluation using combined PIV-PLIF technique. In Proceedings of the 13th International Symposium on Applications of Laser Techniques to Fluid Mechanics, Lisbon, Portugal, 26–29 June 2006; pp. 1–10.
142. Lund, M.T.; Jian, C.Q.; Sojka, P.E.; Gore, J.P.; Panchagnula, M.V. The influence of atomizing gas molecular weight on low mass flow rate effervescent atomization. *J. Fluids Eng.* **1998**, *120*, 750–754. [CrossRef]
143. Broniarz-Press, L.; Ochowiak, M.; Markuszewska, M.; Włodarczak, S. Effect of polyethylene oxide addition on drop si histograms for the Medel Family inhaler. *Chem. Eng. Equip.* **2013**, *52*, 157–158. (In Polish)
144. Mun, R.P.; Young, B.W.; Boger, D.V. Atomisation of dilute polymer solutions in agricultural spray nozzles. *J. Non-Newton. Fluid Mech.* **1999**, *83*, 163–178. [CrossRef]
145. Zhu, H.; Dexter, R.W.; Fox, R.D.; Reichard, D.L.; Brazee, R.D.; Ozkan, H.E. Effect of polymer composition and viscosity on droplet size of recirculated spray solutions. *J. Agric. Eng. Res.* **1997**, *67*, 35–45. [CrossRef]
146. Williams, P.A.; Eeblish, R.J.; Blanchard, R.L.; Rose, S.A.; Lyons, L.; Whitehead, M. The influence of the extensional viscosity of very low concentrations of high molecular mass water-soluble polymers on atomisation and droplet impact. *Pest Manag. Sci.* **2008**, *64*, 497–504. [CrossRef] [PubMed]

energies

MDPI

Article

Particle-Resolved Computational Fluid Dynamics as the Basis for Th Intensification of Fixed-Bed Reactors on Multiple Scales

Nico Jurtz [1,*], Urvashi Srivastava [2], Alireza Attari Moghaddam [3] and Matthias Kraume [1]

1 Chemical & Process Engineering, Technische Universität Berlin, Ackerstr. 76, 13355 Berlin, Germany; matthias.kraume@tu-berlin.de
2 Clariant Corporation, BU Catalysts, 1600 West Hill Street, Louisville, KY 40210, USA; urvashi.srivastava@clariant.com
3 Clariant Produkte (Deutschland) GmbH, Arabellastr. 4a, 81925 Munich, Germany; alireza.moghaddam@clariant.com
* Correspondence: nico.jurtz@tu-berlin.de; Tel.: +49-30-314-70078

Abstract: Process intensification of catalytic fixed-bed reactors is of vital interest and can be conducted on different length scales, ranging from the molecular scale to the pellet scale to the plant scale. Particle-resolved computational fluid dynamics (CFD) is used to characterize different reactor designs regarding optimized heat transport characteristics on the pellet scale. Packings of cylinders, Raschig rings, four-hole cylinders, and spheres were investigated regarding their impact on bed morphology, fluid dynamics, and heat transport, whereby for the latter particle shape, the influence of macroscopic wall structures on the radial heat transport was also studied. Key performance indicators such as the global heat transfer coefficient and the specific pressure drop were evaluated to compare the thermal performance of the different designs. For plant-scale intensification, effective transport parameters that are needed for simplified pseudo-homogeneous two-dimensional plug flow models were determined from the CFD results, and the accuracy of the simplified modeling approach was judged.

Keywords: fixed-bed reactor; wall structures; complex particle shapes; process intensification; heat transfer

Citation: Jurtz, N.; Srivastava, U.; Moghaddam, A.A.; Kraume, M. Particle-Resolved Computational Fluid Dynamics as the Basis for Thermal Process Intensification of Fixed-Bed Reactors on Multiple Scales. *Energies* **2021**, *14*, 2913. https://doi.org/10.3390/en14102913

Academic Editors: Goodarz Ahmadi, Kiao Inthavong and Pouyan Talebizadeh Sardari

Received: 31 March 2021
Accepted: 11 May 2021
Published: 18 May 2021

1. Introduction

Fixed-bed reactors are heavily used in the chemical and process industry, especially in the field of heterogeneous catalysis, where there are thousands of individual catalytic fixed-bed reactors with a low tube-to-particle diameter ratio $N \leq 10$ that are interconnected to tube-bundle reactors. This design decision is the result of optimizing multiple objectives, such as low pressure drop, good radial heat transport, and high active catalytic surface area [1]. Nevertheless, advancing climate change and the shortage of raw materials make more resource- and energy-efficient processes necessary. Here, numerical methods can play a paramount role to develop better designs faster and that are more cost efficient. In the last few years, particle-resolved computational fluid dynamics (CFD) was heavily used by numerous authors to develop process intensification strategies with the focus on the effects on the mesoscopic pellet scale. The range of works extends from investigations of the influence of particle shape on bed morphology and fluid dynamics [2–4], heat transport [5–8], and mass transfer processes [9–12] to the development of novel reactor concepts, such as packed foams [13–15], periodic open-cell structures [16–18], finned reactors [19], or the use of random macroscopic wall structures [20,21]. However, particle-resolved CFD is a numerically very demanding method, and its applicability is currently limited to systems with a few thousand particles [22].

When talking about process intensification, however, this can take place at different spatial scales [23], ranging from the molecular scale to the pellet scale to plant scale. On the plant-scale level, process intensification options are process optimization [24–26], the development of process integration concepts [1,27,28], and applying dynamic operating conditions [29,30]. Due to the restrictions discussed above, particle-resolved CFD cannot directly be applied for the simulation on the largest scale. For this, process simulation software, e.g., *Aspen Plus*, *gPROMS*, or the open-source solution *DWSIM*, is the more efficient choice. Most often, these software packages use two-dimensional pseudo-homogeneous models to describe fluid dynamics and the heat and mass transfer of fixed-beds. For these kinds of models, the knowledge of effective transport parameters, e.g., the effective viscosity and thermal conductivity, the wall heat transfer coefficient, and the axial dispersion coefficient, is necessary to obtain accurate results. Since those parameters need to lump a series of effects that have their fundamentals on micro- and meso-scale fluid dynamic effects, published data varies greatly [31] and can often only be found for certain reactor designs. This is where particle-resolved CFD comes into play, since it has the potential to act as a data source to derive the effective transport parameters that are needed for a reliable process simulation. Recent publications by Dixon [32] and Moghaddam [33] show encouraging results.

In the scope of this work, we investigated different fixed-bed reactor concepts numerically, using particle-resolved CFD. Besides reactors filled with different particle shapes, namely spheres, cylinders, Raschig rings, and four-hole cylinders, additionally, the impact of macroscopic wall structures was studied for packings of spherical particles. The research focus lied on the quantification and qualitative characterization of their heat transport characteristics. For the sake of reduced complexity and to nail the investigations down to the impact of fluid dynamic effects only, no chemical reactions were considered in the investigated cases. The aim of this study was to:

1. understand the effect of particle shape and macroscopic wall structures on the packing morphology and, with this, the fluid dynamics and heat transport in fixed-beds;
2. quantify improvements in the fixed-bed reactor design that can be achieved, only from a fluid dynamics point of view;
3. increase the phenomenological understanding of fluid dynamics and heat transfer in fixed-bed reactors;
4. show how effective transport parameters, such as the effective thermal conductivity and wall heat transfer coefficient, can be extracted from particle-resolved CFD results. Those parameters can then be used in simplified process simulation models for process intensification on the plant scale.

2. Materials and Methods

In this section, the fundamentals of the numerical models are briefly discussed. For a more detailed description, the reader is referred to literature that explains the fundamentals of CFD [34], particle-resolved CFD [22] and simplified fixed-bed reactor modeling [24,35] in more detail.

2.1. Particle-Resolved CFD

Particle-resolved CFD is an established numerical method for the simulation of fixed-bed reactors. This CFD-based modeling approach is characterized by a full three-dimensional spatial resolution of all particles and their interstices. The general procedure consists of four fundamental steps: packing generation, CAD generation, meshing, and the CFD simulation itself. For a more detailed discussion about particle-resolved CFD, the interested reader is referred to comprehensive review articles by Dixon et al. [36,37] and Jurtz et al. [22]. A description of all numerical methods used in the scope of this work can be found in Supplementary Material, Sections S1 and S2 attached to the article. The most important material properties and boundary conditions are summarized in Table 1. All numerical simulations were conducted with the commercial CFD tool *Simcenter STAR-CCM+* provided by *Siemens PLM Software*.

Table 1. Material properties and boundary conditions for the DEM and CFD simulations.

DEM Simulation	
Normal/tangential spring stiffness (N/m)	1×10^6
Static friction coefficient (-)	0.61 / 0.01 (loose/dense bed)
Normal/tangential restitution coefficient (-)	0.5
CFD Simulation	
Fluid density (kg/m^3))	Ideal gas law
Fluid specific heat (J/(kg K))	1006.82
Fluid thermal conductivity (W/(m K))	$0.02414\left(\frac{T}{273.15\,\mathrm{K}}\right)^{3/2}\left(\frac{273.15\,\mathrm{K}+194.0\,\mathrm{K}}{T+194.0\,\mathrm{K}}\right)$
Fluid dynamic viscosity (Pa s)	$1.716 \times 10^{-5}\left(\frac{T}{273.15\,\mathrm{K}}\right)^{3/2}\left(\frac{273.15\,\mathrm{K}+111.0\,\mathrm{K}}{T+111.0\,\mathrm{K}}\right)$
Particle density (kg/m^3)	1500
Particle thermal conductivity (W/(m K))	$0.21 + 1.5 \times 10^{-4} T$ [38]
Particle specific heat (J/(kg K))	1046.7
Inlet velocity (m/s)	0.138; 0.688; 1.376; 2.75
Inlet temperature (K)	293.15
Wall temperature (K)	473.15
Pressure (bar)	1.01325

2.1.1. Numerical Packing Generation Using DEM

In this study, the discrete element method (DEM) was used to numerically generate random packings of spherical and various cylinder-like particle shapes. For the non-spherical particles, the contact detection algorithm of Feng et al. [39] for cylindrical particles was used. The particle beds of Raschig rings and four-hole cylinders are identical to the ones of the cylinders in terms of particle position and orientation, since they are based on the same DEM simulation results. This means that for particles with inner voids, as Raschig rings and multi-hole cylinders, the effect of the inner voids on the particle dynamics during the filling process was neglected. In our previous studies, it was shown that this was a valid assumption [4]. For all filling simulations, the linear spring contact model was used. An overview of all generated packings is given Figure 1.

The aim of this study was to analyze the impact of particle shape and packing mode on the fluid dynamics and heat transfer. Therefore, fixed-beds filled with different particle shapes were generated, whereby the tube-to-particle diameter ratio was held fixed to a value of $N = 5$ by setting a constant volumetric sphere-equivalent particle diameter of $d_{\mathrm{p,v}} = 11\,\mathrm{mm}$ and inner tube diameter of $D = 55\,\mathrm{mm}$. It is known from previous studies [40,41] that even or odd numbered tube-to-particle diameter ratios can lead to additional heterogeneities in the bed morphology. It was expected that additional morphological heterogeneities would lead to an increase of thermal heterogeneities as well. In order to identify the limitations of the pseudo-homogeneous two-dimensional plug flow model, we decided to benchmark the model for such an extreme case against particle-resolved CFD results. Increasing particles thermal conductivity to extreme values, as recently done by Moghaddam et al. [33], is also an option to increase thermal heterogeneities. However, since in most applications, the ceramic-type catalyst support, which is often porous, is used, which is characterized by a low thermal conductivity ($\lambda_s \approx 0.2\,\mathrm{W/(m\,K)}$), we decided to not choose that option. A sketch of each investigated particle shape, including its dimensions, is given in Figure 2. The bed height was set to $h = 100\,d_{\mathrm{p,v}}$. For each particle shape, two different packing modes were created: a rather loose and a dense bed configuration. To achieve different packing densities, the static friction coefficient was used as a tuning parameter during the DEM simulation. A more detailed description of this method can be found in our previous publications [4,42].

The macroscopic wall structure was generated with a Java macro. Along the reactor length of 1.1 m, on 96 axial stages, fifteen spheres were homogeneously distributed on each stage, whereby the center of mass of each sphere was in coincidence with the tube radius. Subsequently, each sphere was move in the outward direction by a factor of rand$(0,1) \cdot d_{p,v}/2$. Thereafter, all spheres of each stage were rotated with a random angle around the reactor axis, using a rotation angle of rand$(0,1) \cdot 2\pi$. In a final step, the macroscopic wall structure was generated by subtracting all spheres from the reactor tube. The structured tube was afterwards filled with spherical particles, whereby the same simulation parameters and boundary conditions were used as for the smooth wall setup.

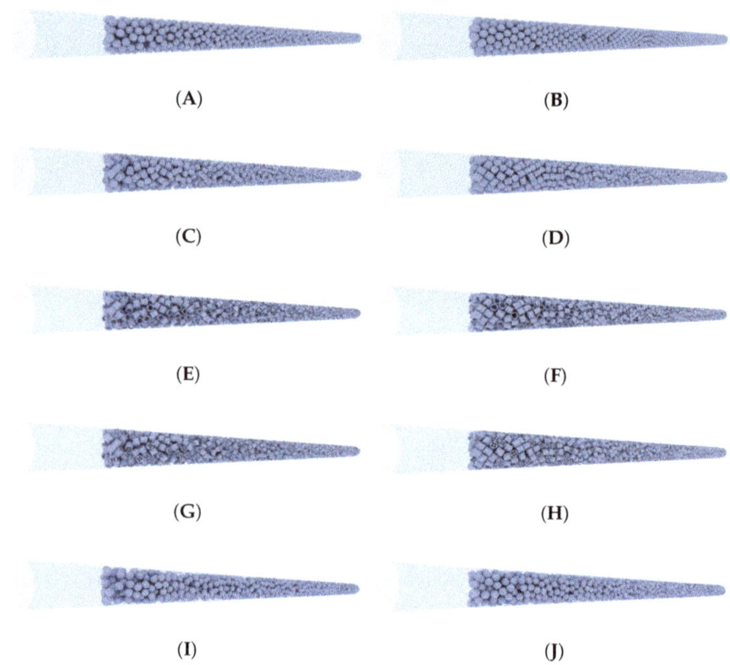

Figure 1. Overview of all generated packings. Left: loose packings of **(A)** spheres, **(C)** cylinders, **(E)** rings, **(G)** 4-hole cylinders, and **(I)** spheres with a macroscopic wall structure. Right: dense packings of **(B)** spheres, **(D)** cylinders, **(F)** rings, **(H)** 4-hole cylinders, and **(J)** spheres with a macroscopic wall structure.

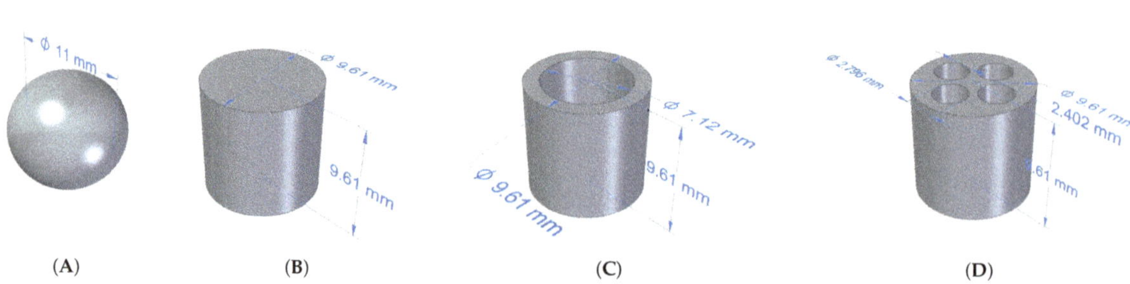

Figure 2. Investigated particle shapes: **(A)** spheres, **(B)** cylinders, **(C)** rings, and **(D)** 4-hole cylinders.

2.1.2. Meshing

After the bed generation was completed, the position and orientation of all particles were extracted, and based on these data, a CAD model of the fixed-bed was generated by placing CAD parts of the respective particle shape. Subsequently, the geometry was meshed, whereby the improved local "caps" approach, developed by Eppinger et al. [8], was used to avoid bad cell quality near particle–particle and particle–wall contacts. This enhanced meshing strategy was based on earlier work of the authors [41]. In one of our previous studies, it was proven that this meshing approach not only worked fine for spherical particles, but also for more complex particle shapes such as cylinders, Raschig rings, and multi-holed cylinders [4,43]. It was found that the mesh settings used led to mesh-independent results regarding fluid dynamics and heat transfer [3,10,44]. Recently, Eppinger and Wehinger [8] investigated the impact of the gaps that were introduced between the particles during the meshing process. They found that the gap size had only a marginal effect on bed voidage and pressure drop. However, the authors found that the fluid in the gaps was no longer stagnant if the gap size was increased above a value of 0.01 $d_{\mathrm{p,v}}$, which was the value that was used in the present work. Therefore, a bigger gap size than the one used in this study could negatively affect the accuracy of heat transfer simulations, since an additional thermal resistance would be introduced for the inter-particle heat transfer.

To avoid unwanted inlet and outlet effects, the inlet and outlet faces were extruded a distance of 1 D and 3 D away from the bed, respectively. Two prism layers with a target thickness of 0.025 $d_{\mathrm{p,v}}$ were used to capture the fluid dynamic and thermal boundary layer at the particles and the tube wall.

2.1.3. CFD Simulation

The momentum, energy, and turbulence transport equations were solved in a segregated manner (see Section S2). For the sake of reduced complexity and to be able to study the convective and conductive heat transfer without any superposing heat transfer mechanism, the heat transfer simulations were conducted under conditions where radiative heat transfer can be neglected, which was, therefore, not accounted for. Because of this, an inlet temperature of $T_0 = 20\,^{\circ}\mathrm{C}$ and a constant wall temperature of $T_{\mathrm{w}} = 200\,^{\circ}\mathrm{C}$ were used along the fixed-bed region.

For all simulations, the SIMPLE-algorithm was used for pressure-velocity coupling, and turbulence was considered through Reynolds-averaged Navier–Stokes (RANS) equations in conjunction with realizable k-ε model-based closures. This turbulence model was successfully used in our previous works [4,8,10,41,43].

2.2. *Simplified Heat Transfer Modeling*

The class of pseudo-homogeneous models is widely used for the simulation of fixed-bed reactors. Here, the particle scale is not resolved. Instead, all effects are lumped into effective transport parameters. In terms of heat transport, the effective transport parameters needed are either a radially invariant effective thermal conductivity $\lambda_{\mathrm{eff,r}}$ and a wall heat transfer coefficient α_{w} or only a radially varying effective radial thermal conductivity $\lambda_{\mathrm{eff,r}}(r)$. The two different concepts are described in the following sections.

2.2.1. Pseudo-Homogeneous $\lambda_{\mathrm{eff,r}}$-α_{w} Model

The pseudo-homogeneous two-dimensional plug flow heat transfer model under steady-state conditions is described by:

$$\rho_{\mathrm{f}} c_{\mathrm{p,f}} u_z \frac{\partial T}{\partial z} = \lambda_{\mathrm{eff,r}} \left(\frac{\partial^2 T}{\partial r^2} + \frac{1}{r} \frac{\partial T}{\partial r} \right) + \lambda_{\mathrm{eff,z}} \frac{\partial^2 T}{\partial z^2}, \tag{1}$$

whereas the following boundary conditions are used:

$$-\lambda_{\mathrm{eff,r}} \frac{\partial T}{\partial r} = \alpha_{\mathrm{w}} (T - T_{\mathrm{w}}) \qquad \text{at } r = R \tag{2}$$

$$\frac{\partial T}{\partial r} = 0 \qquad \text{at } r = 0 \tag{3}$$

$$T = T_0 \qquad \text{at } z = 0 \tag{4}$$

$$\frac{\partial T}{\partial z} = 0 \qquad \text{at } z = L. \tag{5}$$

Here, $u_z = u_0/\varepsilon$ is the constant interstitial velocity, ρ_f the fluid density, and $c_{p,f}$ the specific heat of the fluid. The $\lambda_{eff,r}$-α_w model lumps all radial heat transfer mechanisms into a constant effective radial thermal conductivity $\lambda_{eff,r}$. The steep temperature drop at the tube walls is modeled by the introduction of an artificial wall heat transfer coefficient α_w, using Equation (2). The thermal conductivity in axial direction can be assumed to be equal to the stagnant effective thermal conductivity $\lambda_{eff,z} = \lambda^0_{eff,r}$, or it can be neglected if the system is dominated by convective effects. The model itself has been critically discussed by many authors [31,45,46], but nevertheless widely spread due to its simplistic nature.

It was found by Yagi and Kunii [47] that the radial effective thermal conductivity can be expressed as:

$$\frac{\lambda_{eff,r}}{\lambda_f} = \frac{\lambda^0_{eff,r}}{\lambda_f} + \frac{1}{Pe_{f,r}(\infty)} Pr Re_p. \tag{6}$$

The first term on the right-hand side is the effective radial thermal conductivity of the stagnant bed. A huge number of correlations exists to determine $\lambda^0_{eff,r}$, which have been reviewed by van Antwerpen et al. [48]. Based on a unit cell approach, Zehner and Schlünder [49] derived the following correlation that is widely used:

$$\frac{\lambda^0_{eff,r}}{\lambda_f} = \left(1 - 1\sqrt{1-\varepsilon}\right) + \frac{2\sqrt{1-\varepsilon}}{1-\kappa^{-1}B} \cdot \left(\frac{(1-\kappa^{-1})B}{(1-\kappa^{-1}B)^2} \ln\left(\frac{1}{\kappa^{-1}B}\right) - \frac{B+1}{2} - \frac{B-1}{1-\kappa^{-1}B} \right) \tag{7}$$

Here, κ is the ratio of solid to fluid thermal conductivity and B is the deformation parameter, which is related to the void fraction by $B = 1.25((1-\varepsilon)/\varepsilon)^{10/9}$. The correlation can be further extended by incorporating secondary effects like radiative heat transfer or the effect of particle–particle contacts on the heat transfer. For a more detailed description, the interested reader is referred to the work of Tsotsas [50] and van Antwerpen et al. [48].

For the heat transfer coefficient at the wall, Yagi and Kunii [51] proposed the following correlation for $Nu_w = \alpha_w d_{p,v}/\lambda_f$:

$$Nu_w = Nu^0_w + \frac{1}{(1/Nu^*_w) + (1/Nu_m)}, \tag{8}$$

using:

$$Nu_m = 0.054 Pr Re_p \tag{9}$$

$$Nu^*_w = 0.3 Pr^{1/3} Re_p^{3/4} \tag{10}$$

$$Nu^0_w = \left(1.3 + \frac{5}{N}\right) \frac{\lambda^0_{eff,r}}{\lambda_f}. \tag{11}$$

Nilles and Martin [52,53] developed the following correlation that is widely used:

$$Nu_w = \left(1.3 + \frac{5}{N}\right) \frac{\lambda^0_{eff,r}}{\lambda_f} + 0.19 Pr^{1/3} Re_p^{3/4}. \tag{12}$$

According to Dixon [31] two methods are commonly used to determine $\lambda_{eff,r}$ and α_w. The first option is a parameter estimation done by conducting an optimization study based on the $\lambda_{eff,r}$-α_w model, whereas the objective is to minimize the sum of squared error regarding the radial temperature profile at one or more axial positions. Alternatively, the method described by Wakau and Kaguei [54] can be used. This method is based on the approximate solution of the pseudo-homogeneous $\lambda_{eff,r}$-α_w model and allows determining $\lambda_{eff,r}$ and α_w from the axial temperature profile in the core of the bed and the average outlet temperature. Both can easily be extracted from the particle-resolved simulation results.

The latter method was used in this study and presented in great detail by Wakao and Kaguei [54]. By neglecting the axial thermal conductivity, the analytical solution of Equation (1) is:

$$\frac{T_w - T}{T_w - T_0} = 2\sum_{n=1}^{\infty}\left(\frac{J_0(2a_n r/D)\exp(-a_n^2 y)}{a_n\left(1+(a_n/Bi)^2\right)J_1(a_n)}\right). \tag{13}$$

Here, r is the radial position and Bi the Biot number $Bi = \frac{\alpha_w D}{2\lambda_{\text{eff,r}}}$. a_n is the n-th root of the following equations that include the Bessel function of the first kind and zeroth-order J_0 and the first kind and first-order J_1:

$$Bi J_0(a_n) = a_n J_1(a_n). \tag{14}$$

The parameter y is expressed by:

$$y = \frac{\lambda_{\text{eff,r}} z}{\rho_f u_z c_{p,f}(D/2)^2}, \tag{15}$$

whereas z is the axial position, ρ_f the fluid density, and $c_{p,f}$ its specific heat. Deep in the bed, when $y \geq 0.2$, the first term in the series of Equation (13) becomes predominant, leading to:

$$\frac{T_w - T}{T_w - T_0} = \frac{2J_0(2a_1 r/D)\exp(-a_1^2 y)}{a_1\left(1+(a_1/Bi)^2\right)J_1(a_1)}. \tag{16}$$

with:

$$Bi J_0(a_1) = a_1 J_1(a_1). \tag{17}$$

In the center of the bed ($r = 0$ and $T = T_{\text{core}}$), Equation (16) is reduced to:

$$\frac{T_w - T_{\text{core}}(z)}{T_w - T_0} = \frac{2\exp(-a_1^2 y)}{a_1\left(1+(a_1/Bi)^2\right)J_1(a_1)}. \tag{18}$$

If Equation (18) is logarithmized, it gives:

$$\ln\left(\frac{T_w - T_{\text{core}}(z)}{T_w - T_0}\right) = -a_1^2\left(\frac{\lambda_{\text{eff,r}}}{\rho_f u_z c_{p,f}(D/2)^2}\right)z + \ln\left(\frac{2}{a_1\left(1+(a_1/Bi)^2 J_1(a_1)\right)}\right). \tag{19}$$

It was shown by Wakao and Kaguei [54] that the following relationship for the average outlet temperature T_m is valid for a reasonably large axial position:

$$\frac{T_w - T_m}{T_w - T_{\text{core}}} = \frac{2J_1(a_1)}{a_1}. \tag{20}$$

From Equation (20), a_1 can be solved iteratively, and $\lambda_{\text{eff,r}}$ can be calculated from the slope of Equation (19), subsequently. The wall heat transfer coefficient can either be determined from the intercept of Equation (19) or from Equation (17). The latter method was promoted by Wakao and Kaguei [54], since the authors argued that α_w is very sensitive to slight changes of the intercept.

Both methods were tested during this study. A sensitivity test was conducted based on the particle-resolved CFD results for a packing of spherical particles at $Re_p = 100$. The sensitivity analysis of α_w and $\lambda_{\text{eff,r}}$ towards the accounted temperature range was conducted, whereas the range of $\Theta_{\text{core}} = (T_{\text{core}} - T_0)/(T_w - T_0)$ was varied as follows: $\{\Theta_{\text{core}} \in \mathbb{R} : 0.05 \leq \Theta_{\text{core}} \leq 0.4 \wedge 0.6 \leq \Theta_{\text{core}} \leq 0.95\}$. It was found that $\lambda_{\text{eff,r}}$ had a relative standard deviation (RSD) of $\pm 6\,\%$. Then, from the intercept of Equation (19) calculated, the values of α_w had a very low RSD of $\pm 3\,\%$, while the suggested method of Wakao and Kaguei increased the RSD to ±15 %, which was in contrast to the authors' argumentation. Nevertheless, a huge discrepancy in α_w was found: the values of the intercept-method were up to three times lower in comparison to the values determined from Equation (17). A comparison of particle-resolved CFD results against the results of the two-dimensional plug flow model in terms of axial and radial temperature profiles revealed that the temperature profiles were mispredicted if the intercept method was used, while the method promoted by Wakao and Kaguei led to reasonable results. Therefore, as Wakao and Kaguei did, we also highly recommend calculating α_w from Equation (17) instead of the intercept of Equation (19). To evaluate $\lambda_{eff,r}$ and α_w, the axial temperature profile was limited to $0.2 \leq \Theta \leq 0.8$ in this work.

2.2.2. Pseudo-Homogeneous $\lambda_{eff,r}(r)$-$u_z(r)$ Model

Instead of describing the additional thermal resistance close to the wall with a heat transfer coefficient, a radially varying effective radial thermal conductivity can be introduced. Furthermore, the radial variations of the interstitial velocity and effective axial thermal conductivity can also be considered. With this, Equation (1) is modified as follows:

$$\rho_f c_{p,f} u_z(r) \frac{\partial T}{\partial z} = \frac{1}{r}\frac{\partial}{\partial r}\left(\lambda_{eff,r}(r) r \frac{\partial T}{\partial r}\right) + \lambda_{eff,z}(r)\frac{\partial^2 T}{\partial z^2}. \tag{21}$$

In this case, the artificial boundary condition described in Equation (2) vanishes and is replaced by the following Dirichlet boundary condition:

$$T = T_w \qquad \text{at } r = R. \tag{22}$$

As reviewed by Dixon [31], multiple models exist to determine $\lambda_{eff,r}(r)$. Most often, the reactor is split into two regions to characterize the heat transfer in the near-wall and the bulk region separately. The models reported in the literature vary in their definition of the extent of each region and the description of $\lambda_{eff,r}(r) = f(r)$. Ahmed and Fahien [55] defined the wall region to be $2\,d_{p,v}$ thick and used a cubic dependency for $\lambda_{eff,r}(r)$ in the bulk and a linear decrease in the wall region. They used the correlations of Argo and Smith [56] in combination with correlations for the radial void fraction distribution to obtain the necessary values of $\lambda_{eff,r}$ in the center of the bed, at the tube wall, and at the interface of both regions. Contrary, Gunn et al. [57–59] used a constant value for $\lambda_{eff,r}$ in the bulk region and assumed a quadratic dependency of $T(r)$ in the wall region. They defined the wall region to be $0.3\,d_{p,v}$ thick. Smirnov et al. [60] defined a wall thickness that depended on bed voidage and particle specific surface area. They used a constant effective thermal conductivity in the bulk region and a linear dependency close to the wall. Winterberg et al. [61] proposed a Reynolds number-dependent thickness of the wall region. In the core region, a constant $\lambda_{eff,r}$ was assumed, which decreased in the wall region, using a power-law approach that depends on the Reynolds number, Péclet number, and three more parameters. Recently, Pietschak et al. [62] reviewed several heat transfer correlations and found the correlation of Winterberg et al. [61] to be superior, especially if axial and radial variations of fluid properties were considered. Pietschak et al. [63] proposed a new correlation that accounted for the drop of $\lambda_{eff,r}$ close to the wall, but without the need to introduce a discontinuity at the interface of the near wall and the bulk region. The authors correlated $\lambda_{eff,r}(r) = f\left(\rho_f, c_{p,f}, d_{p,v}, \varepsilon_0, \varepsilon_w, \varepsilon(r), u_0(r)\right)$ and added the cross-mixing factor and an exponent as additional parameters. The radial velocity and void fraction profiles were taken from additional correlation, but the needed data could potentially also be derived from particle-resolved simulations, as recently shown by Dixon [32].

In this work, the correlation of Winterberg et al. [61]:

$$\lambda_{eff,r}(r) = \lambda^0_{eff,r} + \frac{u_{0,c} d_{p,v} \rho_f c_{p,f}}{K_w} \cdot f(R - r), \tag{23}$$

using:

$$f(R-r) = \begin{cases} \left(\frac{R-r}{k_{\mathrm{f,w}} d_{\mathrm{p,v}}}\right)^{n_{\mathrm{f,s}}} & \text{if } 0 < R-r < k_{\mathrm{f,w}} d_{\mathrm{p,v}} \\ 1 & \text{if } k_{\mathrm{f,w}} d_{\mathrm{p,v}} < R-r < R, \end{cases} \tag{24}$$

was used as the basis to determine $\lambda_{\mathrm{eff,r}}(r)$. In Equation (23), K_{w} is the cross-mixing factor that describes the relationship between effective thermal conductivity, particle shape, and flow velocity deep in the bed. The cut-off parameter $k_{\mathrm{f,w}}$ in Equation (24) sets the dimensionless wall distance, after which the constant thermal conductivity, which was assumed in the core region, drops towards the wall. The exponent $n_{\mathrm{f,s}}$ describes the curvature of the damping function close to the wall.

To determine the parameters above, the circumferentially averaged radial temperature profiles in the interval $z = [0.1 : 0.1 : 1.1]$ were extracted from the particle-resolved CFD simulations for the simulation with $Re_{\mathrm{p}} \geq 500$. For the lowest investigated Reynolds number of $Re_{\mathrm{p}} = 100$, the radial temperature gradients flattened out quickly. Therefore, in this case, only the temperature profiles in the range of $z = [0.1 : 0.1 : 0.3]$ were considered. Based on the model described by Equation (21), a parameter optimization study was conducted, using the Nelder–Mead algorithm, while the objective was to minimize the sum of least squares of the difference of the radial temperature profiles between the simplified model and the particle-resolved results. In total, a number of 1100 ($Re_{\mathrm{p}} \geq 500$) and 300 ($Re_{\mathrm{p}} = 100$) data points were available for the optimization task for each operating condition.

3. Heat Transfer Validation

Experimental validation data for axial or radial temperature profiles are scarce and hard to find. Nevertheless, Wehinger et al. [3] and Dong et al. [6] were able to prove the accuracy of the particle-resolved CFD approach, especially in combination with the local "caps" meshing strategy, in terms of axial and radial temperature profiles.

Based on experimental data that were provided by *Clariant International Ltd.*, a validation study was conducted in this work to confirm the reliability of particle-resolved CFD also under industrially relevant conditions (T > 1000 °C). The experimental setup consisted of a hot box, fired with an electrical furnace, and a single reformer tube with an inner diameter of 0.1016 m and a bed height of 1 m. With thermocouples, placed at the outside of the reformer tube, the axial profile of the outer wall temperature was measured. The temperature in the center of one of the packed reformer tubes was measured with a 0.25″ standard 316 SS axial thermowell until an axial distance of approximately 0.5 m. The thermocouples used were of type K, with an accuracy of ±1.5 °C or ±0.4%, whichever was greater. In the scope of this study, two different particle shapes, a tablet-like cylinder with six holes (33 × 18 mm) and an almost equilateral cylinder with ten holes (19 × 16 mm), were investigated.

The experimental setup was replicated numerically, whereas special emphasis was given to meet the particle count that was determined in the experiments. To achieve this, particle static friction coefficients were calibrated, as described by Jurtz et al. [4,42]. Since the tube thickness was relatively big, not only the fluid and the particles, but also the reformer tube itself were spatially discretized. While a fully conformal contact interface was used for the particle-fluid interface, for the sake of reduced cell count, the tube-fluid interface was performed as a non-conformal mapped contact interface. An overview of the investigated setups, including snippets of the resulting meshes, can be seen in Figure 3. Preliminary studies showed that the thermowell not only affected the flow field significantly, as recently discussed by Dixon and Wu [64]. It was found that the heat conduction through the thermowell could not be neglected, since it significantly affected the temperature distribution in the vicinity of the measuring device. To consider for heat conduction in the thermowell, a three-dimensional shell model was used that solved for the lateral conductive heat transport and modeled the heat transport in the face normal direction via the assumption of a constant temperature gradient. The radiative heat transport was considered using a surface-to-surface radiation model as done by Wehinger at al. [7] recently.

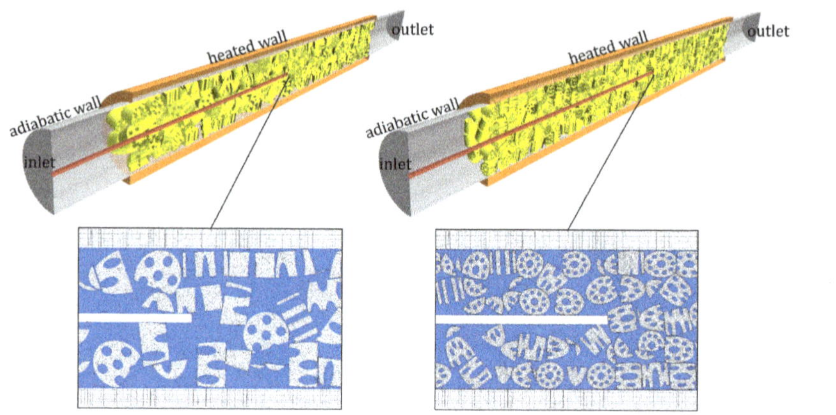

Figure 3. Visualization of the geometry and the mesh for a 6-hole tablet (**left**) and a 10-hole cylinder (**right**).

The experimentally measured temperature distribution on the outer side of the reformer tubes was applied as a spatially varying fixed temperature boundary condition at the outer tube surface. The inlet temperature, according to experimental data, was set to 560 °C and the operating pressure to 1.5 barg. Nitrogen was used as the working fluid, whereas the ideal gas equation of state was used. The fluid viscosity and thermal conductivity were calculated using the Chapman–Enskog model. Inlet flow rates were varied between 15 and 50 Nm^3/h. Particles' thermal conductivity was set to 0.25 W/(m K), whereas for the tube and thermowell, the following function, derived from the spec sheet, was used: $\lambda_s[W/(m\,K)] = 8.195 \cdot \exp(1.188 \cdot 10^{-3} \cdot T)$. The emissivity was set to 0.75 for the particles surface and to 0.6 for the inner reformer tube and the thermowell.

The simulation results are given in Figure 4 in terms of the axial profiles of the dimensionless temperature $\Theta = (T - T_0)/(T_{ref} - T_0)$, whereas T_{ref} is the furnace temperature. The numerical data are presented as a scattered cloud of small symbols to also visualize the temperature variation in the circumferential direction. It can be seen that due to the conductive heat transport within the solid of the thermowell, temperature variations in circumferential direction were low. Without considering this heat transfer mechanism, temperature differences of over 50 K were found (see Section S3), which indicated that the measuring device not only affected the fluid dynamics, but also the measured temperature field significantly. This strengthened the argument that the use of high-fidelity numerical methods can improve the accuracy of determining effective heat transfer parameters significantly. Deep in the bed, an excellent agreement could be found between the predicted and measured temperatures for the six-hole tablets. Only for $z/h \approx 0.1$, some deviations were found. However, if one considers the obvious impact of the heterogeneous bed morphology on the axial temperature profile, the accuracy was still acceptable. For the 10-hole cylinders, the experimental temperature profile was hit almost perfectly for $z/h \leq 0.35$. For a flow rate of 15 N m^3/h, also deep in the bed, a good agreement with the experimental data was found. However, at higher flow rates, for $z/h \approx 0.5$, the simulations results were far off. The reason for this could not be identified, but the fact that the experimental data showed an increase of the axial temperature gradient at higher bed depths for high flow rates was suspicious and may indicate that the temperature sensors were damaged under the harsh operating conditions. Similar problems have been noted by other authors [31] and illustrate the challenges associated with experimental temperature measurements in fixed beds. A possible re-ordering of particles at the tip of the thermowell during operation might also be a possible reason for the deviations observed.

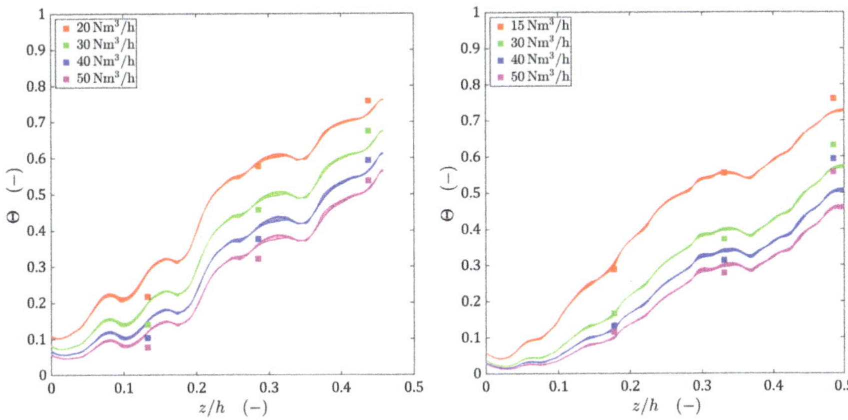

Figure 4. Axial profiles of dimensionless temperature at different flow rates for a 6-hole tablet (**left**) and a 10-hole cylinder (**right**). Comparison of CFD results (scattered cloud) against experimental data (big symbols).

4. Results and Discussion

The heat transport in fixed-bed reactors is strongly coupled to fluid dynamics effects that are induced by the heterogeneous bed morphology. Therefore, in the first part, the bed morphology and fluid dynamics of all generated packings were investigated. Afterwards, the different configurations were compared with regard to their thermal performance. The global heat transfer coefficient $U = \dot{Q}_w / \left(A_w \Delta_{\log} T \right)$, using $\Delta_{\log} T = (T_{out} - T_{in}) / \log((T_w - T_{in})/(T_w - T_{out}))$, was used to compare the different designs, whereby U was evaluated for an axial threshold of the reactor that fulfilled the criterion $0.0 \leq \Theta_{core} \leq 0.8$. In the last part, the simulation results were used to determine effective thermal transport parameters that are commonly needed for the pseudo-homogenous two-dimensional plug-flow model. The results were compared against particle-resolved CFD results to understand the reliability of simplified models. A summary of the most important results and simulation parameters is given in Table 2. A schematic drawing of the setup is presented in Figure 5.

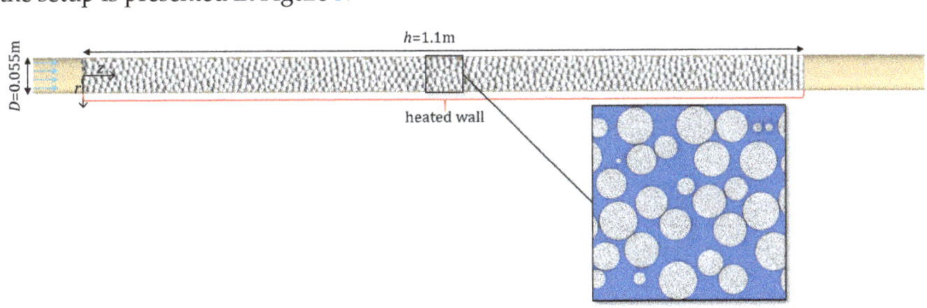

Figure 5. Schematic drawing of the numerical setup, including a snippet of the mesh. Exemplarily shown for the loose packing of spheres.

Table 2. Summary of important results and simulation parameters.

Shape		General Properties					$\lambda_{eff,r}$-α_w-Model Parameters		$\lambda_{eff,r}(r)$-Model Parameters			
		Re (-)	ε (-)	$\Delta p/\Delta z$ (Pa/m)	$\lambda^0_{eff,r}$ (W/(m K))	U (W/(m² K))	α_w (W/(m² K))	$\lambda_{eff,r}$ (W/(m K))	K_n (-)	n_{fw} (-)	k_{fw} (-)	MSE (K²)
spheres	loose	100	0.473	50.0	0.07943	30.86	108.44	0.4048	7.487	1.017	0.2808	1.08
		500	0.473	651.2	0.07943	44.73	172.74	1.4251	10.210	1.242	0.3210	1.40
		1000	0.473	2206.5	0.07943	59.08	236.83	2.3836	10.741	1.246	0.3431	2.41
		2000	0.473	8429.9	0.07943	95.09	369.63	4.9837	11.448	1.310	0.3031	3.26
		100	0.426	65.0	0.08759	20.55	255.61	0.2726	12.051	0.566	0.2711	12.76

4.1. Bed Morphology and Fluid Dynamics

Already, the first visual impression of the generated packings that was given in Figure 1 showed the strong impact that the packing mode had on the particle arrangement. This can best be seen for spherical and cylindrical particles. While for the loose packing configuration, although the confining walls exerted an ordering effect on the particles, to some extent, random arrangement of the particles close to the wall can be seen, the compacted beds were characterized by a high degree of order. Especially the spherical particles tended to build band-like structures at the wall, whereas cylindrical particles built stacked structures and were mostly oriented parallel or perpendicular to the wall. From a fluid dynamics and reaction engineering point of view, the most important effect was the significant reduction in bed voidage that was caused by bed densification. By this, the pressure drop, local flow phenomena, hydraulic residence time, and the active catalytic surface area per reactor volume were significantly affected. The evaluated bed voidage, listed in Table 2, shows that for spherical particles, the bed voidage was reduced by 10%. An extreme reduction of 20% was found for cylindrical particles. For particles with inner voids, like rings and four-hole cylinders, the effect was less pronounced, giving a drop of 10% and 6%, respectively. However, this reduced impact was only a result of the overall higher bed voidage for these particles. For the configuration of spherical particles in the reactor with macroscopic random wall structures, the densification-induced reduction of bed voidage was 11%, which was similar to the reactor with plain walls.

The axial and radial void fraction profiles were good resources to understand the packing morphology of the different designs. Strong and regular oscillations are indicators of ordered particle arrangements and a loss of randomness in the system, whereas low non-regular fluctuations in bed voidage point towards an increasing randomness of the particle arrangement. For an ideally random packing arrangement, the void fraction profile should end in a constant value. Distinct peaks in the void fraction profile are indicators of additional voids that are a result of a non-appropriate filling strategy, which leads to jamming of particles. The axial void fraction profiles of all investigated packings are given in Figure 6. Since the bed rested on a bottom plate, the lowest layers of particles experienced a certain ordering effect, which was induced by the adjacent wall. For spherical particles, only a point contact was possible between the particles and the bottom plate, leading to a value of $\varepsilon = 1$ at $z/d_\mathrm{p} = 0$. Particles of the cylindrical shape type may have a point, line, or face contact with the wall. If face contacts are present, it is possible that $\varepsilon < 1$ at $z/d_\mathrm{p} = 0$. However, for most of the investigated packing, it can be seen that the ordering effect of the bottom plate led to regular oscillations in the void fraction that flattened out after a distance of 3–5 d_p and ended up in random oscillations of lower magnitude, indicating a stochastic axial distribution of the particles. The only exceptions were the compacted packing of spherical particles and the loose packing of spheres in the reactor with macroscopic wall structures. For the dense packing of spheres, regular oscillations were observed between $0 \leq z/d_\mathrm{p} \leq 22$. This indicated a pronounced layer formation in the bottom part of the reactor. In the remaining part of the reactor as well, regular oscillations were observed, albeit to a lesser extent. In the wall structured reactor, high fluctuations were observed that suddenly appeared and flattened out. A probable reason for this was jamming of particles during the filling process that led to additional voids. This hypothesis was strengthened by the fact that this effect vanished for the densified packing.

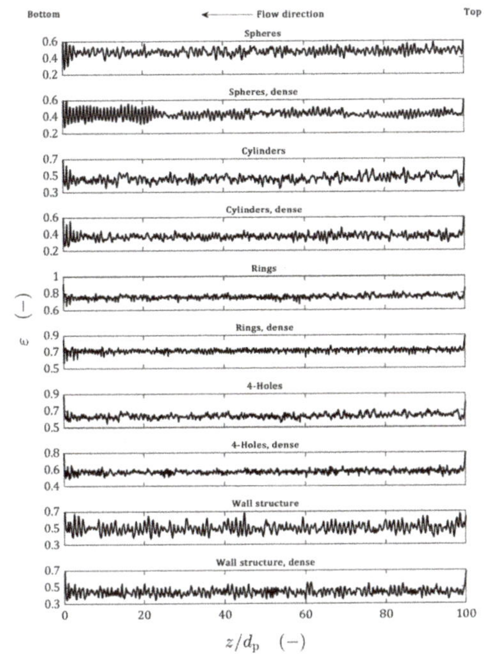

Figure 6. Axial void fraction profiles for all investigated geometries.

Of fundamental interest for the understanding of the fluid dynamics are the radial void fraction profiles and the radial profiles of the circumferentially averaged axial velocity, given in Figure 7. Here, the axial velocity was normalized to the local interstitial velocity $u_0/\varepsilon(r)$. With the exception of the structured wall reactor, for all particle shapes, directly at the wall, a void fraction of $\varepsilon = 1$ was found due to the presence of point and/or line contacts, only. For spherical particles, a first minimum in the void fraction was reached after the distance of one particle radius away from the wall, indicating that the majority of spheres were in direct contact with the wall, forming a closed particle layer. Furthermore, local minima and maxima occurred at positions corresponding to multiples of the particle radius, whereas the oscillations slightly decreased. The global minimum of the void fraction was located in the center of the bed, indicating that an almost stacked arrangement of spheres was present. This was the result of odd tube-to-particle diameter ratios [40,41]. A strong correlation could be found between the void fraction and the velocity profile. Close to the wall, the velocity reached its maximum, known as the wall channeling effect. The position of further minima and maxima corresponded directly to the position where high/low void fractions were found. While the minima of axial velocity did not change with varying Re_p, the maxima increased slightly in the center of the bed if Re_p was lowered. This effect could be attributed to the gas expansion due to heating and to the decreasing wall effect if Re_p was lowered. The center of the bed was almost completely blocked for the flow. The above findings were also valid for the densified packing of spheres; however, the effects were even more pronounced, resulting in a complete blockage of flow paths at $r^* = (R - r)/d_\mathrm{p} = [0.5, 1.5, 2.5]$, and strong channeling was observed at $r^* = [0.1, 1.0, 2.0]$, whereas for $Re_\mathrm{p} \leq 500$, the strongest channeling was not found at the wall, but at $r^* = 2.0$, which is very uncommon.

For cylindrical particles, the trend was similar as for spheres; however, the minima/maxima in the void fraction and velocity were slightly shifted towards the bed center, which indicated that some particles were diagonally aligned. For the densified packing, the minima/maxima were found at multiples of the particle radius, which was a result of the particles' preferred parallel/orthogonal alignment. In contrast to the packings of spheres, where the wall channeling was almost independent of Re_p, for cylindrical particles, the wall channeling effect increased significantly if Re_p was raised. This effect became very dominant for the compacted packing. The void fraction profiles for Raschig rings and four-hole cylinders looked pretty complex; nevertheless, especially for $r^* < 1$, the inner voids of the particles were clearly reflected by corresponding additional maxima in the void fraction. However, no maxima in velocity could be found at void fraction maxima that corresponded to inner voids. This indicated that the flow through the inner particle voids was partially blocked, which might be because of an orthogonal particle alignment. Overall, the void fraction and velocity oscillations were less pronounced for those particle shapes, but heterogeneities increased if the beds were compacted. Similar to cylindrical particles, the wall channeling effect increased with Re_p and became more pronounced for densified packings.

The use of macroscopic random wall structures for packings of spherical particles changed the void fraction and velocity profiles significantly. Due to the presence of the wall structure, the void fraction at the wall fell to a value of $\varepsilon \approx 0.56$. As a result, the wall channeling effect was hindered, and fluctuations in the void fraction and velocity were qualitatively more comparable to the ones of Raschig rings than spheres. The densification of the bed led to slightly more pronounced minima and maxima; however, this effect was not as distinct as for spherical particles in a smooth walled reactor.

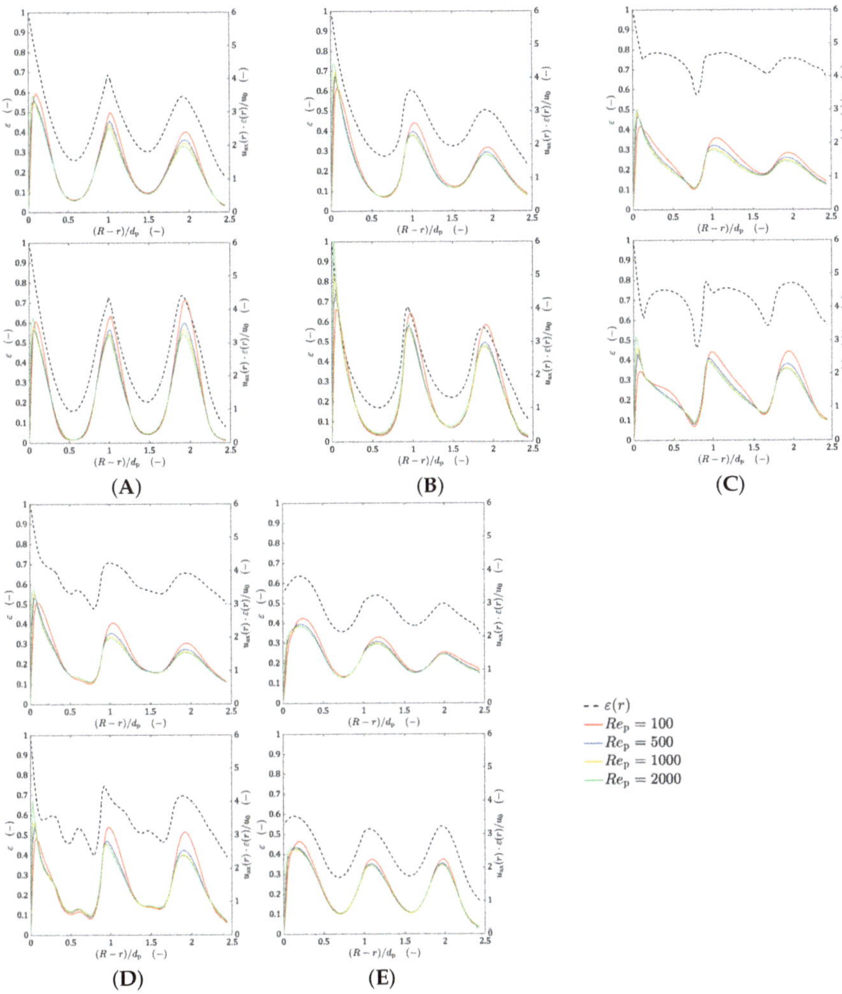

Figure 7. Radial void fraction profiles (dashed line) and radial profiles of the normalized averaged axial velocity (solid lines) for loose (top) and dense (bottom) packings of (**A**) spheres, (**B**) cylinders, (**C**) rings, (**D**) 4-hole cylinders, and (**E**) spheres with the wall structure.

4.2. Heat Transfer Characteristic

A fair comparison of the thermal performance of different reactor concepts always depends on the process boundary conditions that are set. Figure 8 shows the global heat transfer coefficient U as a function of different parameters. Re-fitting of an existing unit that is integrated in a complex production process can lead to the necessity of keeping the throughput constant, which is equivalent to keeping Re_p invariant. In this case, especially at low Re_p, cylindrical particles showed the most beneficial heat transfer characteristic, followed by the wall-structured reactor, Raschig rings, and four-hole cylinders. Spherical particles performed worst over the complete range of investigated Re_p. At high Re_p, cylindrical particles still performed best; however, rings, four-hole cylinders, and the reactor with wall structures were close. For spheres, cylinders, and four-hole cylinders, U decreased if the packings were compacted. This is of special interest, since in industrial applications, most often, densified packings are used to ensure the same pressure drop in the different tubes of the tube bundle reactor. Interestingly the effect was less pronounced for Raschig rings and the wall-structured reactor. At high Re_p, even a slight increase in thermal performance could be seen for those reactor types. In general, the performance gain induced by macroscopic random wall structures was significant.

Another valid process boundary condition can be the necessity of keeping the hydraulic residence time invariant. In this case, Re_p/ε needs to be kept constant. Under this constraint, Raschig rings and four-hole cylinders performed best for moderate to high Re_p, followed by cylinders and the wall-structured reactor. For the lowest investigated Re_p, again, cylindrical particles seemed to perform slightly better than rings.

If a new plant is built and process-driven constraints are low, the most energy efficient particle shape might be an appropriate choice. In this case, the specific pressure drop $\Delta p/\Delta z$ can be one parameter that should be kept constant when comparing different designs. In this case, Raschig rings, the wall-structured reactor, and cylinders performed best. The comparison of the designs from this energetic point of view showed that bed densification led to a less energy-efficient thermal performance, whereas this effect was less pronounced for Raschig rings and the reactor with macroscopic wall structures.

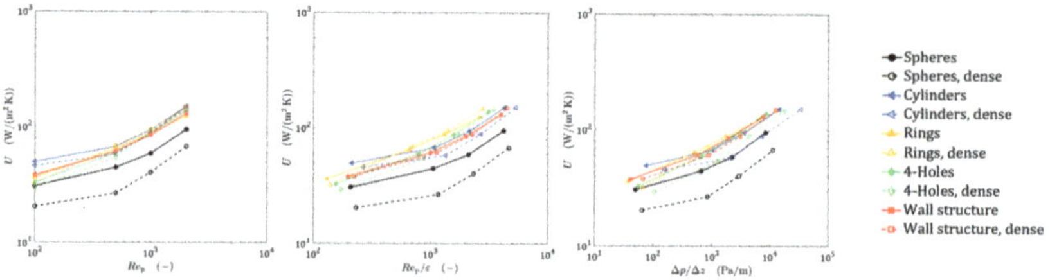

Figure 8. Global heat transfer coefficient as a function of Re_p (**left**), Re_p/ε (**middle**), and specific pressure drop (**right**).

4.3. *Effective Thermal Transport Properties*

As discussed, particle-resolved CFD is a valuable tool to support process intensification on the meso-scale level, e.g., by finding optimized particle shapes [4,10–12] or new reactor concepts, e.g., by applying macroscopic wall structures [20,21] or using internals [19]. However, for process intensification on a macroscopic scale, e.g., by running plants under dynamic operation conditions, developing process integration strategies, or doing plant optimization, different numerical tools are necessary. Process simulation platforms often use pseudo-homogeneous two-dimensional plug flow models. Depending on the class of model used, certain effective transport parameters are needed, which are often not known. In this section, methods are presented for how those parameters can be extracted from particle-resolved CFD results.

4.3.1. $\lambda_{\mathrm{eff,r}}$-α_{w} Model

Although its limitations are well known [31], the $\lambda_{\mathrm{eff,r}}$-α_{w} model is still widely spread, due to its efficiency and simple implementation. Here, the radial heat transport was characterized by the wall heat transfer coefficient α_{w} and the effective radial thermal conductivity $\lambda_{\mathrm{eff,r}}$, which was assumed to be uniform everywhere in the reactor. By extracting the axial core temperature profile and average inlet/outlet temperatures from the CFD simulations, both parameters were determined by using Equations (17), (19), and (20). The results are summarized in Table 2. The parameters were then used to calculate the temperature fields by using the pseudo-homogeneous model described by Equation (1) in conjunction with the boundary condition in Equation (2).

A one-to-one comparison of all investigated cases in terms of radial temperature profiles at different axial positions is provided in Supplementary Material, Section S4. A condensed visualization of the results is given in Figure 9. Here, the deviations of the circumferentially averaged temperature fields, predicted by the pseudo-homogeneous model, are given in relation to the particle-resolved CFD results. Deep red and deep blue colors indicate that the deviation was above or below 10 K. This critical cut-off temperature was chosen, motivated by the rule of van't Hoff, saying that the speed of a chemical reaction doubles to triples itself when the temperature is raised by 10 K [65]. The characteristic temperature drop at the wall that was a result of α_W can only hardly be seen in Figure 9. The reader is referred to the radial temperature profiles given in Section S4. It can be seen that the temperatures close to the wall ($r^* \leq 0.2$–0.4) were systematically underpredicted by the simplified model. This drawback is well known and deeply discussed by many authors [31]. Furthermore, the model was not able to capture morphological and fluid dynamic heterogeneities, which led to step-like temperature profiles, as can be seen best for the radial temperature profiles of the dense spherical packing. Recently, this was also found by Moghaddam et al. [33], who introduced heterogeneities by increasing the solid thermal conductivity. Besides those systematic errors, the deviation in relation to the CFD results was relatively low for the majority of cases. For all investigated designs, the deviation was less than 5 K for the rear part of the reactor ($z/d_p \geq 40$). The threshold of 10 K was mostly exceeded in the entry zone ($z/d_p \leq 20$). Overall, there seemed to be a trend that deviations increased if Re_p was raised. The method seemed to work equally well for loose and densified packings with a slight trend towards less deviations for dense beds. Considering the numerical effort that the simplified model needed in comparison to the particle-resolved CFD simulation, which was ≈10 s compared to ≈24 h, the accuracy was still remarkable.

4.3.2. $\lambda_{eff,r}(r)$-Model

The obvious drawback of the $\lambda_{eff,r}$-α_W model, that the additional near wall thermal resistance is only captured with an artificial temperature drop directly at the reactor wall, can be circumvented by using a radially varying effective radial thermal conductivity. In this work, the correlation of Winterberg [61] (see Equation (23)) was used as the basis to determine the effective radial thermal conductivity. The three necessary parameters of the Winterberg correlation were determined by conducting a parameter optimization study. The basis of this study was the transport equation described by Equation (21). A summary of the optimized model parameters, including the mean squared error MSE, is given in Table 2.

The comparison of the radial temperature profiles for different axial positions can be found in Supplementary Material, Section S5. The spatially resolved deviations between the simplified model and the CFD results are given in Figure 10. It is obvious that in comparison to the results of the $\lambda_{eff,r}$-α_W model, the accuracy was significantly improved. The temperature close to the reactor wall was predicted with a high degree of accuracy by the model. Only sporadically, the temperatures were overestimated by more than 10 K in the vicinity of the wall, whereby the location was mostly limited to the entry zone ($z/d_p \leq 20$). A direct comparison of the $\lambda_{eff,r}$-α_W model and the $\lambda_{eff,r}(r)$ model in relation to the CFD results is given in Figure 11 for the loose packing of Raschig rings at $Re_p = 1000$, showing the superior accuracy of the $\lambda_{eff,r}(r)$ model, especially close to the wall. Deep in the bed, deviations outside of the 10 K threshold were mostly found for packings that were characterized by a higher degree of morphological heterogeneity, like the packings of cylindrical particles, the dense bed of spheres, and the loose packing of spheres in the reactor with a random wall structure. While the former configurations were characterized by strong variations in the radial void fraction distribution, the latter showed big fluctuations in the axial void fraction profile. Bigger deviations were mostly limited to the entry zone, indicating that thermal entrance effects, which were not resolved by an axially invariant $\lambda_{eff,r}(r)$, might be the reason for this. In contrast to the $\lambda_{eff,r}$-α_W model, the deviations did not seem to increase if Re_p was raised. Since the Winterberg correlation did not explicitly consider local variations in the void fraction or axial velocity, it was, similar to the $\lambda_{eff,r}$-α_W model, not able to capture the step-like effects of the temperature profiles.

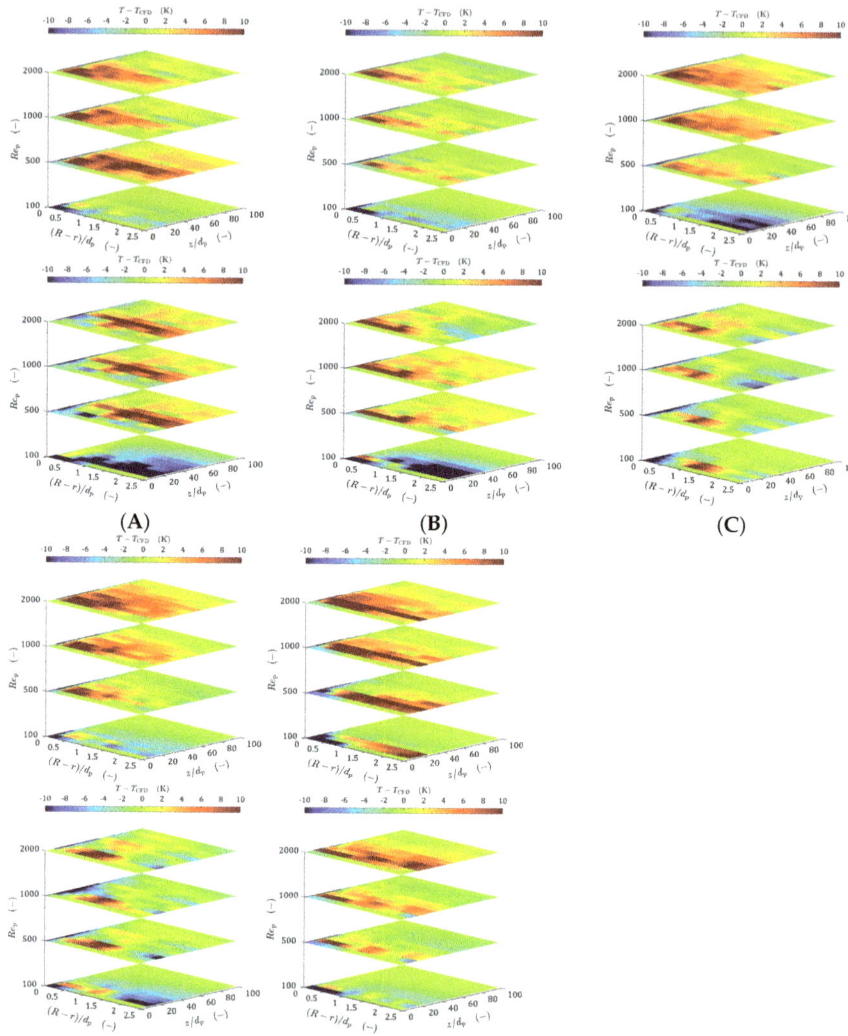

Figure 9. Deviation of the $\lambda_{\text{eff,r}}$-α_{w} model results in comparison to particle-resolved CFD data, for loose (top) and dense (bottom) packings of (**A**) spheres, (**B**) cylinders, (**C**) rings, (**D**) four-hole cylinders, and (**E**) spheres with the wall structure.

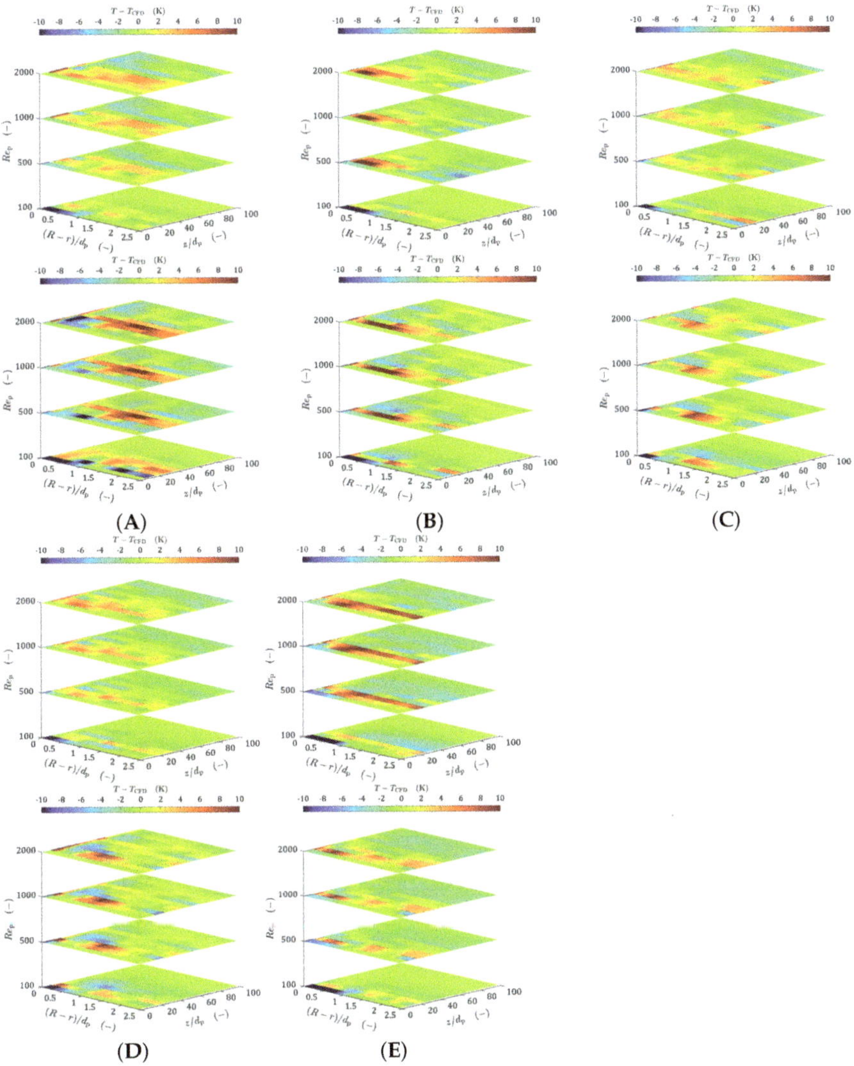

Figure 10. Deviation of the $\lambda_{\text{eff,r}}(r)$ model results (correlation of Winterberg) in comparison to particle-resolved CFD data, for loose (top) and dense (bottom) packings of (**A**) spheres, (**B**) cylinders, (**C**) rings, (**D**) four-hole cylinders, and (**E**) spheres with the wall structure.

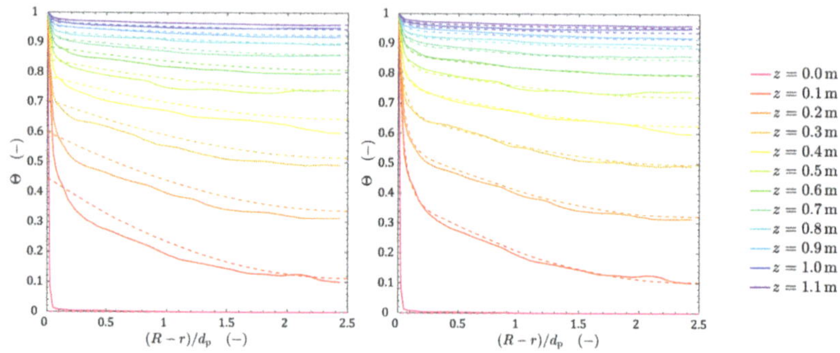

Figure 11. Comparison of the radial temperature profiles derived from a particle-resolved CFD simulation (solid line) and the pseudo-homogeneous model (dashed line) at different radial positions for a loose packing of Raschig rings at $Re_p = 1000$ (left: $\lambda_{\text{eff,r}}$-α_W model; right: $\lambda_{\text{eff,r}}(r)$ model).

5. Conclusions

In this work, it was shown in which way the particle-resolved simulation of fixed-bed reactors can play a central role in the process the intensification of this reactor type. After a brief validation study, showing that also under harsh industrial conditions, particle-resolved CFD was able to predict the temperature field accurately, the heat transfer characteristics of different particle designs were investigated. The studied designs differed in the used particle shape and the bed density. The results showed that heterogeneities in radial void fraction distribution and axial velocity increased, if packings were compacted. As a result, the overall heat transfer coefficient U decreased for most particle shapes. Although the wall channeling effect was most pronounced for the fixed-beds of cylindrical particles, it was found that this particle shape was among the most efficient, with respect to U. Furthermore, a novel reactor tube design that used random macroscopic wall structures was investigated. For packings of spherical particles, it was found that macroscopic random wall structures can significantly decrease morphological heterogeneities, leading to a significantly better heat transfer characteristic. Taking into account various process-related boundary conditions, cylindrical particles, Raschig rings, and wall-structured reactors were identified as the most promising concepts to intensify the radial heat transport.

Methods were presented to determine effective thermal transport parameters, which are needed for simplified pseudo-homogeneous models, from the particle-resolved CFD results. Depending on the degree of morphologically induced heterogeneities, an excellent to fair agreement was found for the $\lambda_{\text{eff,r}}$-α_W model in comparison to the CFD results, whereas deviations became bigger if the morphology was more heterogeneous. The known problem of underestimated temperatures close to the reactor tube, as one of the biggest drawbacks of this model, was confirmed. To circumvent this problem, parameter optimization studies, based on the Winterberg correlation, were performed to predict the radially varying effective radial thermal conductivity, which was needed for the $\lambda_{\text{eff,r}}(r)$ model. A very good agreement regarding the radial temperature profiles was found between the $\lambda_{\text{eff,r}}(r)$ model and the particle-resolved CFD results. In comparison to the $\lambda_{\text{eff,r}}$-α_W model, the $\lambda_{\text{eff,r}}(r)$ model showed its superior accuracy close to the reactor wall. Nevertheless, it was found that both pseudo-homogeneous models became less accurate if step-like temperature profiles, which were a either a result of morphological heterogeneities or a high particle solid thermal conductivity, were present.

In terms of process intensification, this work showed that particle-resolved CFD can either directly be used to study improvements on the meso-scale through:

- studying the impact of particle shape, internals, or reactor tube design on the performance;
- investigating the effect of operating conditions and physical properties;
- testing of novel reactor tube concepts, e.g., reactors with random macroscopic wall structures or heat fins;

- identifying local phenomena as hot/cold spot formation or catalyst poisoning;

or as a reliable source for parameters, and correlations of those, that are needed for process simulation. This allows a more reliable analysis of process intensification by studying:

- dynamic operating conditions;
- process integration concepts;
- conducting process design optimization.

Supplementary Materials: The following are available online at https://www.mdpi.com/article/10.3390/en14102913/s1.

Author Contributions: Conceptualization, N.J. and M.K.; methodology, N.J.; validation, N.J., U.S. and A.A.M.; formal analysis, N.J.; investigation, N.J., U.S. and A.A.M.; writing—original draft preparation, N.J.; writing—review and editing, N.J., U.S., A.A.M. and M.K.; visualization, N.J.; supervision, M.K. All authors read and agreed to the published version of the manuscript.

Funding: This research was funded by Clariant.

Institutional Review Board Statement: Not applicable.

Informed Consent Statement: Not applicable.

Data Availability Statement: The data presented in this study are available on request from the corresponding author. The data are not publicly available due to large file sizes, and, partly, restrictions that might apply.

Conflicts of Interest: The authors declare no conflict of interest.

Abbreviations

The following abbreviations are used in this manuscript:

CAD	Computer-aided design
CFD	Computational fluid dynamics
DEM	Discrete element method
RANS	Reynolds-averaged Navier–Stokes
Nomenclature—Roman	
N	tube-to-particle diameter ratio [-]
$d_{\mathrm{p,v}}$	sphere-equivalent particle diameter [m]
h	bed height [m]
T	temperature [K]
A	area magnitude [m^2]
B	parameter [-]
U	global heat transfer coefficient [W/(m^2 K)]
$c_{\mathrm{p,f}}$	fluid specific heat [J/(kg K)]
r	radial coordinate [m]
z	axial coordinate [m]
y	parameter [-]
a	parameter [-]
K_{w}	parameter [-]
$k_{\mathrm{f,w}}$	parameter [-]
$n_{\mathrm{f,s}}$	parameter [-]
$\dot{Q}$	heat flow rate [W]
u_0	superficial velocity [m/s]
u_{z}	interstitial velocity [m/s]
r^*	dimensionless wall distance [-]
$J_i()$	Bessel function of the first kind and i-th order [-]
Nu_{w}^*	parameter, wall film Nusselt number [-]
Nu_{m}	parameter, mechanical Nusselt number [-]
Nu_{w}^0	parameter, stagnant wall Nusselt number [-]

Nomenclature—Greek

ε	bed voidage [-]
Θ	dimensionless temperature [-]
$\Delta_{\log}T$	logarithmic temperature difference [K]
λ_f	fluid thermal conductivity [W/(m K)]
λ_s	particles' thermal conductivity [W/(m K)]
$\lambda_\mathrm{eff,\,z}$	effective axial thermal conductivity [W/(m K)]
$\lambda_\mathrm{eff,\,r}$	effective radial thermal conductivity [W/(m K)]
$\lambda^0_\mathrm{eff,\,r}$	stagnant bed thermal conductivity [W/(m K)]
κ	ratio of solid to fluid thermal conductivity [-]
α_w	wall heat transfer coefficient [W/(m^2 K)]
ρ_f	fluid density [kg/m^3]
μ	dynamic viscosity [Pa s]

Nomenclature—Indices

f	fluid phase
s	solid particles
w	wall
0	inlet
core	value at $r = 0$

Dimensionless Numbers

$Re_\mathrm{p} = \frac{u_0\rho_\mathrm{f}d_\mathrm{p,v}}{\mu}$	particle Reynolds number
$Pr = \frac{\mu c_\mathrm{p,f}}{\lambda_\mathrm{f}}$	Prandtl number
$Pe_\mathrm{f,r} = \frac{u_0 c_\mathrm{p,f}\rho_\mathrm{f}d_\mathrm{p,v}}{\lambda_\mathrm{eff,r}}$	radial effective P/'eclet number
$Bi = \frac{\alpha_\mathrm{w}D}{2\lambda_\mathrm{eff,r}}$	Biot number
$Nu_\mathrm{w} = \frac{\alpha_\mathrm{w}d_\mathrm{p,v}}{\lambda_\mathrm{f}}$	wall Nusselt number

References

1. Eigenberger, G. Principles of chemical reaction engineering and plant design. In *Ullmann's Encyclopedia of Industrial Chemistry, Principles of Chemical Reaction Engineering and Plant Design*, 5th ed.; Chapter B4: Fixed; Elvers, B., Ed.; Wiley VCH: Weinheim, Germany, 1992; pp. 199–238. [CrossRef]
2. Caulkin, R.; Jia, X.; Fairweather, M.; Williams, R.A. Predictions of porosity and fluid distribution through nonspherical-packed columns. *AIChE J.* **2012**, *58*, 1503–1512. [CrossRef]
3. Wehinger, G.D. *Particle-Resolved CFD Simulations of Catalytic Flow Reactors*; Technische Universitaet: Berlin, Germany, 2016.
4. Jurtz, N.; Wehinger, G.D.; Srivastava, U.; Henkel, T.; Kraume, M. Validation of pressure drop prediction and bed generation of fixed-beds with complex particle shapes using discrete element method and computational fluid dynamics. *AIChE J.* **2020**, *66*, e16967. [CrossRef]
5. Wehinger, G.D.; Kraume, M.; Berg, V.; Korup, O.; Mette, K.; Schlögl, R.; Behrens, M.; Horn, R. Investigating dry reforming of methane with spatial reactor profiles and particle-resolved CFD simulations. *AIChE J.* **2016**, *62*, 4436–4452. [CrossRef]
6. Dong, Y.; Sosna, B.; Korup, O.; Rosowski, F.; Horn, R. Investigation of radial heat transfer in a fixed-bed reactor: CFD simulations and profile measurements. *Chem. Eng. J.* **2017**, *317*, 204–214. [CrossRef]
7. Wehinger, G.D. Radiation Matters in Fixed-Bed CFD Simulations. *Chem. Ing. Tech.* **2019**, *91*, 583–591. [CrossRef]
8. Eppinger, T.; Wehinger, G.D. A Generalized Contact Modification for Fixed-Bed Reactor CFD Simulations. *Chem. Ing. Tech.* **2020**, *93*, 143–153. [CrossRef]
9. Dixon, A.G.; Boudreau, J.; Rocheleau, A.; Troupel, A.; Taskin, M.E.; Nijemeisland, M.; Stitt, E.H. Flow, transport, and reaction interactions in shaped cylindrical particles for steam methane reforming. *Ind. Eng. Chem. Res.* **2012**, *51*, 15839–15854. [CrossRef]
10. Wehinger, G.D.; Eppinger, T.; Kraume, M. Detailed numerical simulations of catalytic fixed-bed reactors: Heterogeneous dry reforming of methane. *Chem. Eng. Sci.* **2015**, *122*, 197–209. [CrossRef]
11. Karthik, G.; Buwa, V.V. A computational approach for the selection of optimal catalyst shape for solid-catalysed gas-phase reactions. *React. Chem. Eng.* **2020**, *5*, 163–182.
12. Partopour, B.; Dixon, A.G. Effect of particle shape on methanol partial oxidation in a fixed bed using CFD reactor modeling. *AIChE J.* **2020**, *66*, e16904, [CrossRef]
13. Fratalocchi, L.; Groppi, G.; Visconti, C.G.; Lietti, L.; Tronconi, E. Adoption of 3D printed highly conductive periodic open cellular structures as an effective solution to enhance the heat transfer performances of compact Fischer-Tropsch fixed-bed reactors. *Chem. Eng. J.* **2020**, *386*, 123988. [CrossRef]

14. Ambrosetti, M.; Bracconi, M.; Maestri, M.; Groppi, G.; Tronconi, E. Packed foams for the intensification of catalytic processes: assessment of packing efficiency and pressure drop using a combined experimental and numerical approach. *Chem. Eng. J.* **2020**, *382*, 122801. [CrossRef]
15. Ambrosetti, M.; Groppi, G.; Schwieger, W.; Tronconi, E.; Freund, H. Packed Periodic Open Cellular Structures—An Option for the Intensification of Non-Adiabatic Catalytic Processes. *Chem. Eng. Process. Process Intensif.* **2020**, *155*, 108057. [CrossRef]
16. Lämmermann, M.; Schwieger, W.; Freund, H. Experimental investigation of gas-liquid distribution in periodic open cellular structures as potential catalyst supports. *Catal. Today* **2016**, *273*, 161–171. [CrossRef]
17. Lämmermann, M.; Horak, G.; Schwieger, W.; Freund, H. Periodic open cellular structures (POCS) for intensification of multiphase reactors: Liquid holdup and two-phase pressure drop. *Chem. Eng. Process. Process Intensif.* **2018**, *126*, 178–189. [CrossRef]
18. Busse, C.; Freund, H.; Schwieger, W. Intensification of heat transfer in catalytic reactors by additively manufactured periodic open cellular structures (POCS). *Chem. Eng. Process. Process Intensif.* **2018**, *124*, 199–214. [CrossRef]
19. Jurtz, N.; Flaischlen, S.; Scherf, S.C.; Kraume, M.; Wehinger, G.D. Enhancing the Thermal Performance of Slender Packed Beds through Internal Heat Fins. *Processes* **2020**, *8*, 1528. [CrossRef]
20. Eppinger, T.; Jurtz, N.; Kraume, M.; Zobel, N.; Frank, B. Influence of the Wall Structure on the Heat Transfer in Packed Beds with Small Tube to Particle Diameter Ratio. In Proceedings of the 2012 AIChE Annual Meeting, David L. Lawrence Convention Center, Pittsburgh, PA, USA, 28 October–2 November 2012; Conference Proceedings; Non-Topical Conferences; American Institute of Chemical Engineers: New York, NY, USA, 2012; p. 9.
21. Eppinger, T.; Jurtz, N.; Kraume, M. Influence of Macroscopic Wall Structures on the Fluid Flow and Heat Transfer in Fixed Bed Reactors with Small Tube to Particle Diameter Ratio. *Processes* **2021**, submitted.
22. Jurtz, N.; Kraume, M.; Wehinger, G.D. Advances in fixed-bed reactor modeling using particle-resolved computational fluid dynamics (CFD). *Rev. Chem. Eng.* **2019**, *35*, 139–190. [CrossRef]
23. Stankiewicz, A.I.; Moulijn, J.A. Process intensification: Transforming chemical engineering. *Chem. Eng. Prog.* **2000**, *96*, 22–34.
24. Elnashaie, S.S. *Modelling, Simulation and Optimization of Industrial Fixed Bed Catalytic Reactors*; CRC Press: Boca Raton, FL, USA, 1994; Volume 7.
25. Mirzaei, A.A.; Shirzadi, B.; Atashi, H.; Mansouri, M. Modeling and operating conditions optimization of Fischer–Tropsch synthesis in a fixed-bed reactor. *J. Ind. Eng. Chem.* **2012**, *18*, 1515–1521. [CrossRef]
26. Eppinger, T.; Wehinger, G.D.; Jurtz, N.; Aglave, R.; Kraume, M. A numerical optimization study on the catalytic dry reforming of methane in a spatially resolved fixed-bed reactor. *Chem. Eng. Res. Des.* **2016**, *115*, 374–381. [CrossRef]
27. Kolios, G.; Frauhammer, J.; Eigenberger, G. Autothermal fixed-bed reactor concepts. *Chem. Eng. Sci.* **2000**, *55*, 5945–5967. [CrossRef]
28. Manenti, F.; Leon-Garzona, A.; Giulia Bozzano, G. Energy-process integration of the gas-cooled/water-cooled fixed-bed reactor network for methanol synthesis. *Chem. Eng. Trans.* **2013**, *35*. [CrossRef]
29. Fischer, K.L.; Freund, H. On the optimal design of load flexible fixed bed reactors: Integration of dynamics into the design problem. *Chem. Eng. J.* **2020**, *393*, 124722. [CrossRef]
30. Fache, A.; Marias, F. Dynamic operation of fixed-bed methanation reactors: Yield control by catalyst dilution profile and magnetic induction. *Renew. Energy* **2020**, *151*, 865–886. [CrossRef]
31. Dixon, A.G. Fixed bed catalytic reactor modelling—the radial heat transfer problem. *Can. J. Chem. Eng.* **2012**, *90*, 507–527. [CrossRef]
32. Dixon, A.G. Local transport and reaction rates in a fixed bed reactor tube: Exothermic partial oxidation of ethylene. *Chem. Eng. Sci.* **2020**, *231*, 116305. [CrossRef]
33. Moghaddam, E.; Foumeny, E.; Stankiewicz, A.; Padding, J. Multiscale Modelling of Wall-to-Bed Heat Transfer in Fixed Beds with Non-Spherical Pellets: From Particle-Resolved CFD to Pseudo-Homogenous Models. *Chem. Eng. Sci.* **2021**, 116532. [CrossRef]
34. Ferziger, J.H.; Perić, M.; Street, R.L. *Computational Methods for Fluid Dynamics*; Springer: Berlin/Heidelberg, Germany, 2002; Volume 3.
35. Levenspiel, O. *Chemical Reaction Engineering*; John Wiley & Sons: Hoboken, NJ, USA, 1999.
36. Dixon, A.G.; Nijemeisland, M.; Stitt, E.H. Packed tubular reactor modeling and catalyst design using computational fluid dynamics. *Adv. Chem. Eng.* **2006**, *31*, 307–389.
37. Dixon, A.G.; Partopour, B. Computational Fluid Dynamics for Fixed Bed Reactor Design. *Annu. Rev. Chem. Biomol. Eng.* **2020**, *11*, 109–130. [CrossRef] [PubMed]
38. Koning, G.W. *Heat and Mass Transport in Tubular Packed Bed Reactors at Reacting and Non-Reacting Conditions: Experiments and Models*; Twente University Press: Enschede, The Netherlands, 2002.
39. Feng, Y.; Han, K.; Owen, D. A generic contact detection framework for cylindrical particles in discrete element modelling. *Comput. Methods Appl. Mech. Eng.* **2017**, *315*, 632–651. [CrossRef]
40. Krischke, A.M. Modellierung und experimentelle Untersuchung von Transportprozessen in durchstromten Schuttungen. In *Fortschritt Berichte-VDI Reihe 3 Verfahrenstechnik*; VDI-Verlag: Dusseldorf, Germany, 2001.
41. Eppinger, T.; Seidler, K.; Kraume, M. DEM-CFD simulations of fixed bed reactors with small tube to particle diameter ratios. *Chem. Eng. J.* **2011**, *166*, 324–331. [CrossRef]
42. Jurtz, N.; Waldherr, P.; Kraume, M. Numerical Analysis of the Impact of Particle Friction on Bed Voidage in Fixed-Beds. *Chem. Ing. Tech.* **2019**, *91*, 1260–1266. [CrossRef]

43. Eppinger, T.; Jurtz, N.; Aglave, R. Automated workflow for spatially resolved packed bed reactors with spherical and non-spherical particles. In Proceedings of the 10th International Conference on CFD in Oil & Gas, Metallurgical and Process Industries SINTEF, Trondheim, Norway, 17–19 June 2014; pp. 17–19.
44. Minhua, Z.; He, D.; Zhongfeng, G. A particle-resolved CFD model coupling reaction-diffusion inside fixed-bed reactor. *Adv. Powder Technol.* **2019**, *30*, 1226–1238. [CrossRef]
45. Vortmeyer, D. Die mathematische Modellierung von Reaktions-und Austauschprozesen in durchströmten Festbetten unter Berücksichtigung von ungleichmäßigen Strömungsverteilungen. *WÄRme-Und StoffÜBertragung* **1987**, *21*, 247–257. [CrossRef]
46. Tsotsas, E.; Schlünder, E.U. Heat transfer in packed beds with fluid flow: Remarks on the meaning and the calculation of a heat transfer coefficient at the wall. *Chem. Eng. Sci.* **1990**, *45*, 819–837. [CrossRef]
47. Yagi, S.; Kunii, D. Studies on effective thermal conductivities in packed beds. *AIChE J.* **1957**, *3*, 373–381. [CrossRef]
48. van Antwerpen, W.d.; Du Toit, C.; Rousseau, P. A review of correlations to model the packing structure and effective thermal conductivity in packed beds of mono-sized spherical particles. *Nucl. Eng. Des.* **2010**, *240*, 1803–1818. [CrossRef]
49. Zehner, P.; Schlünder, E. Wärmeleitfähigkeit von Schüttungen bei mäßigen Temperaturen. *Chem. Ing. Tech.* **1970**, *42*, 933–941. [CrossRef]
50. Tsotsas, E. M11 Wärmeleitfähigkeit von Schüttschichten. In *VDI-Wärmeatlas*; Springer: Berlin/Heidelberg, Germany, 2019; pp. 1831–1843.
51. Yagi, S.; Kunii, D. Studies on heat transfer near wall surface in packed beds. *AIChE J.* **1960**, *6*, 97–104. [CrossRef]
52. Nilles, M. Wärmeübertragung an der Wand durchströmter Schüttungsrohre, Wärmetransport in Schüttungen, Reihe 3, Fortschritt-Berichte VDI. Ph.D. Thesis, Universität Karlsruhe, Düsseldorf, Germany, 1991.
53. Martin, H.; Nilles, M. Radiale wärmeleitung in durchströmten schüttungsrohren. *Chem. Ing. Tech.* **1993**, *65*, 1468–1477. [CrossRef]
54. Wakao, N.; Kagei, S. *Heat and Mass Transfer in Packed Beds*; Taylor & Francis: New York, NY, USA, 1982; Volume 1.
55. Ahmed, M.; Fahien, R. Tubular reactor design—I Two dimensional model. *Chem. Eng. Sci.* **1980**, *35*, 889–895. [CrossRef]
56. Argo, W.; Smith, J. Heat transfer in packed beds-prediction of radial rates in gas-solid beds. *Chem. Eng. Prog.* **1953**, *49*, 443–451.
57. Gunn, D.; Ahmed, M. The characterisation of radial heat transfer in fixed beds. In *IChemE Symp. Ser.*; Elsevier: Leeds, UK, 1984; Volume 86, pp. 513–520.
58. Gunn, D.; Ahmad, M.; Sabri, M. Radial heat transfer to fixed beds of particles. *Chem. Eng. Sci.* **1987**, *42*, 2163–2171. [CrossRef]
59. Gunn, D.; Sabri, M. A distributed model for liquid-phase heat transfer in fixed beds. *Int. J. Heat Mass Transf.* **1987**, *30*, 1693–1702. [CrossRef]
60. Smirnov, E.; Muzykantov, A.; Kuzmin, V.; Kronberg, A.E.; Zolotarskii, I. Radial heat transfer in packed beds of spheres, cylinders and Rashig rings: Verification of model with a linear variation of λer in the vicinity of the wall. *Chem. Eng. J.* **2003**, *91*, 243–248. [CrossRef]
61. Winterberg, M.; Tsotsas, E.; Krischke, A.; Vortmeyer, D. A simple and coherent set of coefficients for modelling of heat and mass transport with and without chemical reaction in tubes filled with spheres. *Chem. Eng. Sci.* **2000**, *55*, 967–979. [CrossRef]
62. Pietschak, A.; Maußner, J.; Dixon, A.G.; Freund, H. Comparative evaluation of heat transfer correlations with different fluid property considerations for fixed-bed reactor modeling. *Int. J. Heat Mass Transf.* **2020**, *148*, 119099. [CrossRef]
63. Pietschak, A.; Dixon, A.G.; Freund, H. A new heat transfer correlation suited for the design of fixed-bed reactors via numerical optimization. *Chem. Eng. Sci.* **2020**, *220*, 115614. [CrossRef]
64. Dixon, A.G.; Wu, Y. Flow and heat transfer in narrow fixed beds with axial thermowells. *Numer. Heat Transf. Part A Appl.* **2019**, *76*, 811–829. [CrossRef]
65. Holleman, A.F. *Lehrbuch der Anorganischen Chemie*; Walter de Gruyter GmbH & Co KG: Berlin, Germany, 2019.

energies

MDPI

Review

Investigation of the Photon to Charge Conversion and Its Implication on Photovoltaic Cell Efficient Operation

Vasileios Kapsalis [1,*], Grigorios Kyriakopoulos [2], Miltiadis Zamparas [3] and Athanasios Tolis [1]

1 Sector of Industrial Management and Operational Research, School of Mechanical Engineering, Zografou Campus, National Technical University of Athens, 157 80 Athens, Greece; atol@central.ntua.gr
2 Photometry Laboratory, Electric Power Division, School of Electrical and Computer Engineering, Zografou Campus, National Technical University of Athens, 157 80 Athens, Greece; gregkyr@chemeng.ntua.gr
3 School of Science and Technology, Hellenic Open University, Parodos Aristotelous 18, 263 35 Patras, Greece; mzamparas@upatras.gr
* Correspondence: bkapsal@mail.ntua.gr

Abstract: Efficient photon to charge (PTC) transfer is considered to be the cornerstone of technological improvements in the photovoltaic (PV) industry, while it constitutes the most common process in nature. This study aims to investigate the parameters that impact efficient PV-cell photon to charge conversion in two ways: (a) providing a brief research analysis to extract the key features which affect the electrical and optical performance of PV cells' operation, and (b) investigating the dependance of these characteristics on the photon to charge mechanisms. The former direction focuses on the latest advances regarding the impacts of the microenvironment climate conditions on the PV module and its operational performance, while the latter examines the fundamental determinants of the cell's efficient operation. The electrical and optical parameters of the bulk PV cells are influenced by both the external microenvironment and the intrinsic photon to charge conversion principles. Light and energy harvesting issues need to be overcome, while nature-inspired interpretation and mimicking of photon to charge and excitation energy transfer are in an infant stage, furthering a better understanding of artificial photosynthesis. A future research orientation is proposed which focuses on scaling up development and making use of the before mentioned challenges.

Keywords: photovoltaic cell efficiency; thermal regulation; energy and light harvesting; irreversibility losses; quantum dynamics; nature-inspired mimicking

Citation: Kapsalis, V.; Kyriakopoulos, G.; Zamparas, M.; Tolis, A. Investigation of the Photon to Charge Conversion and Its Implication on Photovoltaic Cell Efficient Operation. *Energies* **2021**, *14*, 3022. https://doi.org/10.3390/en14113022

Academic Editors: Goodarz Ahmadi, Kiao Inthavong and Pouyan Talebizadeh Sardari

Received: 14 April 2021
Accepted: 18 May 2021
Published: 23 May 2021

1. Introduction

Nowadays, carbon footprint awareness and the necessity of environmental protection have become significant issues in our daily activities. These considerations affect the decisions we make on energy sources and use of alternatives, as well as technological production and consumption patterns, leading to major economic and social consequences. Many efforts are being conducted to model and design efficient carbon emission systems in all industrial sectors, such as metallurgy, construction, shipping, manufacturing, transportation, to name some [1–6]. In most cases, efficient processes go together with an efficient transition to a low carbon economy, however, we have to admit that there are always cases with contradicting and competitive interactions which are worthy of investigation. In this research work, our concern refers to PV systems in order to get a deeper insight into the role of efficiency in operational mode.

Solar photovoltaic (PV) systems that directly convert sunlight into electricity are small scale and highly modular devices. They offer efficient resilience, flexibility, and adaptation to the grid energy supply. They have often been considered as the ideal distributed electricity production since they are derived from solar energy, which is a ubiquitous, inexhaustible, and renewable form of energy and is widely exploited in our society. In recent years, the value of integrated grid-connected PVs has been recognized around the world and many programs have been carried out in many countries to enhance architectural and technical quality in the built environment, storage potential, and the removal of economic and non-technical barriers in order introduce PVs as an energy-significant resource. Energy market reforms and the concept of decentralized energy systems drove research into different hybrid systems with a combined effect on energy production and consumption, therefore, analyzing both demand and supply-side management. The intensive research orientation toward efficiency [7–9] produced the proposal that every house could act as a net energy positive provider, taking into account combined utilizations to increase the benefits of outputs [10]. Electricity production is the main driving force for the installation of integrated PV systems, but, in recent years, installations have also been recognized which establish combined electricity and thermal energy, such as coupled photovoltaic–thermal collectors to enhance heating and cooling demand coverage, to act as refrigerator alternatives [11], or to increase thermal energy storage [12], while there are many reports of the application of standalone PV systems to solar power pumps in irrigation, livestock watering, and solar-powered water purification [13,14]. The entry of these renewable systems to the grid can also affect buildings' energy demand mixture, providing effective management of short-term buffering options in alignment with battery storage [15,16]. Many concerns regarding more efficient ways to implement energy and environmental systems lead to exploring hybrid combinations in order to optimize the path for a low carbon transition towards these systems. Renewable energy sources penetration into the grid-connected power technologies bring new contradicting issues, including the cost of the mismatch between demand and supply and intermittent and unpredictable availability [17], although in many studies they are viewed as a net provider due to the cumulative declining production costs [18–20]. In any case, increased renewable penetration, rational exergy management models, and reduced interaction with the utility grid are of paramount importance and delignate a robust pathway for CO_2 mitigation from the built environment [21], as well as effective management of grid parity [22]. Typically, a PV system depends on cell performance which in turn depends on different mechanisms regarding complex design, fabrication, and operation parameters.

Therefore, investigation of the efficient operation of PV cell electricity output is a matter of continuous interest. This study investigates the PV efficient cell operation from both sides, namely the incoming microenvironmental conditions and the exploitation of the fundamental mechanisms of PTC transition, which affect the solar cell efficient operation. The rest of the paper is organized as follows: in Section 2 the critical parameters are derived under a brief literature review on the microenvironment determinants on PV efficient operation and output. Section 3, following an analysis on PV fundamental concepts, proceeds with a deeper insight on the different technologies of solar cells. In particular, the mechanisms of the photon to energy conversion are explored in terms of design, fabrication, and materials. In Section 4 the results are analyzed and a discussion is developed. The main findings with a proposed future research orientation are concluded in Section 5.

2. Materials and Methods

Many efforts have been made in the field of the cooling effect of photovoltaic cells [23], increasing efficiency by reducing the cell temperature [24] or expanding thermal regulation time during the charging and discharging process in sensible, latent, or thermochemical storage mechanisms [12,25–32]. A review on PV cooling technologies regarding the three-heat transfer mechanism (convection, conduction, and radiation) can be found in literature [33]. The optimization of operating temperature increases in the lifetime of the system by decreasing the thermo-mechanical fatigue for typical failure mechanisms [34] and has a significant influence on PV efficiency output. The correlation between the examined PV cell temperature and the environmental independent variables is found by a regression analysis [34] and for fixed period measurements. The model was implemented at various periods within the year. Sensitivity analysis for different weather conditions was conducted for longer and seasonal periods. It was found that the coefficients of the above method have to be taken with care during other periods since they refer to a local solution. Indeed, different declined angles of the independent variables fitted different periods, however the regression model conserved its linearity over the whole year. It is of great importance that the PV array temperature is controlled because it has been found that the power output decreases about 0.5% for every degree of Celsius increase, depending on the environmental conditions [34]. Reduced cell temperature is associated with increased efficiency which in turn leads to reduced thermal stresses and degradation of the modules. Therefore, it is essential to prevent the cell from overheating. The fill factor (*FF*) is considered as a measurement of exergy degradation of the PV at a maximum power point output (P_{max} *at mpp*), as in Figure 1, and it is related to the voltage and current in that point (V_{mpp} *and* I_{mpp}), as well as the open circuit current voltage and the short circuit current values (V_{oc} *and* I_{sc} *correspondingly*). It is derived from the following equation [35].

$$FF = \frac{V_{mpp} I_{mpp}}{V_{oc} I_{sc}} = \frac{P_{max}}{V_{oc} I_{sc}} \quad (1)$$

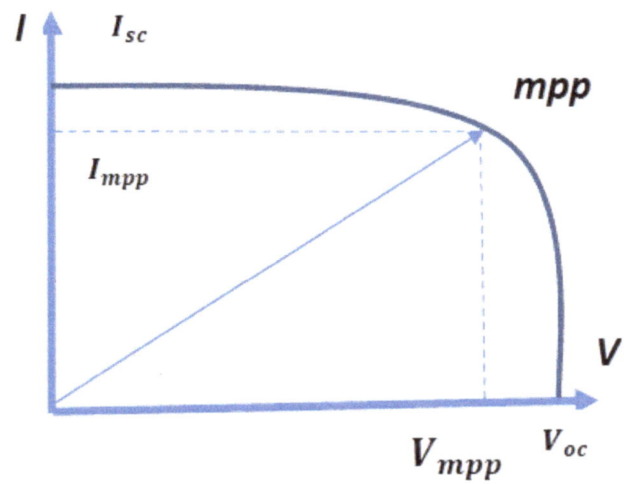

Figure 1. Load characteristics in solar cells [Authors own study].

In addition, it measures the reduced quality of a diode due to recombination and space-charges formation due to unbalanced transport.

Later advanced data analysis compared different algorithms to extract the parameters which contribute to the PV array model output [36]. The numerical Levenberg–Marquardt algorithm and metaheuristics such as Differential Evolution (DE), Genetic (GA), Particle Swarm Optimization (PSO), Ant Bee Colony (ABC) algorithms are used to extract and fit the parameters of two PV models, the five parameters model (5PM) and the Sandia Array Performance model (SAPM). The error analysis is made by two metrics, the Route Mean Square Error (RMSE) and the Normalized Mean Absolute Error (NMAE), where it is found that, in general, the metaheuristics depicted fewer errors [36].

Another research report conducted a detailed correlation coefficient analysis regarding PV output and environmental variables. It was noted that, when included in the model, the module characteristics, ambient temperature, air speed, perpendicular irradiance on the array, and relative humidity contributed to less error in the PV prediction of the maximum output [36].

It is worth mentioning that, from the aforementioned analysis of the relationship between the I–V characteristics and the cell temperature, the fundamental energy gap in the different band levels and the irradiation has already been established as a nonlinear relationship [37], besides the impacts of defects and unstable materials behavior on cell degradation—which, in turn, limit the design's geometrical parameters [38,39]—or the Staebler Wronski (SW) defects [40]. In Figure 2, the loss mechanisms in a c-silicon cell are depicted [39].

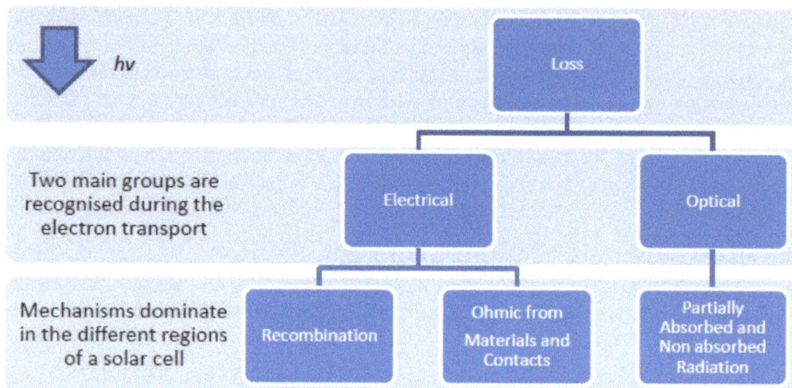

Figure 2. Main photon to charge loss mechanisms.

3. Photovoltaic Cell Energy Conversion

3.1. Fundamental Aspects

Photovoltaic solar to energy conversion is based on the electron behavior of semiconductors which originates from the existence of two electron energy bands: the valence and conduction bands. The energy difference between the bottom of conduction and the top of the valence band is the energy gap, E_g. It is well known that the incoming energy of a photon (h stands for Plank's constant and v for frequency), hv, when is greater than E_g, is absorbed and may create bound pairs of electrons-holes, the excitons. This disruption of a covalent bond transfers electrons from the valence to conduction band, leaves a hole behind, changes the conductivity, and becomes the carrier of electricity. The doping of certain impurities within the material dominates different sites of donors and acceptors in the lattice, corresponding to positive and negative regions. Therefore, the diffusion of electron and holes develops a contact potential, about 1 V under room temperature and certain doping. The potential across the p-n junction is constant and the electric field is limited to a narrow transition region. The ability of the electrons to drift into that field immediately or with delay due to recombination depends on their distance from that field and other interactions. Thus, the separation of the excitons makes the electrons serve as an external current.

3.2. Photogenerated Current

The necessity of absorbing as many photons as possible obligates the use of materials with a low band gap, while the connections between cells are of utmost importance for the optimization of the solar energy yield. In PV systems, since the majority of photodiodes were exposed to photons within differentiated energy streams, the efficiency is prevailed by several mechanisms to exceed beyond certain values or approach the Shockley–Queissier (S–Q) limit [41]. The portion of the unabsorbed photons, the thermalization effect, and the time dependent recombination contribute to electricity conversion, with losses leading to the partial utilization of the spectrum and photon energy. These interactions are usually contradicted and affect the tradeoff for the critical properties of the solar cell design decisions.

Moreover, energy production is simultaneously happening with the fundamental principle of time micro-reversibility, namely that the solar photons from the sun are converted into electricity within solar cells but also reemitted as thermal radiation. The emitted radiation produced by the electro–hole excess energy in the cell is luminescent, which means that the electrochemical potential of photon differs by zero, obeying the modified Planck law, and the thermodynamic efficiency of a solar converter is limited by the Carnot efficiency between the working source (sun temperature at about 6000 K) and the heat sink at the cell temperature [42].

3.3. Recombination Limits

The significance of the effective lifetime of the charge carriers to generate current and their dependance on the recombination processes has already been noted. There are three reasons for this association: (a) the doping level, (b) the irradiance of the cell, and (c) the nature and quality of the semiconductor. Accordingly, we recognize the following recombination processes which are interrelated with the abovementioned reasons: the surface density recombination, the Shockley-Read-Hall (SRH) recombination through undesirable light traps, the radiative recombination, and the Auger recombination—which has to do with the probability of a conduction band electron to transfer the excess energy to a valence band hole or to another conduction band electron. The latter denotes a three-particle process of the electron hole concentration under illumination and increases with the cube of the carrier concentration, making a great contribution as a limited open voltage (V_{oc}) and efficiency factor, depending on the materials used. Typical mechanisms of recombination in solar cells are related to luminescence, SRH defects, SRH on impurities, Auger, and surface. In general, the total recombination rate, τ_{total}, is related to the other ones, namely the radiation ($\tau_{radiation}$), Auger (τ_{Auger}), and trapping (τ_{trap}) recombination by the following equation:

$$\frac{1}{\tau_{total}} = \frac{1}{\tau_{radiation}} + \frac{1}{\tau_{Auger}} + \frac{1}{\tau_{trap}} \tag{2}$$

For example, in crystalline silicon solar cells the Auger recombination dominates and the distribution is: Auger 82%, radiative 9%, SRH 7% and surface 2% [43,44].

3.3.1. Single Junction Cells

In inorganics semiconductors, unlike organics, the Boltzmann energy, kT, (k is the Boltzmann's constant and T the temperature) is much larger than the exciton binding energy atroom temperature and therefore the free charges are easily created while the material is excited. The increasing interest in the solar energy has led to the research and advanced alternatives to enhance the PV output. The inorganic semiconductors have been intensively investigated, and reached high efficiencies with the Tungsten Diselenide (WSe_2) absorbers, while there is further research to investigate the photonic crystals and the trapping mechanisms in certain silicon thin films, potentially beyond the Lambertian limits and conversion efficiency of 30%, close to the S–Q limit [45,46].

On the other hand, in the last decade the organic PVs have doubled in efficiency, reaching about 12% [47,48]. The heterojunction architecture consists of a blend of electron donating (p-type) and electron accepting (n-type) organic semiconductors which drive photogenerated excitons, bound electro-hole pairs, separate free charges, and generate current.

Further developments are made with the dye sensitized cells, which exploit the challenges of liquid electrolytes with little transport between monolayers. The efficiency reached more than 13% and recently this concept also considered inorganic nanoparticles to act as sensitizers to replace the dye [49,50].

An extraordinary incremental increase in efficiency (expected more than 20%) came from the family of hybrid organic–inorganic materials of AMX3, namely perovskites ((CH3NH3)PbI3), where the A stands for the inorganic site, M is the metal site, and X is a halogen [51,52].

Similar research is being made in the emerging field of nanotechnology, since the manipulation of solar cells nanostructures has become of utmost importance in terms of energy gap tunability, transport decoupling, and interfacial interactions which prove very different from the bulk material. The representatives of those structures, quantum dots, still remain at low levels, about 10% [53,54].

3.3.2. Multi Cells Approaches

The utilization of different band gap materials is a relatively new concept and depicts many challenges to be resolved due to the small efficiencies. Practically, the principles are based on the doping of large E_g materials to formulate defect band in the gap [55,56] or the creation of many separated states according to the quantum size effect [57,58]. The development of multi cell devices suppresses the unabsorbed photons fraction and the thermal effects but also increase the recombined carriers due to the photogeneration extraction.

3.3.3. Spectrum Oriented Optimization

Spectral manipulation PV systems include different strategies to maximize the overall solar radiation utilization of the incident broadband spectrum. They employ technologies shifting the solar spectrum to match more efficient wavelengths conversion, different band-gap materials in series or parallel, and splitting the spectrum in certain long and short-wave ranges to run efficiently with the combined thermal and photovoltaic processes [59].

The spectrum splitting systems are composed of two parts: the optical one to split the spectrum and concentrate the light and the other one to harvest the energy. With the creation of intermediate bands and recent technologies, potentially more than 50% efficiency may be obtained [60–62].

In a similar approach, cascade splitting by superposing is the simplest way to separate two photo diodes with different band-gaps and threshold. The problem to be overcome here is the interconnection of the diodes due to the non-illuminated face of the device which is metalized to act as a current collector, but also this makes it opaque and useless in a cascade configuration. Advance solutions are to build the used diodes from materials with the almost the same lattice constants but different band-gaps [63], tuning it by varying the stoichiometry of quaternary compounds. Another way is to filter concentrated sunlight through cells that absorb part of the energy (e.g., cobalt sulfate) while transmits the rest to underlying silicon diodes. Interference filters consist of a number of layers of materials with different refractive indices with band reflect and ripple regions.

Holographic concentrator operation is an attractive beam splitting technique and prepared to act as an extremely dispersive cylindrical lens. The far infrared is directed to a region to dissipate heat and to not affect the photodiode which has to operate at low temperatures to avoid loss of efficiency.

Different refractive and diffractive-spectral-beam-splitters techniques use (Table 1) holograms in common axis [64], non-uniform diffractive grafting or micro prism arrays as spectral beam splitters in multiple optical axis, and integrated diffractive/refractive arrays (Figure 3) in zig-zag axis [65].

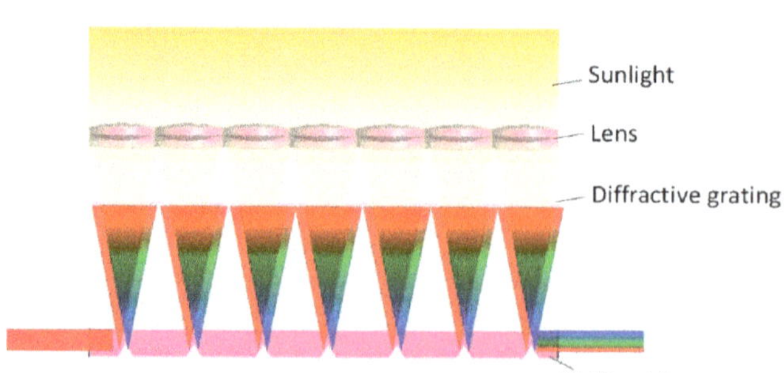

Figure 3. Spectral-beam splitting scheme in side absorption concentrated modules. Reproduced from [65], mdpi:2020.

Table 1. Diffractive and spectral beam splitters (SBS) techniques. Reproduced from [65], mdpi:2020.

Category	Characteristics of Technique	Efficiency (Module or Systems in Some Cases)
Common axis	SBS is formed by a reflective hologram and a quadratic surface	N/A
	A broadband receiver combined with an opening in the center which includes another spectrally selective receiver, all together mounted in a above hologram	21.4% (PV/T system)
Multiple optical axis	A nonuniform diffractive-grating as a SBS	34.7% (Dual-cell (InGaP/GaAs) system) In: Indium, Ga: Gallium, As: Arsenide
	Micro-prism arrays as a SBS	46.05% (Triple-cell (InGaP/GaAs/InGaAs) system) P: Phosphorus
Zig-zag axis	Integrated diffractive / refractive optical element as a SBS	≤55% (Module)
	A waveguide with engraved microstructures beneath the SBS	
	Combined condenser, SBS and out-coupling adapter in lenses, diffractive grating and a light-guide, respectively	34.8% (module) 29.5% (Dual-cell silicon/germanium (Si/Ge) system)

Carrier multiplication may occur when ionization of an atom by impact creates another exciton. This occurs with low efficiency, e.g., <1%, in bulk Si, while it can proceed with nearly 100% efficiency if PbSe is in the form of nanocrystals of about 5 nm size. The excess photon energy is partitioned between the created electron and hole according to their respective masses [66,67]. The generation of multiple excitons before relaxation, and the very fast photo current collection or slowing down the relaxation are researched in bulk and quantum sized systems resulting in carrier multiplication (CM) and hot carrier transportation [68–70].

Another option is the design of specific optical nonlinear systems to manipulate as broad a spectrum as possible ranging from far infra-red into ultraviolet to reduce the thermal losses. The aforementioned concepts are enabled by some nano materials like graphene or carbon nanotubes (CNT) [71]. Broader light harvesting and increased current density have been successful due to the use of new sensitized dyes containing two carboxylic acids or two cyanoacrylic acids as linker groups to bind the metal oxide photo anode, instead of the conventional single linker group of the dye sensitized solar cell (DSSC) [72].

The thermophotovoltaic (TPV) systems use optical concentration devices to distinguish the heat generated from the photovoltaic conversion and generate more electricity. Potentially, they may provide high levels of efficiency, reporting a focus on the PV photothermal conversion to electricity and the thermal regulation of cell temperature (Figure 4). The basic parameters that are affected by the cell temperature are the open circuit voltage, V_{oc}, the short circuit current, I_{sc}, the fill factor, FF, and the energy conversion efficiency value (η) [73].

A promising expansion of the PV concept concerns graphene-based thermionic and thermo radiative converters. The thermionic efficiency may improve the efficiency by including various irreversibility losses, such as space-charge effects and non-radiative recombination or suppression [68], mainly due to graphene's excellent conductivity, mobility, and linear band gap structure [69–72]. Moreover, the thermo radiative cell is able to produce electric current from the coldness of the outer space due to the exploitation of the negative illumination effect [73–80].

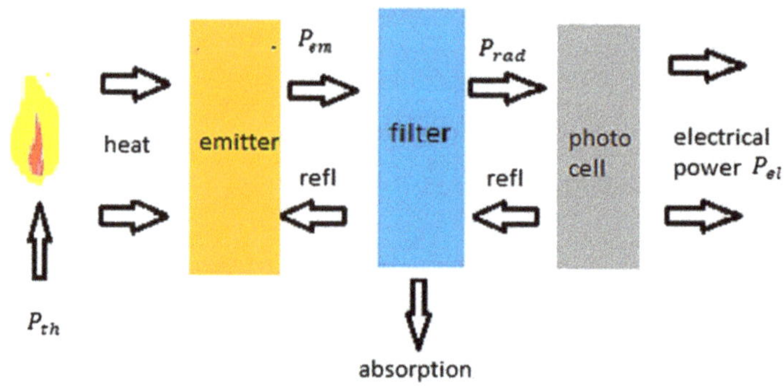

Figure 4. General structure of TPV system. Reproduced from [73], Oxford: 2020.

3.3.4. Design Optimization

The main performance parameters of the output are influenced by the physical and technological parameters of the design and manufacturing process, which, in turn, characterize the electrical, thermal, and optical properties on the operation. This optimization facilitates more efficient structures, regarding an integrated view of the sustainable development in PV fabrication. For example, in the most mature silicon-based PV industry, it is well known that the SW effect limits the architecture of the cells within certain boundaries (e.g., thickness in thin films) due to the hydrogen bonding configurations and the diffusion behavior under elevated and specified range of temperatures regarding the materials [40].

There are also reports of interesting interpretations which evolved the two side junctions, namely the npn structure, by adding a top pn junction in order to vertically absorb the light generated excitons and laterally collect them [81]. Moreover, the wafer-based silicon solar cells are challenged with the threshold between an expensive dense array of radial-orientated nanorods which are doped in pn junctions and the compensated cheap high-density defects and impurities of substrate absorbers, in order to optimize the demand. The former enhances the efficiency while the latter reduces the diffusion length of the carriers [81]. The main parameters that are analyzed refer to the dimensions (length, width, thickness), the doping quantity and quality, and the photon flux. The modelling displayed the causal effects of these parameters to the electrical characteristics (the open circuit voltage, short current density, the filling factor and the conversion efficiency), the optical performance, the quantum efficiency, and the spectral response.

3.3.5. Nature-Inspired Solar to Energy Conversion

In nature, the ideal photosynthesis efficiency differs, due to the debate on light energy definitions and also the irreversibility losses, such as the partially active light wavelength, probabilistic light trapping mechanisms, the recombination effects, the space-state, and the quantum considerations of the transitions. [82,83]. Yet, it has been reported that some molecular systems are optimized in photosynthetic processes and specific photon to charge transformations take place up to 1 conversion ratio within certain conditions. The intermediate functionality of the excitation energy transfer from the electronic state of the antenna pigments to the reaction centers has been observed to be accelerated by some microbial organisms playing the role of an active molecular pipeline, but also depicts quantum oscillations to be considered during the energy transfer [84–86]. The latter opened new pathways to the fabrication of artificial solar cells based on nature-inspired mimicking photosynthesis [87]. The competitive mechanisms between incoherence hopping and coherent tunneling are revealed as the dominant mechanism in the charge transfer.

4. Results and Discussion

4.1. Impacts of Material Properties and Fabrication Processes

The main materials that impact on fabrication processes are linked to the intrinsic defects at the front and back interface of PV window layers. Such defects in these bulk materials can easily form photo-active alloys due to high defect density, thus restricting the current intensity of the device and determining the absorption of photons [88]. Subsequently, higher efficiency can be achieved according to the enhancement of the current intensity, necessitating control of fabrication methods of window layers. Besides, the increase in the carrier concentration results in challenging doping materials with different band gaps and wavelength absorption. These materials are designed with preferred band bending and reduced rear barrier heights of window layers, thus, avoiding the hole transportation resistance that limits the performance of Schottky junction [89,90]. It is noteworthy that, while the interaction of the window layer materials with the deposited and diffused doping atoms reduces the rear contact, it may also increase the potential barrier between them, restricting the photo generated charge carriers, e.g., Cadmium Telluride (CdT) cell technology, when doped with Cu/Au (Copper/Gold), interacts with gold (Au) atoms, while increasing the potential. Carbon nanotubes (CNT), nanocomposites and nanocrystals with suitable valence band edges are also materials that can be used to overcome the contradicting effects [91–93]. Thermal evaporation, magnetron sputtering and chemical etching [94–98], electrostatic spray assisted vapor deposition [99], electrospinning, and annealing indium tin oxide (ITO) processes [100,101] are preparation methods that have influential roles in the electrical and optical parameters. Enhanced operational characteristics of the examined devices, such as the short circuit current, *Isc*, the open circuit voltage, *Voc* and the fill factor, *FF*, can relate surface/layer treatment with the photoconversion efficiency, *n*, via the output power derived, *P*, following equation:

$$n = V_{oc} \times I_{sc} \times FF/P \tag{3}$$

Besides, it can be noted that the minimization of charge recombination can increase the Fill Factor of the PV cells, thus increasing the efficiency of organic PV. Other ways of enhancing the charge transport can be implemented by the incorporation of cascaded atom number (Z) chalcogens, namely the small molecular donors (SMD), in the solubilizing side chains of the active layers which, in turn, can both promote the intermolecular interactions of atoms and further improve the connectivity. From an elementary viewpoint, oxygen (O), sulfur (S), and Selenium (Se) atoms are capable of affecting the donor–acceptor phase aggregation. In particular, the O atoms promote tighter π–π tighter stacking and the atoms of S and Se support a greater crystalline order in thin films [102]. Research efforts are also linked to manufacturing improvements, targeting the following:

- The smaller size heteroatoms, higher electronegativity of the heteroatoms, and larger moments of furan conjugated polymers or fullerene-based heterojunctions [103,104].
- Utilization of certain physical and chemicals treatments to manipulate the active layer absorption [105].
- Thermal stability of the inverted cell structure, thus improving the exciton transportation and efficient separation [106].

4.2. Impact of Energy Harvesting on Energy Conversion Value

When considering the impact energy harvesting on energy conversion value it is of utmost importance to note that the excess photon energy that becomes thermal losses accounts for more than half of the losses, leading towards possible energy harvesting exploitation. There are also reports of vivid interest in devices exploiting the unlimited dissipated thermal energy regarding solar to energy processes, resulting in manufacturing of small autonomous electronic devices with no need for power supply and maintenance, as well as increased conversion efficiency. Such a conversion efficiency increase can be attributed to the broadening of recycling and exploitation of thermal energy waste in many ways that commonly follow the principles of the thermoelectric conversion.

In a similar study [107] a power synthetic inductance circuit was created from the heat generated by the bearing, while other researches [108] proceeded to optimize the figure of merit of pyroelectric materials due to the improved properties of crystallinity, density, and the reduced permittivity derived from energy harvesting. The analysis supported advanced waste to energy applications with composite materials, due to the enhanced phonon transferring properties of the conductive networks.

In another recent study [109] a day-and-night combined operation of electricity and latent energy storage was investigated. Specifically, a Bismuth telluride (Bi2Te3) based thermoelectric generator (TEG), firstly, harvests the concentrated solar energy directed from Fresnel lens, while aluminum fins and mixed nanocomposite materials PCMs are functionally charging and discharging energy. Moreover, the creation of active layers of (QDs) and lithium chloride (LiCl) on top, being sandwiched between a substate and an aluminum contact, can develop a multi-step photon absorption mechanism that enabled the so-called mid gap states mechanism of the incorporated nanocrystals, allowing for harvesting of human body radiation [110]. Another critical research consideration is the exploitation of the second order phase transition above the Curie temperature from the ferromagnetic to paramagnetic phase to increase the cooling rate [111].

4.3. Impacts of Light Harvesting on Photon to Charge Transfer

The enhancement of light to current conversion, correspondingly, induces a wider range of power output, and may in the future yield over 35%. Naturally occurring dipole–dipole interactions between well oriented molecular excited states, generating quantum interference effects, may drive the future research. It is reported [112] that, in lasing, the quantum coherence breaks the balance in a cavity light trap, suppressing the absorption process, and, consequently, the emission dominates. The idea behind the reverse engineered phenomenon in the photovoltaic conversion is an optical pump device which could suppress the emission rate, hence, promoting the absorption rate, e.g., at ground level. In quantum mechanics, this can be achieved by adding a two-level relative transition amplitude which can coherently result in a destructive interference of the undesirable process, namely the emission [112]. Recent research on thin films, copper indium gallium diselenide (CIGS), towards the passivation of rear surface and light trapping in a nanocavity array showed the superior behavior of aluminum oxide (Al2O3) substrate and *Voc* and *Jsc* improvements about 10% [113].

Energy Transfer Interactions

The interactive behavior of photons with solids comprises ionic and electronic oscillations and drives our focus on lattice vibrations. The dynamic response of the dielectric function on electromagnetic radiation can be comprehended by elementary oscillators, while yielding strong interaction of photons and transverse optical (TO) phonons with the accompanying large Reststrahl absorption in the infrared (IR) range [114]. The dispersion is described by a phonon-polariton, which is observed in inelastic scattering processes. In addition, Brillouin scattering at acoustic phonons along with Raman scattering at optical phonons can disseminate useful direct information about the spectrum and symmetry of vibrations in a semiconductor [115].

Charge and excitation energy transfer complexes are influenced by interactive mechanisms between the relative interactions of photon, phonon, electron, and excitons, to name a few, as in Figure 5, but the exact mechanisms have not yet been understood. Charge transfer is perturbatively understood by electron–electron interactions and the Dexter consideration of the exponentially decaying wave functions which is dominant within the field of some Angstroms.

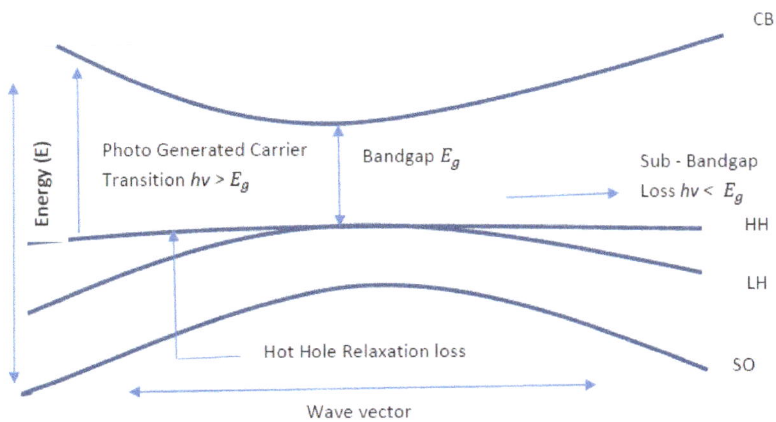

Figure 5. Photogenerated carrier transition, hot carrier relaxation, and sub-bandgap loss depicted in band gap GaAs (Gallium Arsenide). CB: The lowest conduction band., LH is the light hole valence, SO is the split-off hole valence band. [Authors own study] [116,117].

The interaction with lattice phonons leads to the conversion of excess kinetic energy into thermal energy. A hot electron may interact multiple times with longitudinal optical (LO) phonons and holes primarily interact with the transverse optical (TO) phonons. The latter relaxes faster than electrons due to the larger electronic density of states in the valence bands [116]. Moreover, the transversal and longitudinal photons intervene in excitation transfer and dipole–dipole interactions are mainly affected by the Coulombic participation beyond the Dexter domain, in the far field on some nanometers [117]. It is reported that efficiency strongly depends on the intrinsic coherence time of the coupled systems, the short and the long domain, which characterize the decoherent and coherent systems. However, it seems that the electro-electron interactions play a dominant role even in the excitation transfer [117], while, in another view, in organic solar cells, the electron–phonon interactions revealed substantial influence on open circuit voltage and limited the efficiency performance [118]. Further analysis is needed to define the limitations between the quantum coherence and the incoherence hopping competition, which seems to be a main driver for the charge transfer efficiency, and the corresponding learning parameters for nature-adaptive optimization of the photosynthetic efficiency, from 1% to 100%.

5. Conclusions

In the solar PV supply industry, the efficiency of the individual components of a PV system still remains at the cutting edge of technological interest and knowledge diffusion in niche markets. Among these technological challenges is PV-cell operating regulation, since it results in the efficient dissipation of thermal energy surplus. Furthermore, the concurring interaction between the PV system and its micro-environment leads to a capacity factor increase for the whole system while the PV components are more efficient and durable. Therefore, this study stressed the need for future research work towards module degradation, materials deterioration, and proper machinery engagement in producing PV systems at the desired cooling effect. The key findings are summarized here, and also define the limitations of our study:

- Key technology determinants on photon to charge efficiency are examined.
- Design, fabrication, and material impacts are presented.
- Efficient thermal energy dissipation and light harvesting improve performance.
- Nature-inspired photosynthetic optimization is discussed.
- Quantum dynamics are emphasized.

The technological progress in efficiency of energy and materials stock and flows processes may play a key role, but also the analysis showed that there are multi-parameter long-run factors with positive or negative casualties to investigate further.

In summary, the electrical and optical parameters of the bulk PV cells are influenced both from the external microenvironment and the intrinsic fundamental principles of the energy gap difference between the two-level bands. A hierarchical taxonomy of the modelling and design critical parameters, fabrication and material impacts, recombination effects, and spectrum optimization technologies are analyzed. Light and energy harvesting issues need to be overcome, while nature-inspired mimicking and interpretation of photon to charge and excitation energy transfer are in an infant stage, leading towards a better understanding of artificial photosynthesis. Distinguishing between conventional diffusion behavior and the dynamics of the quantum walks on coherence and incoherence hopping excitations also opens a new route of research, due to competitive interactions with other fields, such as lasing, quantum thermal machines, or quantum supercomputing, to name a few.

Further research should include an integrated approach to quantifying the performance of PV systems to include the aforementioned wide spectrum of key-technology determinants, including photon to charge efficiency investigation, thermal regulation, exergy efficiency, irreversibility of losses, quantum coherence speedup limitations, as well as energy and light harvesting interpretation of natural photosynthetic processes.

Author Contributions: Conceptualization, V.K. and G.K.; methodology, V.K., G.K. and M.Z.; software, A.T.; validation, V.K., G.K., A.T. and M.Z.; formal analysis, V.K.; investigation, V.K and G.K. and M.Z.; resources, G.K.; data curation, V.K and A.T.; writing—original draft preparation, V.K.; writing—review and editing, V.K., G.K. and M.Z.; visualization, V.K.; supervision, V.K. and A.T.; project administration, V.K.; funding acquisition, V.K. All authors have read and agreed to the published version of the manuscript.

Funding: This research received no external funding.

Conflicts of Interest: The authors declare no conflict of interest.

References

1. Lin, B.; Xu, M. Regional differences on CO_2 emission efficiency in metallurgical industry of China. *Energy Policy* **2018**, *120*, 302–311. [CrossRef]
2. Adom, P.K.; Kwakwa, P.A.; Amankwaa, A. The long-run effects of economic, demographic, and political indices on actual and potential CO_2 emissions. *J. Environ. Manag.* **2018**, *218*, 516–526. [CrossRef] [PubMed]
3. Dulebenets, M.A. A comprehensive multi-objective optimization model for the vessel scheduling problem in liner shipping. *Int. J. Prod. Econ.* **2018**, *196*, 293–318. [CrossRef]
4. Andersson, F.N.; Opper, S.; Khalid, U. Are capitalists green? Firm ownership and provincial CO2 emissions in China. *Energy Policy* **2018**, *123*, 349–359. [CrossRef]
5. Abioye, O.F.; Dulebenets, M.A.; Pasha, J.; Kavoosi, M. A vessel schedule recovery problem at the liner shipping route with Emission Control Areas. *Energies* **2019**, *12*, 2380. [CrossRef]
6. Lo, P.L.; Martini, G.; Porta, F.; Scotti, D. The determinants of CO2 emissions of air transport passenger traffic: An analysis of Lombardy (Italy). *Transp. Policy* **2018**, *91*, 108–119. [CrossRef]
7. Wiginton, L.K. Quantifying rooftop solar photovoltaic potential for regional renewable energy policy. *Comput. Environ. Urban. Syst.* **2010**, *34*, 345–357. [CrossRef]
8. Ko, L. Evaluation of the development potential of rooftop solar photovoltaic in Taiwan. *Renew. Energy* **2015**, *76*, 582–595. [CrossRef]
9. Santos, I.P.; Ruther, R. The potential of building-integrated (BIPV) and building-applied photovoltaics (BAPV) in single-family, urban residences at low latitudes in Brazil. *Energy Build.* **2012**, *50*, 290–297. [CrossRef]
10. Zhao, X.; Zhang, X.; Riffat, S.B.; Su, Y. Theoretical study of the performance of a novel PV/e roof module for heat pump operation. *Energy Convers. Manag.* **2011**, *52*, 603–614. [CrossRef]
11. Ramos, A.; Chatzopoulou, M.A.; Guarracino, I.; Freeman, J.; Markides, C.N. Hybrid photovoltaic-thermal solar systems for combined heating, cooling and power provision in the urban environment. *Energy Convers. Manag.* **2017**, *150*, 838–850. [CrossRef]
12. Aravossis, K.G.; Kapsalis, V.C. Power Engineering: Advances and Challenges Part A. In *Thermal Energy Storage Technologies*; CRC Press: Boca Raton, FL, USA; Taylor and Francis Group: Abingdon, UK, 2018; Chapter 13; pp. 83–420.

13. Qazi, S. *Fixed Standalone PV Systems for Disaster Relief and Remote Areas. Chapter 5, In: Standalone Photovoltaic (PV) Systems for Disaster Relief and Remote Areas*; Elsevier: Amsterdam, The Netherlands, 2017; Chapter 5; pp. 139–175. [CrossRef]
14. Murty, A.S.R.; Rajanish, Y.P.D. Enhanced energy harvesting and analysis of a high concentration photovoltaic / thermal system with support of cooling fluid and increased mass flow rates. *Int. J. Eng. Technol.* **2016**, *8*, 1077–1085.
15. Maranda, W. Analysis of self-consumption of energy from grid-connected photovoltaic system for various load scenarios with short-term buffering. *SN Appl. Sci.* **2019**, *1*, 406. [CrossRef]
16. Parida, A.; Choudhury, S.; Chatterjee, D. Optimized Solar PV Based Distributed Generation System Suitable for Cost-Effective Energy Supply. In Proceedings of the 9th Power India International Conference, Delhi, India, 28 February–1 March 2020; pp. 1–6. [CrossRef]
17. Ban-Weiss, G.; Wray, C.; Woody, W.; Ly, P.; Akbari, H.; Levinson, R. Electricity production and cooling energy savings from installation of a building-integrated photovoltaic roof on an office building. *Energy Build.* **2013**, *56*, 210–220. [CrossRef]
18. Dale, M.; Benson, S.M. Energy Balance of the Global Photovoltaic (PV) Industry–Is the PV Industry a Net Electricity Producer? *Environ. Sci. Technol.* **2013**, *47*, 3482–3489. [CrossRef] [PubMed]
19. Koo, C.; Hong, T.; Park, H.S.; Yun, G. Framework for the analysis of the potential of the rooftop photovoltaic system to achieve the net-zero energy solar buildings. *Prog. Photovolt. Res. Appl.* **2014**, *22*, 462–478. [CrossRef]
20. Ciulla, G.; Brano, V.L.; Di Dio, V.; Cipriani, G. A comparison of different one-diode models for the representation of I–V characteristic of a PV cell. *Renew. Sustain. Energy Rev.* **2014**, *32*, 684–696. [CrossRef]
21. Kilkis, S. A Rational Exergy Management Model for Curbing Building CO2 Emissions (LB-07-013), American Society of Heating, Refrigerating and Air-Conditioning Engineers (ASHRAE). *Transactions* **2007**, *113*, 76–86.
22. Onu, P.; Mbohwa, C. Advances in Solar Photovoltaic Grid Parity. In Proceedings of the 7th International Renewable and Sustainable Energy Conference (IRSEC), Agadir, Morocco, 27–30 November 2019; pp. 1–6. [CrossRef]
23. Drabiniok, E.; Neyer, A. Bionic micro porous evaporation foil for photovoltaic cell cooling. *Microelectron. Eng.* **2014**, *119*, 65–69. [CrossRef]
24. Ciulla, G.; Brano, V.L.; Cellura, M.; Franzotta, V.; Milone, D. A Finite Difference Model of a PV-PCM System. *Energy Procedia* **2012**, *30*, 198–206. [CrossRef]
25. Kichou, S.; Skandalos, N.; Wolf, P. Energy performance enhancement of a researchcentre based on solar potential analysis and energy management. *Energy* **2019**, *183*, 1195–1210. [CrossRef]
26. Park, J.; Kim, T.; Leigh, S.-B. Application of a phase-change material to improve the electrical performance of vertical-building-added photovoltaics considering the annual weather conditions. *Solar Energy* **2014**, *105*, 561–574. [CrossRef]
27. Browne, M.C.; Norton, B.; McCormack, S.J. Phase change materials for photovoltaic thermal management. *Renew. Sustain. Energy Rev.* **2015**, *47*, 762–782. [CrossRef]
28. Hasan, A.; McCormack, S.J.; Huang, M.J.; Norton, B. Characterization of phase change materials for thermal control of photovoltaics using Differential Scanning Calorimetry and Temperature History Method. *Energy Convers. Manag.* **2014**, *81*, 322–329. [CrossRef]
29. Atkin, P.; Farid, M.M. Improving the efficiency of photovoltaic cells using PCM infused graphite and aluminum fins. *Solar Energy* **2015**, *114*, 217–228. [CrossRef]
30. Chandrasekar, M.; Rajkumar, S.; Valavan, D. A review on the thermal regulation techniques for nonintegrated flat PV modules mounted on building top. *Energy Build.* **2015**, *86*, 692–697. [CrossRef]
31. Hasanuzzaman, M.; Malek, A.B.M.A.; Islam, M.M.; Pandey, A.K.; Rahim, N.A. Global advancement of cooling technologies for PV systems: A review. *Solar Energy* **2016**, *137*, 25–45. [CrossRef]
32. Cui, Y.; Wang, Y.; Huang, Q.; Wei, S. Effect of radiation and convection heat transfer on cooling performance of radiative panel. *Renew. Energy* **2016**, *99*, 10–17. [CrossRef]
33. Kandeal, A.W.; Thakur, A.M.; Elkadeem, M.R.; Elmorshedy, M.F.; Ullah, Z.; Sathyamurthy, R.; Sharshir, S.W. Photovoltaics performance improvement using different cooling methodologies: A State-of-Art Review. *J. Clean. Prod.* **2020**, *273*, 122772. [CrossRef]
34. Hrica, J.; Chatterjee, S.; TamizhMani, G. BAPV array: Thermal modeling and cooling effect of exhaust fan. In Proceedings of the 37th IEEE Photovoltaic Specialists Conference, Seattle, DC, USA, 19–24 June 2011; pp. 3144–3149. [CrossRef]
35. Kapsalis, V.C.; Karamanis, D. On the effect of roof added photovoltaics on building's energy performance. *Energy Build.* **2015**, *108*, 195–204. [CrossRef]
36. Kichou, S.; Silvestre, S.; Guglielminotti, L.; Mora-López, L.; Muñoz-Cerón, E. Comparison of two PV array models for the simulation of PV systems using five different algorithms for the parameters identification. *Renew. Energy* **2016**, *99*, 270–279. [CrossRef]
37. Kim, G.G.; Choi, J.H.; Park, S.Y.; Bhang, B.G.; Nam, W.J.; Cha, H.L.; Park, N.; Ahn, H.-K. Prediction Model for PV Performance With Correlation Analysis of Environmental Variables. *IEEE J. Photovolt.* **2019**, *9*, 832–841. [CrossRef]
38. Kichou, S.; Silvestre, S.; Nofuentes, G.; Torres-Ramírez, M.; Chouder, A.; Guasch, D. Characterization of degradation and evaluation of model parameters of amorphous silicon photovoltaic modules under outdoor long-term exposure. *Energy* **2016**, *96*, 231–241. [CrossRef]
39. Shen, L.; Li, Z.; Ma, T. Analysis of the power loss and quantification of the energy distribution in PV module. *Appl. Energy* **2020**, *260*, 114333. [CrossRef]

40. Stradins, P. Staebler-Wronski defects: Creation efficiency, stability, and effect on a-Si:H solar cell degradation. In Proceedings of the 35th IEEE Photovoltaic Specialists Conference, Honolulu, HI, USA, 20–25 June 2010; pp. 142–145. [CrossRef]
41. Shockley, W.; Queisser, H. Detailed Balance Limit of Efficiency of p-n Junction Solar Cells. *J. Appl. Phys.* **1961**, *32*, 510–519. [CrossRef]
42. Luque, A. Will we exceed 50% efficiency in photovoltaics. *J. Appl. Phys.* **2011**, *110*, 031301. [CrossRef]
43. Kerr, J.M.; Cuevas, A. General parameterization of Auger recombination in crystalline silicon. *J. Appl. Phys.* **2002**, *91*, 2473. [CrossRef]
44. Augusto, A.; Herasimenk, Y.S.; King, R.R.; Bowden, G.S.; Honsberg, C. Analysis of the recombination mechanisms of a silicon solar cell with low bandgap-voltage offset. *J. Appl. Phys.* **2017**, *121*, 205704. [CrossRef]
45. Hsieh, M.; Kaiser, A.; Bhattacharya, S.; Sajeev, J.; Lin, S.-Y. Experimental demonstration of broadband solar absorption beyond the lambertian limit in certain thin silicon photonic crystals. *Sci. Rep.* **2020**, *10*, 11857. [CrossRef] [PubMed]
46. Bhattacharya, S.; Sajeev, J. Photonic crystal light trapping: Beyond 30% conversion efficiency for silicon photovoltaic. *APL Photon.* **2020**, *5*, 20902. [CrossRef]
47. Carle, J.E.; Krebs, F.C. Technological status of organic photovoltaics (opv). *Sol. Energy Mater. Solar Cells* **2013**, *119*, 309–310. [CrossRef]
48. Service, R.F. Outlook brightness for plastic solar cells. *Science* **2011**, *293*. [CrossRef]
49. Kamat, P.V. Quantum dot solar cells. Semiconductor nanocrystals as light harvesters. *J. Phys. Chem. C* **2008**, *112*, 18737–18753. [CrossRef]
50. Skandalos, N.; Karamanis, D. PV glazing technologies. *Renew. Sustain. Energy Rev.* **2015**, *49*, 306–322. [CrossRef]
51. McGehee, M.D. Materials Science: Fast—Truck solar cells. *Nature* **2013**, *501*, 323–325. [CrossRef]
52. Snaith, H.J. Perovskites: A emergence of new era for low cost, high efficiency solar cells. *Phys. Chem. Lett.* **2013**, *4*, 3623–3630. [CrossRef]
53. Bozyigit, D.; Volk, S.; Yarema, O.; Wood, V. Quantification of deep traps in nanocrystal solids, their electronic properties and the influence on device behavior. *Nano Lett.* **2013**, *13*, 5284–5288. [CrossRef] [PubMed]
54. Chuang, C.H.M.; Brown, P.R.; Bulovic, V.; Bawendi, M.G. Improved performance and stability in quantum dot solar cells through band alignment engineering. *Nat. Mat.* **2014**, *13*, 796–801. [CrossRef]
55. Wang, W.; Lin, A.S.; Phillips, J.D. Intermediate-band photovoltaic solar cell based on ZnTe: O. *Appl. Phys. Lett.* **2009**, *95*, 11103. [CrossRef]
56. Marti, A.; Tablero, C.; Antolin, E.; Luque, A.; Campion, R.; Novikov, S.; Foxon, C.T. Potential of mn doped ln1−xgaxn for implementing intermediate band solar cells. *Sol. Energy Mater. Solar Cells* **2009**, *93*, 641–644. [CrossRef]
57. Luque, A.; Marti, A.; Lopez, N.; Antolin, E.; Canovas, E.; Stanley, C.; Farmer, C.; Caballero, L.J.; Cuadra, L.; Balenzategui, J.L. Experimental analysis of the quasi-Fermi level split in quantum dot intermediate-band solar cells. *Appl. Phys. Lett.* **2005**, *87*, 83505. [CrossRef]
58. Shao, Q.; Balandin, A.; Fedoseyev, A.; Turowski, M. Intermediate-band solar cells based on quantum dot supracrystals. *Appl. Phys. Lett.* **2007**, *91*, 163503. [CrossRef]
59. Imenes, A.G.; Mills, D.R. Spectral beam splitting technology for increased conversion efficiency in solar concentrating systems: A review. *Solar Energy Mater. Solar Cells* **2004**, *84*, 19–69. [CrossRef]
60. Hossain, M.I.; Bousselham, A.; Alharbi, F.H. Optical concentration effects on conversion efficiency of a split-spectrum solar cell system. *J. Phys. D Appl. Phys.* **2014**, *47*, 075101. [CrossRef]
61. Abrams, Z.R.; Gharghi, M.; Niv, A.; Gladden, C.; Zhang, X. Theoretical efficiency of 3rd generation solar cells: Comparison between carrier multiplication and down-conversion. *Sol. Energy Mater. Sol. Cells* **2012**, *99*, 308–315. [CrossRef]
62. Bett, A.W.; Dimroth, F.; Stollwerck, G.; Sulima, O.V. III-V compounds for solar cell applications. *Appl. Phys. A* **1999**, *69*, 119–129. [CrossRef]
63. Meckler, M. Fixed Solar Concentrator-Collector-Satellite Receiver and Co-Generator. U.S. Patent 4 490-981, 1 January 1985.
64. Erim, M.N.; Erim, N.; Kurt, H. Spectral splitting for an InGaP/GaAs parallel junction solar cell. *Appl. Opt.* **2019**, *58*, 4265–4270. [CrossRef] [PubMed]
65. Wei, A.C.; Chang, W.J.; Sze, J.R. A Side-Absorption Concentrated Module with a Diffractive Optical Element as a Spectral-Beam-Splitter for a Hybrid-Collecting Solar System. *Energies* **2020**, *13*, 192. [CrossRef]
66. Shaller, R.D.; Klimov, V.I. High efficiency carrier multiplication in PbSe nanocrystals: Implications for solar energy conversion. *Phys. Rev. Lett.* **2004**, *92*, 186601–186604. [CrossRef]
67. Takeda, Y.; Motohiro, T. Requisites to realize high conversion efficiency of solar cells utilizing carrier multiplication. *Sol. Energy Mater. Sol. Cells* **2010**, *94*, 1399–1405. [CrossRef]
68. Green, M.A. Third generation photovoltaics: Solar cells for 2020 and beyond. *Phys. E Low Dimens. Syst. Nanostruct.* **2002**, *14*, 65–70. [CrossRef]
69. Conibeer, G.; König, D.; Green, M.; Guillemoles, J. Slowing of carrier cooling in hot carrier solar cells. *Thin Solid Film.* **2008**, *516*, 6948–6953. [CrossRef]
70. Sun, D.; Aivazian, G.; Jones, A.M.; Ross, J.S.; Yao, W.; Cobden, D. Ultrafast hot-carrier-dominated photocurrent in graphene. *Nat. Nanotechnol.* **2012**, *7*, 114–118. [CrossRef] [PubMed]

71. Gabor, N.M.; Song, J.C.; Ma, Q.; Nair, N.L.; Taychatanapat, T.; Watanabe, K.; Taniguchi, T.; Levitov, L.S.; Jarillo-Herrero, P. Hot carrier-assisted intrinsic photoresponse in grapheme. *Science* **2011**, *334*, 648–652. [CrossRef] [PubMed]
72. Holliman, J.P.; Mohsen, M.; Connell, A.; Kershaw, P.C.; Meza-Rojas, D.; Jones, W.E.; Geatches, D.; Sen, K.; Hsiao, Y.W. Double Linker Triphenylamine Dyes for Dye-Sensitized Solar Cells. *Energies* **2020**, *13*, 4637. [CrossRef]
73. Utlu, Z. Thermophotovoltaic applications in waste heat recovery systems: Example of GaSb cell. *Int. J. Low Carbon Technol.* **2020**, *15*, 277–286. [CrossRef]
74. Pidgeon, C.R.; Ciesle, C.M.; Murdin, B.N. Suppression of non-radiativeprocesses in semiconductor mid infrared emitters and detectors. *Prog. Quantum Electron.* **1997**, *21*, 361–419. [CrossRef]
75. Ang, Y.S.; Yang, H.Y.; Ang, L.K. Universal scaling laws in Schottky heterostructures based on two dimensional materials. *Phys. Rev. Lett.* **2018**, *121*, 56802. [CrossRef]
76. Xiao, G.; Zheng, G.; Qiu, M.; Li, Q.; Li, D.; Ni, M. Thermionic energy conversion for concentrating solar power. *Appl. Energy* **2017**, *208*, 1318–1342. [CrossRef]
77. Liao, T. Improved design of a photon enhanced thermionic energy converter. *IEEE Electron. Device Lett.* **2019**, *40*, 115–118. [CrossRef]
78. Santhanam, P.; Fan, S. Thermal-to-electric energy conversion by diodes under negative illumination. *Phys. Rev. B* **2016**, *93*, 161410. [CrossRef]
79. Ono, M.; Santhanam, P.; Li, W.; Zhao, B.; Fan, S. Experimental demonstration of energy harvesting from sky using the negative illumination effect of a semiconductor photodiode. *Appl. Phys. Lett.* **2014**, *114*, 161102. [CrossRef]
80. Salem, M.S.; Alzahrani, A.J.; Ramadan, R.A.; Alanazi, A.; Shaker, A.; Abouelatta, M.; Gontrand, C.; Elbanna, M.; Zekry, A. Physically Based Analytical Model of Heavily Doped Silicon Wafers Based Proposed Solar Cell Microstructure. *IEEE Access* **2020**, *8*, 138898–138906. [CrossRef]
81. Ross, R.T.; Hsiao, T.L. Limits on the yield of photochemical solar energy conversion. *J. Appl. Phys.* **1977**, *48*, 4783–4785. [CrossRef]
82. Bolton, R.; Hall, D.O. Photochemical conversion and storage of solar energy. *Annu. Rev. Energy* **1979**, *4*, 353–401. [CrossRef]
83. Engel, G.S.; Calhoun, T.R.; Read, E.L.; Ahn, T.K.; ManCal, T.; Cheng, Y.C.; Blankenship, R.E.; Fleming, G.R. Evidence for wavelike energy transfer through quantum coherence in photosynthetic systems. *Nature* **2007**, *446*, 782–786. [CrossRef]
84. Scholes, G.D. Quantum-coherent electronic energy transfer: Did nature think of it first? *J. Phys. Chem. Lett.* **2010**, *1*, 2–8. [CrossRef]
85. Vlaming, S.M.; Silbey, R.J. Correlated intermolecular coupling fluctuations in photosynthetic complexes. *J. Chem. Phys.* **2012**, *136*, 055102. [CrossRef] [PubMed]
86. Mohseni, M.; Shabani, A.; Lloyd, S.; Rabitz, H. Energy-scales convergence for optimal and robust quantum transport in photosynthetic complexes. *J. Chem. Phys.* **2014**, *140*, 035102. [CrossRef] [PubMed]
87. Creatore, C.; Parker, M.; Emmott, S.; Chin, A. Efficient biologically inspired photocell enhanced by delocalized quantum states. *Phys. Rev. Lett.* **2013**, *111*, 253601. [CrossRef] [PubMed]
88. Heo, J.H.; Im, S.H.; Noh, J.H.; Mandal, T.N.; Lim, C.S.; Chang, J.A.; Lee, Y.H.; Kim, H.; Sarkar, A.; Nazeeruddin, M.K.; et al. Efficient inorganic–organic hybrid heterojunction solar cells containing perovskite compound and polymeric hole conductors. *Nat. Photonics* **2013**, *7*, 486–491. [CrossRef]
89. Jeon, N.J.; Lee, J.; Noh, J.H.; Nazeeruddin, M.K.; Gratzel, M.; Seok, S.I. Efficient inorganic–organic hybrid perovskite solar cells based on pyrene arylamine derivatives as hole-transporting materials. *J. Am. Chem. Soc.* **2013**, *135*, 19087–19090. [CrossRef]
90. Li, G.; Govind, N.; Ratner, A.M.; Cramer, J.C.; Gagliardi, L. Influence of Coherent Tunneling and Incoherent Hopping on the Charge Transfer Mechanism in Linear Donor–ridge–Acceptor Systems. *J. Phys. Chem. Lett.* **2015**, *6*, 4889–4897. [CrossRef] [PubMed]
91. Bastola, E.; Alfadhili, F.K.; Phillips, A.B.; Heben, M.J.; Ellingson, R.J. Wet chemical etching of cadmium telluride photovoltaics for enhanced open-circuit voltage, fill factor, and power conversion efficiency. *J. Mater. Res.* **2019**, *34*, 3988–3997. [CrossRef]
92. Munshi, A.H.; Kephart, J.; Abbas, A.; Raguse, J.; Beaudry, J.N.; Barth, K.; Sites, J.; Walls, J.; Sampath, W. Polycrystalline CdSeTe/CdTe absorber cells with 28 mA/cm2 short-circuit current. *IEEE J. Photovolt.* **2018**, *8*, 310. [CrossRef]
93. Paudel, N.R.; Yan, Y. Enhancing the photo-currents of CdTe thin-film solar cells in both short and long wavelength regions. *Appl. Phys. Lett.* **2014**, *105*, 183510. [CrossRef]
94. Li, J.; Diercks, D.R.; Ohno, T.R.; Warren, C.W.; Lonergan, M.C.; Beach, J.D.; Wolden, C.A. Controlled activation of ZnTe:Cu contacted CdTe solar cells using rapid thermal processing. *Sol. Energy Mater. Sol. Cells* **2015**, *133*, 208. [CrossRef]
95. Bastola, E.; Bhandari, K.P.; Subedi, I.; Podraza, N.J.; Ellingson, R.J. Structural, optical, and hole transport properties of earth-abundant chalcopyrite (CuFeS2) nanocrystals. *MRS Commun.* **2018**, *8*, 970. [CrossRef]
96. Subedi, K.K.; Bastola, E.; Subedi, I.; Song, Z.; Bhandari, K.P.; Phillips, A.B.; Podraza, N.J.; Heben, M.J.; Ellingson, R.J. Nanocomposite (CuS)x(ZnS)1x thin film back contact for CdTe solar cells: Toward a bifacial device. *Sol. Energy Mater. Sol. Cells* **2018**, *186*, 227. [CrossRef]
97. Ochoa-Landin, R.; Vigil-Galan, O.; Vorobiev, Y.V.; Ramírez-Bon, R. Chemically-deposited Te layers improving the parameters of back contacts for CdTe solar cells. *Sol. Energy* **2009**, *83*, 134. [CrossRef]
98. Song, T.; Moore, A.; Sites, J.R. Te Layer to Reduce the CdTe Back-Contact Barrier. *IEEE J. Photovolt.* **2018**, *8*, 293–298. [CrossRef]
99. Watthage, S.C.; Phillips, A.B.; Liyanage, G.K.; Song, Z.; Gibbs, J.M.; Alfadhili, F.K.; Alkhayat, R.B.; Ahangharnejhad, R.H.; Almutawah, Z.S.; Bhandari, K.P. Selective Cd removal from CdTe for high-efficiency Te back-contact formation. *IEEE J. Photovolt.* **2018**, *8*, 1125. [CrossRef]

100. Li, M.; Chang, F.; Li, C.; Xia, C.; Wang, T.; Wang, J.; Sun, M. CIS and CIGS thin films prepared by magnetron sputtering. *Procedia Eng.* **2012**, *27*, 12–19. [CrossRef]
101. Gulkowski, S.; Krawczak, E. RF/DC Magnetron Sputtering Deposition of Thin Layers for Solar Cell Fabrication. *Coatings* **2020**, *10*, 791. [CrossRef]
102. Wang, M.; Hossain, M.; Choy, K. Effect of Sodium Treatment on the Performance of Electrostatic Spray Assisted Vapour Deposited Copper-poor Cu(In,Ga)(S,Se) 2 solar cells. *Sci. Rep.* **2017**, *7*, 1–10. [CrossRef] [PubMed]
103. Wagner, J.; Gruber, M.; Hinderhofer, A.; Wilke, A.; Bröker, B.; Frisch, J.; Amsalem, P.; Vollmer, A.; Opitz, A.; Koch, N.; et al. High fill factor and open circuit voltage in organic photovoltaic cells with diindenoperylene as donor material. *Adv. Funct. Mater.* **2010**, *20*, 4295–4303. [CrossRef]
104. Du, J.; Fortney, A.; Washington, E.K.; Biewer, M.C.; Kowalewski, T.; Stefan, M.C. Benzo[1,2-b:4,5-b0]difuran and furan substituted diketopyrrolopyrrole alternating copolymer for organic photovoltaics with high fill factor. *J. Mater. Chem. A* **2017**, *5*, 15591. [CrossRef]
105. Galagan, Y.; Fledderus, H.; Gorter, H.; Mannetje, H.H.; Shanmugam, S.; Mandamparambil, R.; Bosman, J.; Rubingh, J.M.; Teunissen, J.P.; Salem, A.; et al. Roll-to-Roll Slot–Die Coated Organic Photovoltaic (OPV) Modules with High Geometrical Fill Factors. *Energy Technol.* **2015**, *3*, 834–842. [CrossRef]
106. Wang, Z.; Hong, Z.; Zhuang, T.; Chen, G.; Sasabe, H.; Yokoyama, D.; Kido, J. High fill factor and thermal stability of bilayer organic photovoltaic cells with an inverted structure. *Appl. Phys. Lett.* **2015**, *106*, 53305. [CrossRef]
107. Lubieniecki, M.; Uhl, T. Integration of Thermal Energy Harvesting in Semi-Active Piezoelectric Shunt-Damping Systems. *J. Electron. Mater.* **2015**, *44*, 341–347. [CrossRef]
108. Wang, Q.; Bowen, C.R.; Lei, W.; Zhang, H.; Xie, B.; Qiu, S.; Li, M.Y.; Jiang, S. Improved heat transfer for pyroelectric energy harvesting applications using a thermal conductive network of aluminum nitride in PMN-PMS-PZT ceramics. *J. Mater. Chem. A* **2018**, *6*, 5040–5051. [CrossRef]
109. Jeyashree, Y.; Sukhi, Y.; Vimala, J.A.; Lourdu, J.S.; Indirani, S. Concentrated solar thermal energy harvesting using Bi2Te3based thermoelectric generator. *Mater. Sci. Semicond. Process.* **2020**, *107*, 104782. [CrossRef]
110. Ghomian, T.; Kizilkaya, O.; Choi, J.W. Lead sulfide colloidal quantum dot photovoltaic cell for energy harvesting from human body thermal radiation. *Appl. Energy* **2018**, *230*, 761–768. [CrossRef]
111. Chun, J.; Song, H.; Kang, M.; Kang, H.; Kishore, R.A.; Priya, S. Thermo-Magneto-Electric Generator Arrays for Active Heat Recovery System. *Sci. Rep.* **2017**, *7*, 41383. [CrossRef]
112. Scully, M.O. Quantum photocell: Using quantum coherence to reduce radiative recombination and increase efficiency. *Phys. Rev. Lett.* **2010**, *104*, 207701. [CrossRef]
113. Dorfman, K.E.; Voronine, D.V.; Mukamel, S.; Scully, M.O. Photosynthetic reaction center as a quantum heat engine. *Proc. Natl. Acad. Sci. USA* **2010**, *110*, 2746–2751. [CrossRef]
114. Wang, Y.C.; Chen, C.W.; Su, T.Y.; Yang, T.Y.; Liu, W.W.; Cheng, F.; Wang, M.Z.; Chueh, Y.L. Design of suppressing optical and recombination losses in ultrathin CuInGaSe2 solar cells by Voronoi nanocavity arrays. *Nano Energy* **2020**, *78*, 2211–2855. [CrossRef]
115. Boer, K.W.; Pohl, U.W. Photon–Phonon Interaction. In *Semiconductor Physics*; Springer: Cham, Switzerland, 2018; pp. 389–424.
116. Kirk, A.P. Advancing solar cell to the limit of energy cascading. In Proceedings of the 39th Photovoltaic Specialists Conference, Tampra Bay, FL, USA, 16–21 June 2013; pp. 782–787.
117. Schafer, C.; Ruggenthaller, M.; Appel, H.; Rubio, A. Modification of excitation and charge transfer in cavity quantum— Electrodynamical chemistry. *Proc. Natl. Acad. Sci. USA* **2019**, *116*, 4883–4892. [CrossRef]
118. Panhans, M.; Hutsch, S.; Benduhn, J.; Schellhammer, K.S.; Nikolis, V.C.; Vangerven, T.; Vandewal, K.; Ortmann, F. Molecular vibrations reduce the maximum achievable photovoltage in organic solar cells. *Nat. Commun.* **2020**, *11*, 1488. [CrossRef]

energies

MDPI

Article

Numerical Study of Heat Transfer Intensification in a Circular Tube Using a Thin, Radiation-Absorbing Insert. Part 1: Thermo-Hydraulic Characteristics

Piotr Bogusław Jasiński

Institute of Turbomachinery, Lodz University of Technology, 90-924 Lodz, Poland; piotr.jasinski@p.lodz.pl; Tel.: +48-42-631-23-49

Abstract: The presented paper, which is the first of two parts, shows the results of numerical investigations of a heat exchanger channel in the form of a cylindrical tube with a thin insert. The insert, placed concentrically in the pipe, uses the phenomenon of thermal radiation absorption to intensify the heat transfer between the pipe wall and the gas. Eight geometric configurations of the insert size were numerically investigated using CFD software, varying its diameter from 20% to 90% of the pipe diameter and obtaining the thermal-flow characteristics for each case. The tests were conducted for a range of numbers Re = 5000–100,000 and a constant temperature difference between the channel wall and the average gas temperature of ΔT = 100 °C. The results show that the highest increase in the Nu number was observed for the inserts with diameters of 0.3 and 0.4 of the channel diameter, while the highest flow resistance was noted for the inserts with diameters of 0.6–0.7 of the channel diameter. The $f/f_s(Re)$ and $Nu/Nu_s(Re)$ ratios are shown on graphs indicating how much the flow resistance and heat transfer increased compared to the pipe without an insert. Two methods of calculating the Nu number are also presented and analysed. In the first one, the average fluid temperature of the entire pipe volume was used to calculate the Nu number, and in the second, only the average fluid temperature of the annular portion formed by the insert was used. The second one gives much larger Nu/Nu_s ratio values, reaching up to 8–9 for small Re numbers.

Keywords: heat transfer enhancement; radiation insert; numerical simulations; friction factor; Nu number

Citation: Jasiński, P.B. Numerical Study of Heat Transfer Intensification in a Circular Tube Using a Thin, Radiation-Absorbing Insert. Part 1: Thermo-Hydraulic Characteristics. *Energies* **2021**, *14*, 4596. https://doi.org/10.3390/en14154596

Academic Editor: Pouyan Talebizadeh Sardari

Received: 15 June 2021
Accepted: 21 July 2021
Published: 29 July 2021

1. Introduction

There are many ways to intensify the heat transfer in heat exchanger channels, e.g., by using ribs of different shapes or special tabs on the inner wall [1–3]. They aim to increase the fluid's turbulence in the laminar boundary layer and increase the heat transfer surface. Another, also quite a popular method to improve heat transfer is to disturb the flow in the whole pipe cross-section, which can be achieved by, e.g., introducing special disturbing inserts, twisted tapes, or wire coils into the duct. A common feature of this type of turbulisers, regardless of their shapes, is the strong influence of some geometrical parameters on the heat-flow characteristics (e.g., the rib angle, its height or thickness) [4–8]. However, any such turbuliser, in addition to a local increase in heat transfer intensity, simultaneously causes a significant increase in flow resistance, which involves an obvious power increase to pump the working fluid.

This paper presents testing channels with an insert that does not turbulise the flow as most such elements do, but instead uses an additional heat transport mechanism, thermal radiation. Obviously, due to the necessary radiation transmittance of the working fluid, it is only applicable to gases.

The insert presented in this article is made in the form of a thin, smooth pipe and placed parallel and concentrically to the exchanger channel's walls (Figure 1). Apart from the standard, convective method of heat transport between the wall and the fluid, the heat transfer intensification mechanism additionally uses the phenomenon of thermal radiation between the pipe wall and the insert. During the fully developed fluid flow, the insert placed in the centre of the tube does not mix the fluid and has a different temperature than the wall. Due to the temperature difference, heat is transferred by radiation between the wall surface and the insert. Assuming that the working gas is completely transparent and does not absorb radiation, the total radiant heat flux is transferred to the insert, increasing its temperature and becoming an additional heat transfer surface while transferring heat to the fluid on both sides by convection.

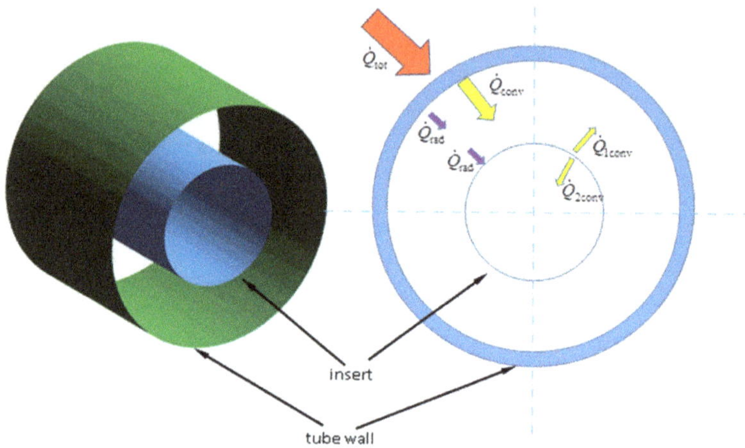

Figure 1. A fragment of a pipe with an insert and a schematic view of heat fluxes.

Unfortunately, no papers describe this type of insert in the available literature, making it impossible to refer directly to the presented research. Therefore, the articles that describe the most similar cases using turbulence inserts and with radiation heat transfer were analysed.

There is quite a lot of research on solar collectors in the literature, where the phenomena of radiation and convection occur together. In such a collector, solar radiation penetrates through a transparent glass tube and, by heating the absorber, will transfer heat to the fluid by convection. However, it is only in the absorber (inside the channel) that the heat transfer is intensified by placing various inserts that turbulise the flow [9–11].

The intensification of the heat transfer in the channel, resulting from the turbulence of the fluid through various methods, has been analysed by many researchers. The variable value was often some geometrical feature of the insert, influencing the flow resistance and heat transfer. For example, Eiamsa-ard and Promvonge [12] studied diamond-shaped turbulence inserts with their ends touching each other. The variable parameters were the length of a single element and the angle of inclination of the diamond face. The highest Nu numbers were observed for the shortest elements and those with the highest inclination angles. Kongkaitpaiboon et al. [13] experimented with an insert consisting of round rings attached to a pipe wall. He analysed two geometric variables: the rings' inner diameters and the distance between them (three diameters and three longitudinal distances) in a small range of Reynolds numbers (4000–20,000). The study showed that the best heat transfer occurred with the ring's smallest inner diameter and the smallest distances between the rings. Durums [14], on the other hand, in their study, presents the influence of the conical nozzle angle on the pressure drop and heat transfer intensity. The presented results show that as the turbuliser nozzle angle increases, the Nusselt number and friction coefficient also increase nonlinearly. In papers [15–17], Jasiński presented a comprehensive experimental and numerical study of a ball insert, with varying ball diameters and longitudinal distances between them, at a fairly large range of *Re* numbers 5000–300,000. It turned out that the largest coefficient of thermal efficiency was obtained by the inserts with the smallest diameter and largest longitudinal distance, and additionally, at large *Re* numbers. One example of twisted tape inserts research was presented in [18] by Bas and Ozceyhan. The highest efficiencies were obtained for the smallest Re numbers 5000–10,000 with a small gap between the pipe wall and the insert for all the insert geometries they studied.

This work's originality is based on the use of an additional heat transport mechanism, which is radiation from the pipe wall to the absorption insert. The presented results were obtained for a relatively small temperature difference ΔT = 100 °C, as is standard for radiation processes. Further work will focus on research for developing such inserts at greater temperature differences to improve the thermal efficiency of various devices, such as solar air ducts, heaters, and heat exchangers. The concept of this type of insert may be of great importance, especially for high-temperature combustion and heat recovery processes, where thermal radiation is the dominant method of heat transfer.

2. Geometry of Insert and Principle of Operation

The principle of operation of such an insert is based on the fact that the total heat flux supplied to the outer wall of the channel is partially transferred by convection to the fluid, and at the same time, a part of this flux is transferred to the insert by radiation, Figure 1. Because the insert does not cause turbulence and mixing of the fluid, its temperature mainly depends on the radius on which it is located. Since the temperature field in the cross-section changes as a function of the radius during a fully developed flow in the channel, there is some temperature difference between the pipe wall and the insert. If the flowing gas is transparent and going through convection, there is also a mechanism of heat transport through radiation $\dot{Q}_{rad}$ between these surfaces. The absorption of radiation by the insert causes its slight heating, and thus the temperature difference between the wall and the insert is decreased, which in turn reduces the radiative heat flux. Simultaneously, the insert is washed on both sides by a working gas that flows both in the annular space and in the insert's centre space, which also causes convective heat transfer to the gas flowing on both sides. The heat receipt from the insert causes a decrease in its temperature, i.e., an increase in ΔT (between the pipe wall and the insert). The convective heat fluxes $\dot{Q}_{1conv}$ and $\dot{Q}_{2conv}$ of the insert's outer and inner surfaces are not equal due to different flow and thermal conditions. As you can see, the mechanism of this process is quite complex and depends on many parameters.

The thickness of such an insert should be as small as possible, and the insert itself should be made of a material that conducts heat well. The thickness can then be neglected, and the temperature can be assumed to be the same on both surfaces of the insert. In order to obtain the highest possible radiative heat flux, the emissivity of both surfaces involved in the radiative heat transfer, i.e., the pipe wall and the insert, should also be as high as possible.

Due to the fact that a boundary condition of constant temperature is assumed on the inner surface of the pipe, we are not really interested in "what happens" in the wall itself. Therefore, the pipe wall was can also be assumed to be negligible in the numerical calculations. Of course, a more realistic case would be with a pipe of a certain thickness, e.g., 1 mm and made of copper, but then the heat conduction resistance through the wall would be so small that the wall temperature would practically be the same anyways.

The use of this type of insert can be justified, especially in high-temperature exchangers, where due to a significant temperature difference, the radiation fraction in heat transfer is significant compared to convection, e.g., heat recovery from exhaust gases in gas boilers, cars, etc.

Nevertheless, the research presented in this article has shown that the use of such an insert, even with small differences in temperatures between the wall and the gas of approx. 100 °C, gives a noticeable effect of intensifying heat transfer. As the gas flow velocity increases, the convective heat flux also increases, and thus the proportion of the radiation flux decreases. Therefore, the channel efficiency with inserts absorbing thermal radiation is the greatest for small Reynolds numbers.

3. Numerical Model

Numerical tests were conducted for various insert diameters d_i (Figure 2c) and a constant temperature difference $\Delta T = 100\ ^\circ\mathrm{C}$ between the wall and the average gas temperature in the pipe cross-section. On the outer wall of the channel, the heat transfer boundary condition of the first kind was applied, i.e., a constant wall temperature equal to $100\ ^\circ\mathrm{C}$, while for gas, its mean volume temperature equal to $0\ ^\circ\mathrm{C}$ was a constant value. The working gas was the air with physical properties that were dependent on temperature.

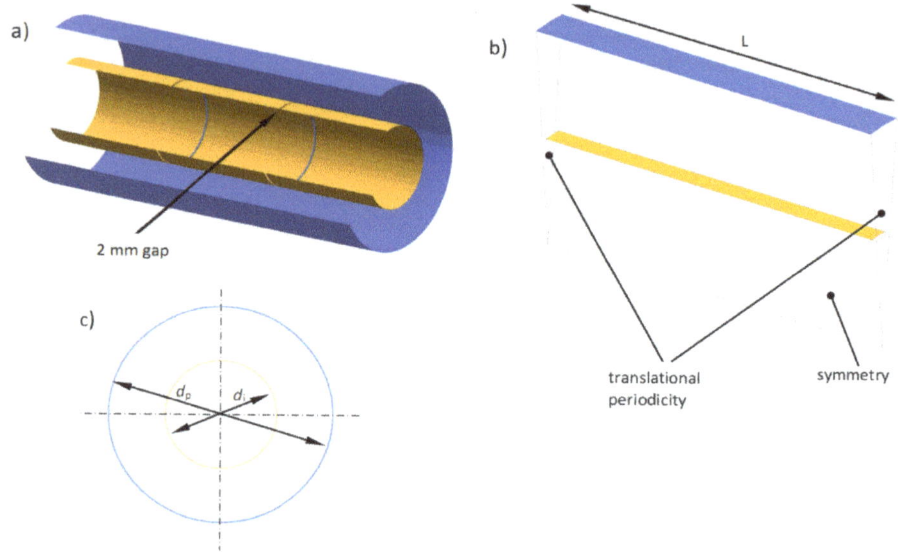

Figure 2. Repeatable insert segment: (**a**) 3D view of the pipe section with insert, (**b**) computational domain, (**c**) channel cross-section diagram.

In order to simplify the notation and the possibility of referencing other geometrical dimensions, the insert diameters d_i are presented in dimensionless form $D = d_\mathrm{p}/d_\mathrm{i}$, with reference to the pipe diameter $d_\mathrm{p} = 200$ mm. The values of d_i and D, along with their final dimensions, are given in Table 1.

Table 1. Dimensions of the tested insert and the corresponding dimensionless diameters.

d_i [mm]	40	60	80	100	120	140	160	180
D [—]	0.2	0.3	0.4	0.5	0.6	0.7	0.8	0.9

Due to the axial nature of the flow and the lack of a rotational velocity component in the flow, it is assumed that each tested insert can be considered as a two-dimensional case. Therefore, a computational domain in the shape of a longitudinal cylindrical slice with an aperture angle of 10° and length L = 152 mm was used in the simulations (Figure 2b), and the axial symmetry as the boundary conditions on the sides of the domain was set.

The appropriate method of conducting numerical simulations allowed the obtaining of a fully developed flow in this relatively short domain. In order to achieve this effect, the translational periodicity (as boundary conditions) at the inlet and outlet of the domain was set, and the fluid flow was forced with a pressure gradient. Thus, the computational domain has been reduced to a repeatable, periodic, and axisymmetric geometry representative of the entire channel. By reducing the domain size and the number of computational mesh nodes (up to 2D), it was possible to significantly shorten the computation time while maintaining the mesh's high quality. This method of study has also been presented in [4,19–21], while other, less important details related to this method of numerical modelling are also presented in [15–17]. In order to ensure the same static pressure on both sides of the insert, slight gaps of 2 mm length in its continuity (Figure 2a) were applied, constituting about 1.5% of the total length of the insert.

For each insert, nine simulations were performed with different Reynolds numbers, resulting in 72 numerically tested cases, not counting the trial simulations. Based on the obtained results, thermal-flow characteristics of the tested inserts were prepared. Numerical simulations were performed with the ANSYS-CFX v18.2 calculation code.

4. Grid, Turbulence Model and Radiation Model

4.1. Governing Equations

Numerical calculations of the frictional resistance and the heat transfer were performed using the ANSYS-CFX code. During the calculations the basic equations of conservation of mass (1), momentum (2) and energy (3), which have the form [22], were solved:

$$\frac{\partial \rho}{\partial t} + \nabla \cdot (\rho \overline{U}) = 0 \tag{1}$$

$$\frac{\partial (\rho \overline{U})}{\partial t} + \nabla \cdot (\rho \overline{U} \times \overline{U}) = -\nabla p + \nabla \cdot \mu_e (\nabla \overline{U} + (\nabla \overline{U})^T - \frac{2}{3} \delta \nabla \cdot \overline{U}) + (\rho - \rho_{ref}) g \tag{2}$$

$$\frac{\partial (i_{tot} \rho)}{\partial t} = \frac{\partial \rho}{\partial t} + \nabla \cdot (\rho \overline{U} i_{tot}) = \nabla \cdot (k \nabla T) + \nabla \cdot (\overline{U} \tau_w) + S_E \tag{3}$$

where the term $\nabla \cdot (\overline{U} \tau_w)$ represents the work of viscous forces, S_E is a term of an energy source, and the total enthalpy i_{tot} is expressed as:$i + \frac{1}{2}\overline{U}^2$. For the turbulence model used, the governing equations are presented in detail in [16].

4.2. Grid Independence

Before the actual simulations, validation calculations for several geometries at different mesh densities were performed. Table 2 gives an example of the grid under test for one of the insert diameters, D = 0.4 and Re = 18,650. It was observed that the deviation between the grid elements of 25,440 and 49,820 is only 0.9% for the friction factor and 0.22% for the Nu number. Thus, for further calculations, a structural and hexagonal mesh (Figure 3) with 25,440 elements was chosen, which is the quality at which its further densification gives results differing by less than 1.5%.

Table 2. Grid independent test for Re = 18,650 and insert diameter D = 0.4.

No. of Elements	f	Dev. %	*Nu*	Dev. %
6140	0.0461	-	73.23	-
11,960	0.0508	9.38	78.05	6.18
18,230	0.0543	6.32	80.13	2.60
25,440	0.0551	1.45	80.62	0.61
49,820	0.0556	0.9	80.80	0.22

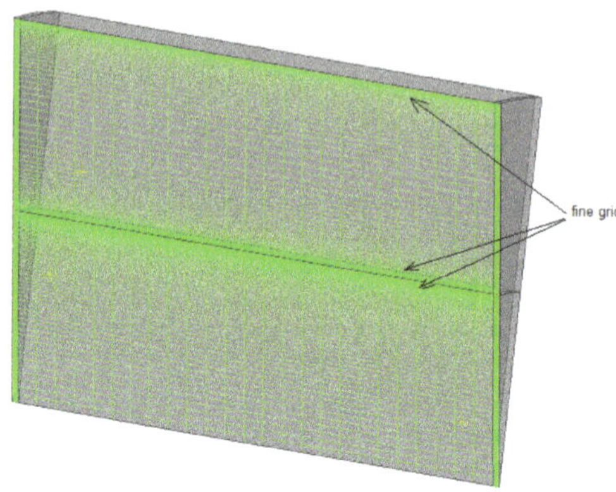

Figure 3. Computational mesh with densification areas at the walls.

For further calculations, a structured and hexagonal mesh, shown in Figure 3, was chosen with a quality at which further densification of the mesh gave results varying less than approximately1.5% for the most important flow parameters.

Of particular importance for flow and heat transfer phenomena is the hydraulic boundary layer. In the studied geometries, these layers form at the channel wall and on both sides of the insert, i.e., where the fluid contacts the solid. In these areas, the mesh was further compacted to obtain an appropriate y+ value for the turbulence model that was used [15–17,22,23].

4.3. Turbulence Model

Due to the assumed turbulent fluid flow, the SST k-ω (Shear Stress Transport) turbulence model was used in all of the numerical simulations. It is one of the most frequently used models in CFD applications due to its much better mapping of the flow-thermal phenomenon in calculations than the standard k-ε model [16,22]. The SST model's main advantage is taking the viscous boundary sublayer into account by applying the k-ω model near the wall and using the standard k-ε model in the turbulent core region. A special function (the so-called blending function) implemented in the SST model [22] is responsible for selecting an appropriate model during the calculations.

One of the criteria for the numerical solution's uniqueness was to achieve the appropriate convergence for the residues: momentum, energy, and turbulence. In all of the simulations, a convergence of 10×10^{-4} for the maximum residues and an order of magnitude lower (10×10^{-5}) for mean residues (RMS—Root Mean Square) were obtained, both for the solutions of hydraulic equations and the energy equations. The second criterion for the solution's uniqueness was stabilizing the flow-thermal parameters, such as velocity, pressure, and temperature, which were monitored both as average values and in several selected points in the computational domain. The computation process was terminated if the above parameters did not change for several consecutive iterations.

4.4. Radiation Model

For the radiative heat transfer calculations, shown in Figure 2c, the geometry of the tube with the insert can be treated as two infinitely long concentric cylinders with surfaces A_1 and A_2, emissivities ε_1 and ε_2, and temperatures T_1 and T_2 (where index 1 is referred to the smaller surface, i.e., the insert, and index 2 to the bigger surface, i.e., the pipe wall). For the tested geometries, where the outer cylinder completely surrounds the inner cylinder, it is assumed that the outer surface will intercept all radiation from the inner surface. Hence, considering the Stefan–Boltzman law for grey bodies, the net radiative heat flux transferred between these surfaces is described by Equation (4).

$$\dot{Q}_{rad} = \frac{\sigma \cdot \left(T_2{}^4 - T_1{}^4\right)}{\frac{1}{\varepsilon_1} + \frac{A_1}{A_2} \cdot \left(\frac{1}{\varepsilon_2} - 1\right)} = \varepsilon_{12} \cdot \sigma \cdot \left(T_2{}^4 - T_1{}^4\right) \quad (4)$$

where:$\varepsilon_{12} = \frac{1}{\frac{1}{\varepsilon_1} + \frac{A_1}{A_2} \cdot \left(\frac{1}{\varepsilon_2} - 1\right)}$—interchange factor.

The above equation is the basic formula describing the radiative heat transfer for the studied geometry, and its derivation can be found in many literature sources, e.g., [24,25].

Using (4), a validation test for the selected insert geometry was performed, comparing the results obtained from the analytical calculations and numerical tests. The investigated quantity was the radiation heat flux exchanged between these surfaces for different pipe wall temperatures. The inner surface (Figure 4a) was kept at a constant temperature of 0 °C, while the outer surface temperature was changed every 50 °C. Both surfaces' emissivities, both for this test and further numerical simulations, were the same and amounted to $\varepsilon_1 = \varepsilon_2 = 0.9$, while there was a vacuum between the surfaces to eliminate the effect of convection on heat transfer.

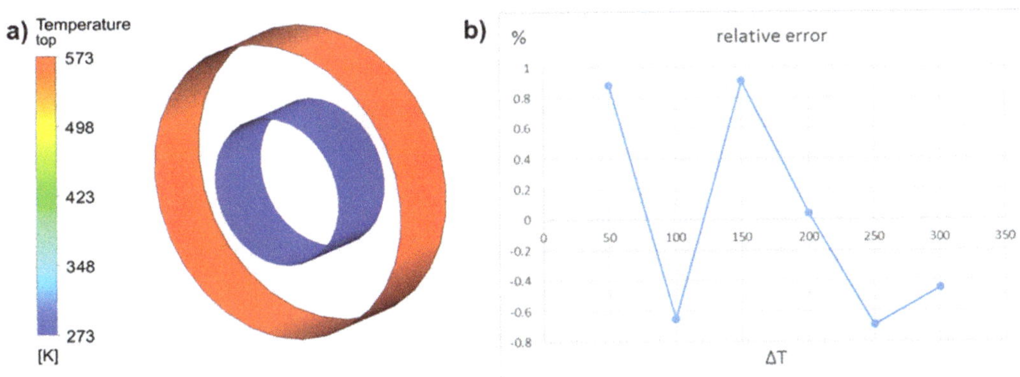

Figure 4. (**a**) View of the tested geometry, 1-insert, 2-pipe; (**b**) relative error of radiation fluxes.

Due to very similar results, which would not be visible in the ordinary comparative chart, in Figure 4b, only the percentage difference of the heat fluxes in the form of the relative error calculated from (5) is shown.

$$\delta = \frac{\dot{Q}_{theory} - \dot{Q}_{numerical}}{\dot{Q}_{theory}} \cdot 100\% \quad (5)$$

As shown in the diagram, the discrepancy in the value of the radiative heat flux transferred between the surfaces is within the limits of ±1% for the six temperature differences tested.

Several radiation models are available in the ANYS-CFX calculation program, such as Rosseland, P1, Discrete Transfer, and Monte Carlo. In a situation where the thermal energy of radiation is transferred between two surfaces, and the medium is transparent to radiation with wavelengths, which is where most of the heat transfer takes place, the Monte Carlo model is the only one that applies—and onlythis model was used in the simulations [20].

5. Hydraulic Characteristics—Friction Factor

5.1. Data Processing

The use of any insert in the flow channel always results in an increase in flow resistance. The situation is similar in the tested insert case, where the resistance increase is caused by the friction of the fluid against its two additional surfaces—outer and inner. As mentioned before, this kind of insert does not cause swirling and turbulisation of the fluid, which means that the flow resistances are much lower than in the case of inserts with a different construction and operation. These resistances mainly depend on the average gas velocity flow in the channel for a given insert diameter.

Figure 5 shows the velocity profiles in the symmetry plane (according to the diagram in Figure 2b) for an insert with a dimensionless diameter $D = 0.2$ and several Re numbers. Similarly, Figure 6 shows the velocity profiles for different dimensionless insert diameters D, but with the same number, Re = 13,500. By analysing the velocity fields shown, it can be concluded that the fluid velocities differ significantly in the inner and outer cross-section of the insert. These differences are due to the flow resistance and, as mentioned earlier, depend on the insert diameter and the Re number. As the flow resistance on both sides of the insert is the same, the gas flows in two streams, "adjusting" the velocity to the flow resistance. It can be assumed that in this way, there is a self-regulation of the amount of flowing medium outside and inside the insert.

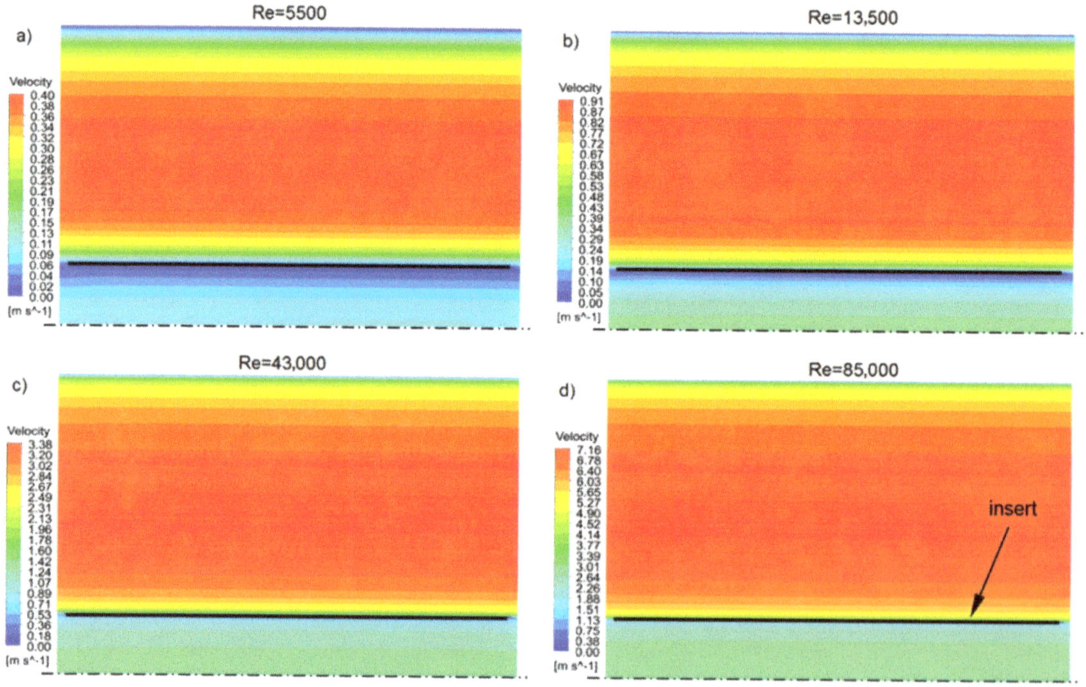

Figure 5. Velocity profiles in the symmetry plane for dimensionless insert diameter $D = 0.2$ and different Re numbers: (**a**) Re = 5500, (**b**) Re = 13,500, (**c**) Re = 43,000 and (**d**) Re = 85,000.

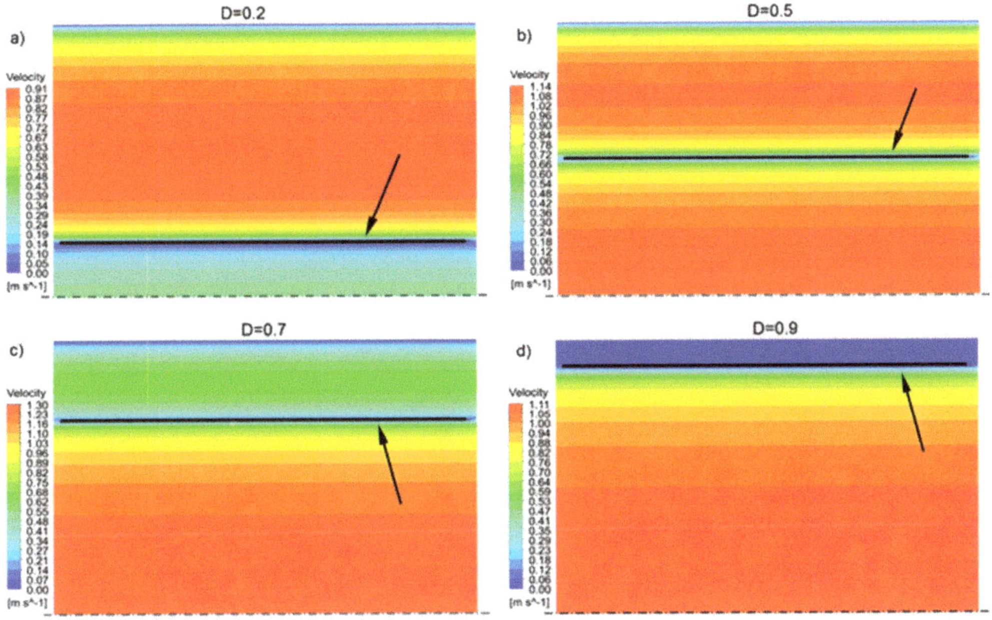

Figure 6. Velocity profiles in the symmetry plane for several dimensionless diameters D of the insert and the same number Re = 13,500. (**a**) $D = 0.2$, (**b**) $D = 0.5$, (**c**) $D = 0.7$, (**d**) $D = 0.9$.

To calculate the friction factor obtained from the numerical results, the Darcy–Weisbach Equation (6) was used:

$$f = \frac{2 \cdot d_{\mathrm{p}}}{\rho \cdot u^2} \cdot \frac{dp}{dx} \tag{6}$$

where the Re number is a function of the average velocity u in the pipe cross-section and the pipe diameter d_{p}:

$$Re = \frac{u \cdot d_{\mathrm{p}}}{\nu} \tag{7}$$

The set value, forcing the fluid flow, was the pressure gradient dp/dx, while the average velocity in the pipe u resulted from numerical calculations. The friction factor for a smooth pipe was calculated using the Blassius formula (8):

$$f_{\mathrm{s}} = 0.3164 \cdot Re^{-0.25} \tag{8}$$

5.2. Results and Discussion—Friction Factor

A total of nine numerical simulations for each geometry and different Reynolds numbers were performed. As mentioned earlier, the set amount forcing the flow was the pressure gradient dp/dx, and the average velocity in the pipe u was the result of the numerical calculations. In this way, the flow characteristics $f(Re)$ for each insert geometry were prepared, while the friction factor was calculated from Equation (6). Figure 7 shows the characteristics of the friction factor as a function of Re for each tested insert diameter. Logarithmic scales were used on both axes, and the smooth pipe characteristics for comparison, calculated from (8), are also shown. As expected, for all insert sizes, the friction factor decreases as the Re number increases. The characteristics shown in the graph have a similar trend as the smooth pipe, but obviously different values.

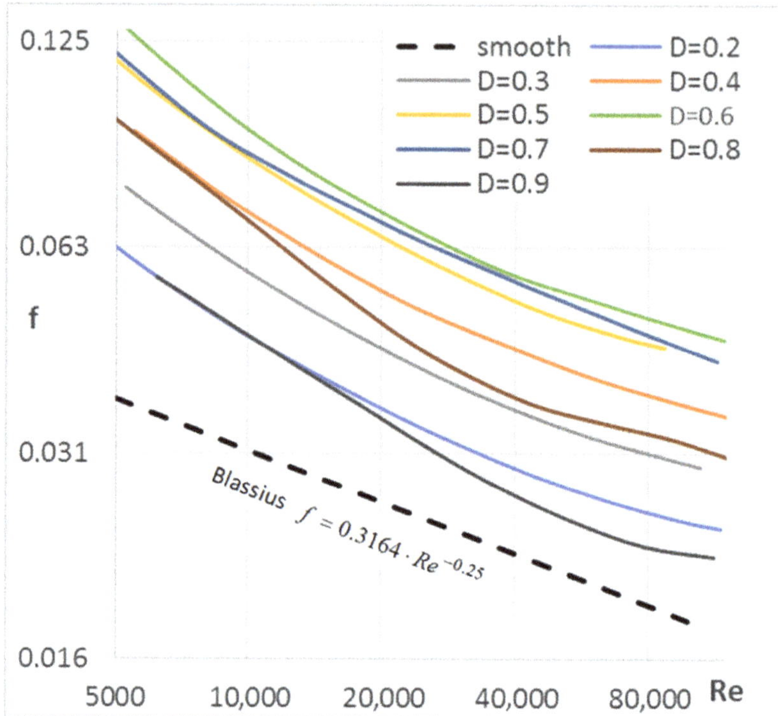

Figure 7. Friction factor $f(Re)$ for dimensionless insert diameters D.

Figure 8a presents the ratio of f/f_s as a Re number function, showing how many times the friction factor increased for a given insert diameter compared to the smooth pipe. As it can be seen, for all geometries, the largest increase occurs for the smallest Re numbers.

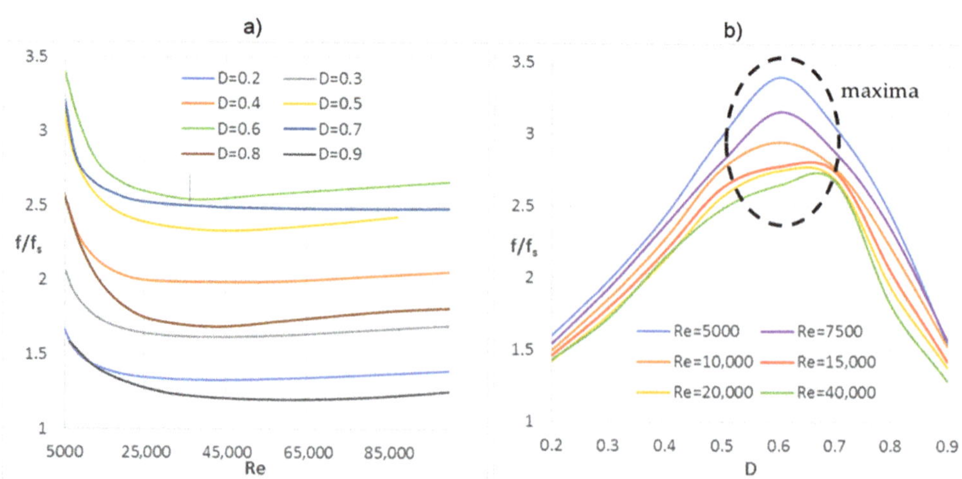

Figure 8. (**a**) $f/f_s(Re)$ ratio for dimensionless insert diameters D; (**b**) $f/f_s(D)$ ratio for several Re numbers.

However, for larger *Re* numbers, from about 20,000 onwards, the characteristics align more or less horizontally or increase minimally, which means that these functions' trends are similar to those of the smooth pipe. On the other hand, in Figure 8b, which shows the f/f_s ratio as a function of the dimensionless insert diameter *D*, one can observe its effect on the friction factor value, according to several selected *Re* numbers. You can clearly see the maxima, exhibited by all of the characteristics in the *D* range from 0.5 to 0.7, which means that pipes with inserts of these diameters have the highest flow resistance.

However, compared to the turbulising flow inserts of other designs, e.g., those described in [26,27], the increase of the friction factor is several times smaller for the same *Re* numbers, which also translates to much smaller pressure drops.

In the aforementioned works, flow turbulising inserts were studied to improve heat transfer. Such inserts, in the form of twisted tape with cutouts that generate additional vortices during flow, specifically disrupt the flow and thus increase convective heat transfer, but at the same time, the flow resistance increases significantly.

The inserts studied in this paper do not disturb the flow because they have negligible thickness, are not equipped with vortex generators, and are arranged parallel to the flow direction. This design and the arrangement of the insert in the pipe result in much lower flow resistance and pressure drops than other inserts, while heat transfer enhancement takes place only through radiation.

6. Thermal Characteristics—Nusselt Number

6.1. Data Processing

In the simulation, the constant temperature of the pipes' outer walls was assumed to be 100 °C (as the thermal boundary conditions), and the average gas temperature was assumed to be in the computational domain of 0 °C. This temperature difference was kept for all calculation cases in order to perform the tests under the same conditions. Due to the different flow conditions for each case, a different heat flux transferred to the fluid was obtained.

As mentioned earlier, the presented computational model obtained a fully developed, periodic flow at the inlet and outlet of the domain (forced by a pressure gradient). In order to obtain a fully developed thermal layer, and at the same time to maintain the energy balance in the system, the heat flux supplied by the wall to the gas domain had to be compensated by the use of an additional, equal in value, negative volumetric energy source that was different from the entire domain flow [16]. The working gas properties (air), such as thermal conductivity, viscosity, and density were functions of the temperature.

In Figure 9, in accordance with the diagram in Figure 2b, temperature profiles and several *Re* numbers, for example, the insert with dimensionless diameter D = 0.2, are shown. Similarly, Figure 10 shows the temperature profiles for several dimensionless diameters *D* of the inserts, but for the same number, *Re* = 13,500. As it can be seen, the temperature profiles in the inner and outer cross-sections of the insert differ significantly, highlighting the effects of radiation and the separation of the gas streams. The radiative heat flux transferred to the insert causes its heating, which allows this heat to be transferred to the fluid during convective heat transfer on both sides of the insert. To calculate the *Nu* number from the numerical analysis, the following relationship was used:

$$Nu = \frac{\dot{Q}_{\mathrm{tot}} \cdot d_{\mathrm{p}}}{k \cdot (T_{\mathrm{w}} - T_{\mathrm{b}}) \cdot A_2} \quad (9)$$

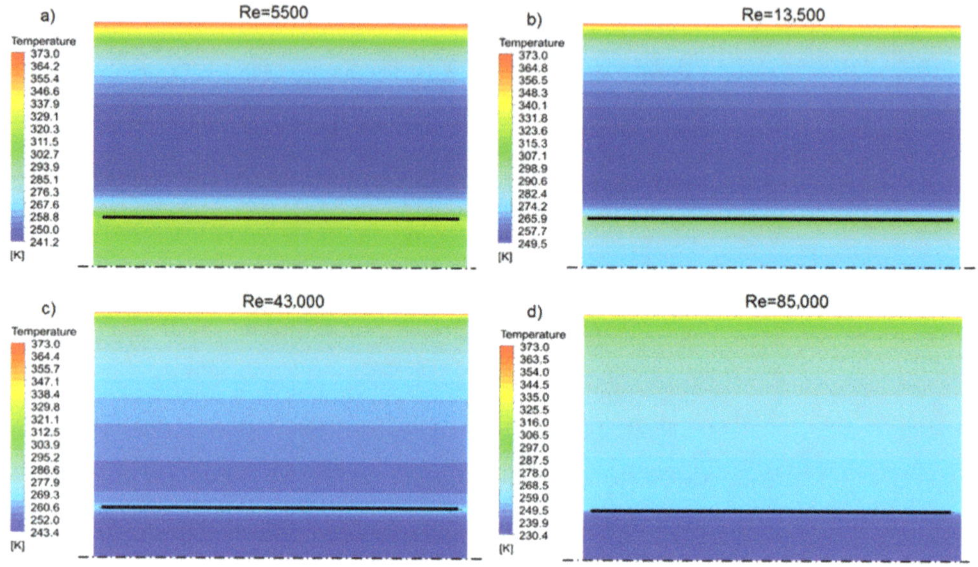

Figure 9. Temperature profiles in the symmetry plane for a dimensionless insert diameter $D = 0.2$ and several Re numbers: (**a**) $Re = 5500$, (**b**) $Re = 13{,}500$, (**c**) $Re = 43{,}000$ and (**d**) $Re = 85{,}000$.

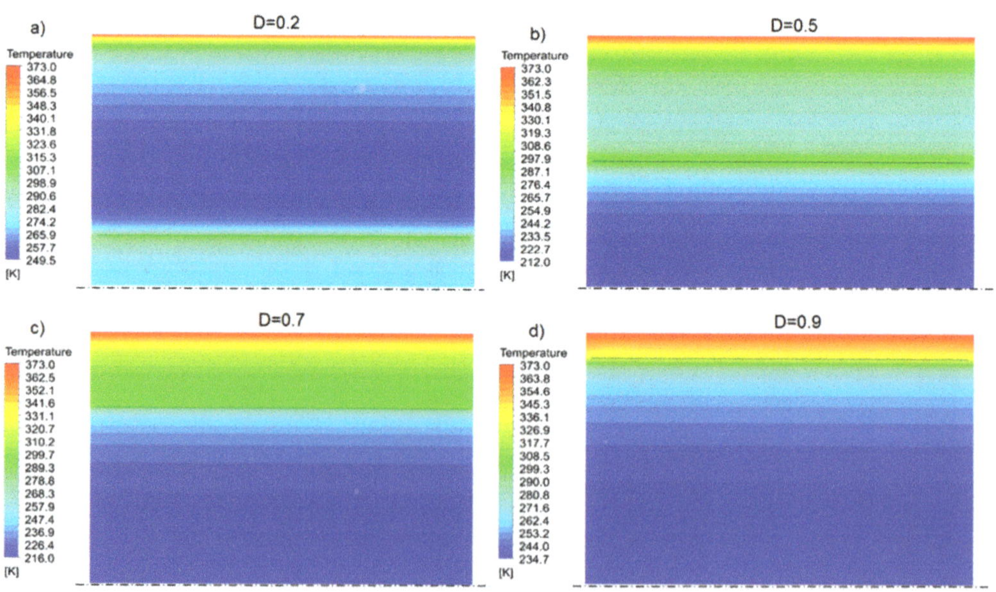

Figure 10. Schemes follow the same formatting. Temperature profiles in the symmetry plane for several dimensionless insert diameters D and the same number $Re = 13{,}500$. (**a**) $D = 0.2$, (**b**) $D = 0.5$, (**c**) $D = 0.7$, (**d**) $D = 0.9$.

While the Nu number for the smooth pipe without insert was calculated using the Dittus–Boelter correlation (10):

$$Nu_s = 0.023 \cdot Re^{0.8} \cdot Pr^{0.4} \tag{10}$$

As mentioned earlier, the insert divides the channel into two hydraulically independent spaces, except for small gaps for equalising static pressures (Figure 2a). This means that the fluids from the annular (outside the insert) and tubular (inside the insert) cross-section do not mix, but instead exchange heat energy among themselves through the insert's thin wall. This situation significantly complicates the heat transfer calculation, i.e., the Nu number calculation, due to the problem's two possible approaches.

The Nu number, calculated according to (9), characterises the heat transfer intensity on the flow channel wall—in this case, a round pipe. Therefore, calculating the temperature Tb can be related to either (a) the entire volume of the channel or (b) only the volume of the annular portion that actually contacts the channel wall. Both approaches have their justification.

Ad. (a) T_b *calculated as the volume average temperature in the whole duct*. This method of calculation is the most commonly used by most researchers, both for empty ducts and those equipped with inserts or other turbulators, e.g., [28–34]. The heat flux is transferred to the fluid that is in contact with the duct wall, and its average temperature is the result of the heat transfer conditions and the fluid velocity on both sides of the insert. One disadvantage of this approach with the considered insert is the lack of fluid mixing from the annular and tubular spaces. Thus, there are different velocity and temperature fields in these areas, and the temperature gradient in the duct wall layer is quite different from what it would be with a classical insert.

Ad. (b) T_b *calculated as the volume average temperature from the annular part only*. This approach is justified because the heat flux is directly transferred to the fluid in contact with the channel wall. Obviously, its average temperature is also affected by the amount of heat supplied to the insert by radiation and returned by the fluid's insert due to convection to the annular and tubular parts.

6.2. Results and Discussion

According to the author, both ways of calculating the Nu number can be considered for this type of insert, even though they give different results. As it will be shown later in the paper, smaller values of Nu number are obtained using calculation method (a), i.e., taking the temperature in the whole volume of the pipe into account. In comparison, larger values are obtained using calculation method (b), i.e., only taking the annular part's temperature into account.

Figure 11 shows the characteristics of the $Nu(Re)$ number calculated for methods (a) and (b), which additionally include the characteristics for a smooth pipe, calculated according to (7). It is clear that these charts differ quite significantly, both in terms of the value and the position of the curves. A characteristic feature is the values of the Nu numbers smaller than those of the smooth pipe, calculated using method (a), mainly for the largest diameters D = 0.9, 0.8, 0.7 and 0.6 (shown in Figure 11a), and calculated by method (b) in the range of large Re numbers and large insert diameters D = 0.9 and 0.8 (shown in Figure 11b). With normal turbulising inserts, such a phenomenon is unlikely to occur because any insert put into the pipe disrupts the flow by intensifying the convective heat flux while increasing the flow resistance.

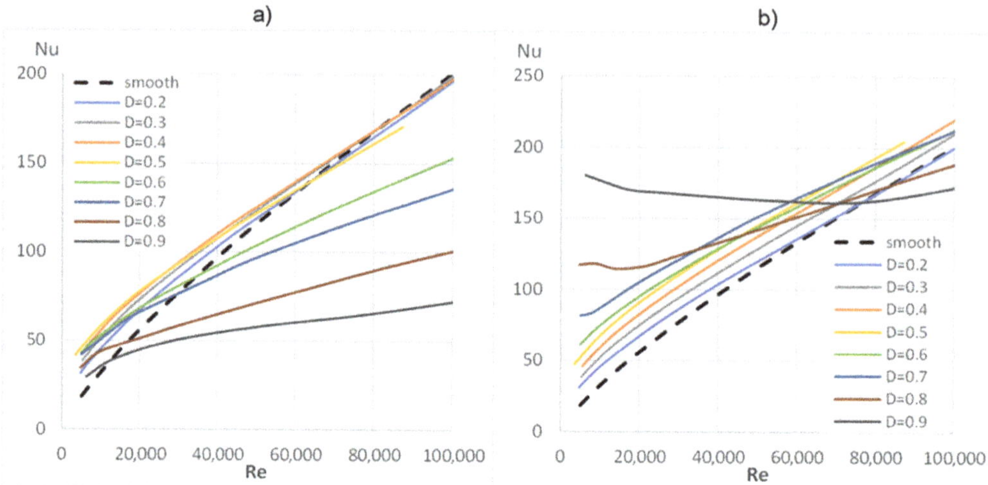

Figure 11. Nusselt number $Nu(Re)$ as a function of Re number, calculated on the basis of the average temperature: (**a**) from the entire pipe; (**b**) from the annular section of pipe.

Figure 11 shows the characteristics of the $Nu(Re)$ number calculated for methods (a) and (b), which additionally include the characteristic for a smooth pipe, calculated according to (7). It is clear that these charts differ quite significantly, both in terms of the value and the position of the curves. A characteristic feature is the values of the Nu numbers smaller than those of the smooth pipe, calculated using method (a), mainly for the largest diameters D = 0.9, 0.8, 0.7 and 0.6 (shown in Figure 11a), and calculated by method (b) in the range of large Re numbers and large insert diameters D = 0.9 and 0.8 (shown in Figure 11b). With normal turbulising inserts, such a phenomenon is unlikely to occur because any insert put into the pipe would disrupt the flow by intensifying the convective heat flux while increasing the flow resistance.

In the case of the tested inserts, the appearance of Nu numbers smaller than the reference level, which would be for a smooth pipe, can be somehow explained by the creation of additional space between the pipe wall and the insert (annular cross-section). The larger the diameter of the insert, the smaller this space becomes, while at the same time, the average temperature of the gas in it increases, and its velocity decreases. This space gap can be treated as a kind of insulation layer between the pipe wall and the main flowing gas stream.

At small diameters, the insert is more or less evenly washed by the fluid on both sides. At large diameters, however, the mass flux of the flowing gas in the annular cross-section is much smaller than inside the insert, so the convective heat transfer decreases significantly, and the heat resistance from the conduction of the gas increases.

Figure 12a,b show the ratio of the Nu number of the tested inserts to the Nu number for a smooth pipe for the two above-mentioned calculation methods. Based on the average temperature T_b in the entire pipe (case a), it can be seen that for several inserts with diameters D = 0.9, 0.8, 0.7, 0.6, the values of the Nu/Nu_s characteristics are less than one. On the other hand, when only taking the temperature T_b for the calculations from the annular cross-section (case b), the curves are arranged in an "ideal" order, i.e., the intensification of heat transfer increases proportionally with the increase of the insert diameter.

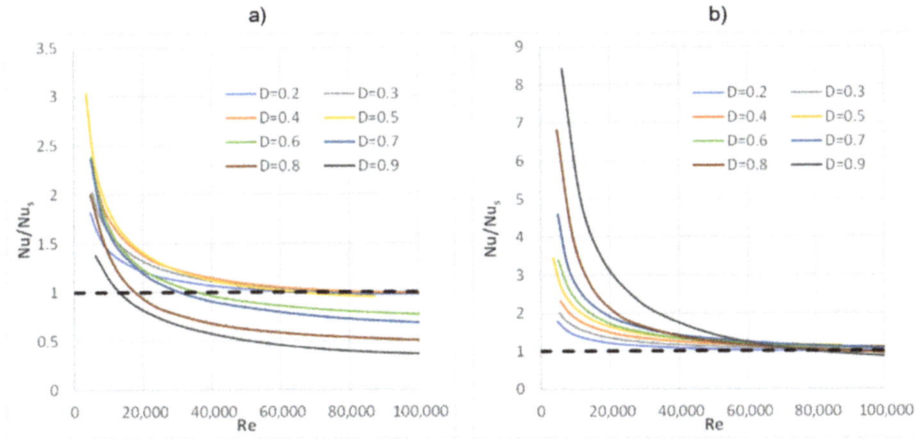

Figure 12. The ratio of the *Nu* number obtained from simulation to the *Nu* number for a smooth pipe $Nu/Nu_s(Re)$ as a function of the *Re* number, calculated on the basis of the average temperature: (**a**) from the entire pipe; (**b**) from the annular cross-section.

Figure 13 shows the Nu/Nu_s ratio as a function of the diameter *D* for several selected *Re* numbers, i.e., it illustrates the effect of the insert geometry on the heat transfer intensification. In Figure 13a, we can see a clear maximum that shows all of the characteristics at $D = 0.5$, which means that this is the largest increase in *Nu* number for this insert diameter in relation to the smooth pipe. These characteristics are very similar to the friction factors shown in Figure 8b.

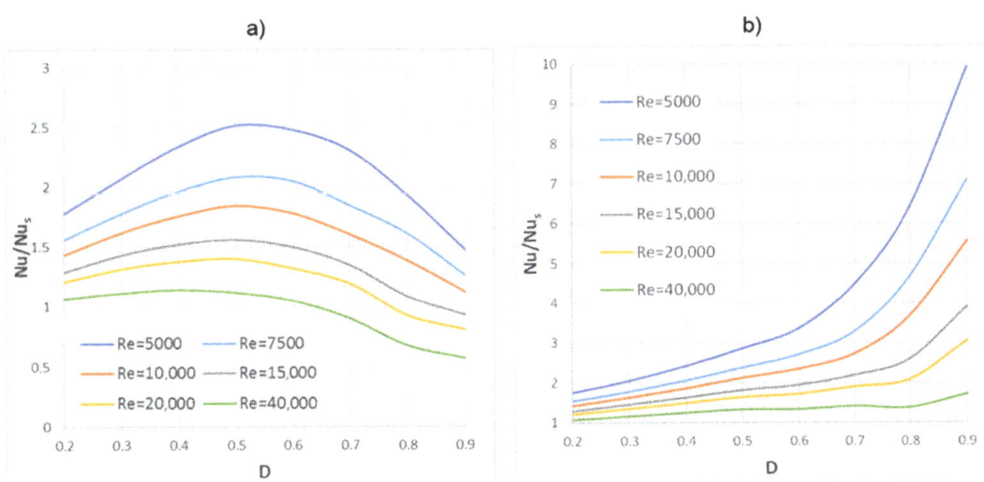

Figure 13. $Nu/Nu_s(D)$ ratio for several *Re* numbers, calculated from the average temperature: (**a**) from the whole pipe; (**b**) from the annular section.

When analysing Figure 13b, it was observed that along with increasing the insert diameter, the value of ratio Nu/Nu_s also increases, which is similar to what can be observed in Figure 12b.

7. Summary and Conclusions

This work presents numerical investigations for different diameters of an insert that absorbs radiative heat flux while simultaneously not turbulising the fluid. Based on these results, the characteristics of $f(Re)$ and $Nu(Re)$ were prepared and presented in graphs (Figures 7 and 11). The characteristics of $f/f_s(Re)$ and $Nu/Nu_s(Re)$ are also shown (in Figures 8 and 12), i.e., the ratio of the friction factor and Nu numbers obtained from simulations, compared to a smooth pipe.

The velocity and temperature fields in the channel's symmetry plane are also presented for a few exemplary insert geometries (Figures 6 and 10) and a few Re numbers and constant diameters D (Figures 5 and 9).

The analysis of the results allowed for the formulation of the following conclusions:

(1) The highest increase in friction resistance (the friction factor ratio f/f_s) was observed for inserts with diameters D = 0.5–0.7, while the lowest values were recorded for inserts with the smallest diameter D = 0.2 and the largest diameter D = 0.9;

(2) The Nu number, which is the intensity of heat transfer, can be calculated in two ways that give different results. As mentioned earlier, there are two methods to determine the temperature Tb in formula (6). At the moment, the author suggests the "classical" way, (a), i.e., calculating the temperature Tb as an average over the entire volume channel. This method gives smaller, or one may say, "safer", values of the Nu number than method (b). The Nu values obtained using method b) are quite large, up to about nine times larger than for a smooth pipe (for D = 0.9 and small Re numbers), but at relatively small velocities and pressure drops. Comparing the obtained results to the results of other insert geometries described in the available literature, it can be concluded that these are very high Nu numbers. Therefore, these values should be approached with caution, and the issue should still be thoroughly investigated and analysed;

(3) The influence of radiation on heat transfer is shown in Figures 11 and 12. Compared to a smooth pipe, the highest Nu numbers are noted for their small Re numbers. As the Re number increases, the difference between the Nu number of the tested inserts and that of a smooth pipe decreases quite quickly, and with large Re numbers, the Nu numbers of the values are comparable to those of a smooth pipe. This is a correct phenomenon because as the Re number and the gas flow velocity increase, the convective heat flux also increases, and the fraction of radiative heat flux decreases in the total heat flux, which is transferred from the wall to the fluid;

(4) For the largest diameters D = 0.9, 0.8, 0.7, 0.6 (Figure 11a), a decrease in Nu numbers below the reference level, which is a smooth pipe, was observed. As previously mentioned, the tested insert does not turbulise the flow and therefore does not increase the convective heat flux. On the other hand, at larger insert diameters, the mainstream of fluid flows inside the insert, and the outer (annular) space becomes thinner, which causes a significant reduction in gas velocity in this region. As a result, the annular part, instead of participating in the heat transport to the gas, due to its low velocities, actually becomes an insulating layer inside the pipe. As it can be seen from the research, this phenomenon causes a deterioration of the overall heat transfer, even when compared to a smooth pipe.

Funding: This research received no external funding.

Institutional Review Board Statement: Not applicable.

Informed Consent Statement: Not applicable.

Data Availability Statement: Not applicable.

Conflicts of Interest: Author declares no conflict of interest.

Nomenclature

A	heat transfer area (m^2)
d_i	insert diameter (m)
d_p	pipe diameter (m)
D	diameter ratio, (d_i/d_p)
dp/dx	pressure gradient (Pa/m)
f	friction factor
k	thermal conductivity (W/mK)
L	domain length (m)
Nu	Nusselt number
$\dot{Q}_{tot}$	total heat flux (W/m^2)
$\dot{Q}_{rad}$	radiative heat flux (W/m^2)
$\dot{Q}_{conv}$	convective heat flux (W/m^2)
Pr	Prandtl number
Re	Reynolds number
T_b	bulk temperature (K)
T_w	wall temperature (K)
u	average velocity (m/s)
ν	kinematic viscosity (m^2/s)
ΔT	temperature difference (K)
δ	relative error (%)
ε	emissivity
ε_{12}	interchange factor
ρ	density (kg/m^3)
	indexes:
s	smooth tube
1	for smaller heat transfer area
2	for bigger heat transfer area

References

1. Ozceyhan, V.; Gunes, S.; Buyukalaca, O.; Altuntop, N. Heat transfer enhancement in a tube using circular cross sectional rings separated from wall. *Appl. Energy* **2008**, *85*, 988–1001. [CrossRef]
2. Min, C.; Qi, C.; Wang, E.; Tian, L.; Qin, Y. Numerical investigation of turbulent flow and heat transfer in a channel with novel longitudinal vortex generators. *Int. J. Heat Mass Transf.* **2012**, *55*, 7268–7277. [CrossRef]
3. Lotfi, B.; Sundén, B.; Wang, Q. An investigation of the thermo-hydraulic performance of the smooth wavy fin-and-elliptical tube heat exchangers utilizing new type vortex generators. *Appl. Energy* **2016**, *162*, 1282–1302. [CrossRef]
4. Jasiński, P. Numerical Study of Friction Factor and Heat Transfer Characteristics for Single-Phase Turbulent Flow in Tubes with Helical Micro-Fins. *Arch. Mech. Eng.* **2012**, *59*, 469–485. [CrossRef]
5. Yeom, T.; Simon, T.; Zhang, T.; Zhang, M.; North, M.; Cui, T. Enhanced heat transfer of heat sink channels with micro pin fin roughened walls. *Int. J. Heat Mass Transf.* **2016**, *92*, 617–627. [CrossRef]
6. Liu, G.; Yang, C.; Zhang, J.; Zong, H.; Xu, B.; Qian, J.-Y. Internal Flow Analysis of a Heat Transfer Enhanced Tube with a Segmented Twisted Tape Insert. *Energies* **2020**, *13*, 207. [CrossRef]
7. Wijayanta, A.T.; Mirmanto, P.; Kristiawan, B.; Aziz, M. Internal Flow in an Enhanced Tube Having Square-cut Twisted Tape Insert. *Energies* **2019**, *12*, 306. [CrossRef]
8. Bellos, E.; Tzivanidis, C. Enhancing the Performance of Evacuated and Non-Evacuated Parabolic Trough Collectors Using Twisted Tape Inserts, Perforated Plate Inserts and Internally Finned Absorber. *Energies* **2018**, *11*, 1129. [CrossRef]
9. Mwesigye, A.; Bello-Ochende, T.; Meyer, J.P. Heat transfer and thermodynamic performance of a parabolic trough receiver with centrally placed perforated plate inserts. *Appl. Energy* **2014**, *136*, 989–1003. [CrossRef]
10. Cerón, J.F.; Pérez-García, J.; Solano, J.P.; García, A.; Herrero-Martín, R. A coupled numerical model for tube-on-sheet flat-plate solar liquid collectors. Analysis and validation of the heat transfer mechanisms. *Appl. Energy* **2015**, *140*, 275–287. [CrossRef]
11. Luo, L.; Wen, F.; Wang, L.; Sundén, B.; Wang, S. Thermal enhancement by using grooves and ribs combined with delta-winglet vortex generator in a solar receiver heat exchanger. *Appl. Energy* **2016**, *183*, 1317–1332. [CrossRef]
12. Eiamsa-Ard, S.; Promvonge, P. Thermal characterization of turbulent tube flows over diamond-shaped elements in tandem. *Int. J. Therm. Sci.* **2010**, *49*, 1051–1062. [CrossRef]
13. Kongkaitpaiboon, V.; Nanan, K.; Eiamsa-Ard, S. Experimental investigation of convective heat transfer and pressure loss in a round tube fitted with circular-ring turbulators. *Int. Commun. Heat Mass Transf.* **2010**, *37*, 568–574. [CrossRef]
14. Aydin, D. Heat transfer and exergy loss in cut out conical turbulators. *Energy Convers. Manag.* **2004**, *45*, 785–796.

15. Jasiński, P.B. Numerical study of thermal-hydraulic characteristics in a circular tube with ball turbulators. Part 1: PIV experiment and pressure drop. *Int. J. Heat Mass Transf.* **2014**, *74*, 48–59. [CrossRef]
16. Jasiński, P.B. Numerical study of thermal-hydraulic characteristics in a circular tube with ball turbulators. Part 2: Heat transfer. *Int. J. Heat Mass Transf.* **2014**, *74*, 473–483. [CrossRef]
17. Jasiński, P. Numerical study of thermo-hydraulic characteristics in a circular tube with ball turbulators. Part 3: Thermal performance analysis. *Int. J. Heat Mass Transf.* **2017**, *107*, 1138–1147. [CrossRef]
18. Bas, H.; Ozceyhan, V. Heat transfer enhancement in a tube with twisted tape inserts placed separately from the tube wall. *Exp. Therm. Fluid Sci.* **2012**, *41*, 51–58. [CrossRef]
19. Jasiński, P. Numerical optimization of flow-heat ducts with helical micro-fins, using Entropy Generation Minimization (EGM) method. In Proceedings of the WSEAS International Conferences (HTE'11), Florence, Italy, 23–25 August 2011; pp. 47–54.
20. Fodemski, T.; Górecki, G.; Jasiński, P. Corrugated channels heat transfer efficiency Analysis based on velocity fields resulting from computer simulation and PIV Measurements. In Proceedings of the 8th International Conference on Heat Transfer, Fluid Mechanics and Thermodynamics, HEFAT 2011, Pointe Aux Piments, Mauritius, 11–13 July 2011.
21. Li, X.-W.; Meng, J.-A.; Guo, Z.-Y. Turbulent flow and heat transfer in discrete double inclined ribs tube. *Int. J. Heat Mass Transf.* **2009**, *52*, 962–970. [CrossRef]
22. ANSYS-CFX Manual v. 18.1. Available online: http://www.ansys.com (accessed on 25 July 2021).
23. Gutkowski, A.N.; Łęcki, M.; Jasiński, P.; Jędrowiak, B. Flame behavior during propagation in small isothermal tubes characterized by different degrees of the end opening. *Combust. Sci. Technol.* **2018**, *191*, 711–725. [CrossRef]
24. Karwa, R. *Heat and Mass Transfer*; Springer: Berlin/Heidelberg, Germany, 2016.
25. Fakheri, A. *Intermediate Heat Transfer*; Taylor & Francis: Abingdon, UK, 2014.
26. Arjmandi, H.; Amiri, P.; Pour, M.S. Geometric optimization of a double pipe heat exchanger with combined vortex generator and twisted tape: A CFD and response surface methodology (RSM) study. *Therm. Sci. Eng. Prog.* **2020**, *18*, 100514. [CrossRef]
27. Wijayanta, A.T.; Yaningsih, I.; Aziz, M.; Miyazaki, T.; Koyama, S. Double-sided delta-wing tape inserts to enhance convective heat transfer and fluid flow characteristics of a double-pipe heat exchanger. *Appl. Therm. Eng.* **2018**, *145*, 27–37. [CrossRef]
28. Zheng, N.; Liu, P.; Wang, X.; Shan, F.; Liu, Z.; Liu, W. Numerical simulation and optimization of heat transfer enhancement in a heat exchanger tube fitted with vortex rod inserts. *Appl. Therm. Eng.* **2017**, *123*, 471–484. [CrossRef]
29. Huanga, S.; Chena, H.; Zhang, X. Experimental evaluation of thermal performance in a circular tube with Y-branch insert. *Int. Commun. Heat Mass Transf.* **2019**, *106*, 15–21. [CrossRef]
30. Bhuiyaa, M.; Ahamed, J. Heat transfer enhancement and development of correlation for turbulent flow through a tube with triple helical tape inserts. *Int. Commun. Heat Mass Transf.* **2012**, *39*, 94–101. [CrossRef]
31. Hasanpour, A.; Farhadi, M.; Sedighi, K. A review study on twisted tape inserts on turbulent flow heat exchangers: The overall enhancement ratio criteria. *Int. Commun. Heat Mass Transf.* **2014**, *55*, 53–62. [CrossRef]
32. Jasiński, P.; Kowalczyk, M.; Romaniak, A.; Warwas, B.; Obidowski, D.; Gutkowski, A. Investigation of Thermal-Flow Characteristics of Pipes with Helical Micro-Fins of Variable Height. *Energies* **2021**, *14*, 2048. [CrossRef]
33. Ho, C.; Liu, Y.-C.; Yang, T.-F.; Ghalambaz, M.; Yan, W.-M. Convective heat transfer of nano-encapsulated phase change material suspension in a divergent minichannel heatsink. *Int. J. Heat Mass Transf.* **2021**, *165*, 120717. [CrossRef]
34. Ho, C.; Liu, Y.-C.; Ghalambaz, M.; Yan, W.-M. Forced convection heat transfer of Nano-Encapsulated Phase Change Material (NEPCM) suspension in a mini-channel heatsink. *Int. J. Heat Mass Transf.* **2020**, *155*, 119858. [CrossRef]

energies

MDPI

Article

Numerical Study of Heat Transfer Intensification in a Circular Tube Using a Thin, Radiation-Absorbing Insert. Part 2: Thermal Performance

Piotr Bogusław Jasiński

Institute of Turbomachinery, Lodz University of Technology, 90-924 Lodz, Poland; piotr.jasinski@p.lodz.pl; Tel.: +48-42-631-23-49

Abstract: This article is the second part of the work under the same title, which is based on the results of the research presented in the previous article: "Numerical study of heat transfer intensification in a circular tube using a thin, radiation-absorbing insert. Part 1: Thermo-hydraulic characteristics". Part 1 presents an analysis of pressure drops and heat transfer intensification in a round tube with an insert, using the phenomenon of radiation absorption. In this paper, an analysis of the tested insert's thermal performance (PEC) is presented, taking into account the criterion of equal pumping power. The tests were carried out for the range of Re = 5000–100,000 numbers, for various insert diameters (from 20% to 90% of the pipe diameter) and a constant temperature difference between the wall and the gas $\Delta T = 100\ ^\circ C$. The highest Nu numbers were observed for inserts with dimensionless diameters of 0.3 and 0.4, while the highest flow resistance was observed for inserts with diameters of 0.6 and 0.7 of the channel diameter. The thermal efficiency was calculated in two ways, as was the associated Nu number. These results significantly differed from each other: the maximum PEC values for method (I) reached 2, and for method (II) to 8. The common feature for both calculation methods was the fact that the maximum values of the Nu number and the thermal efficiency were observed for small Re numbers; however, as the Re number increases, PEC and Nu number decrease strongly.

Keywords: heat transfer enhancement; radiation insert; numerical simulations; performance evaluation criteria; thermal efficiency

Citation: Jasiński, P.B. Numerical Study of Heat Transfer Intensification in a Circular Tube Using a Thin, Radiation-Absorbing Insert. Part 2: Thermal Performance. *Energies* **2021**, *14*, 4533. https://doi.org/10.3390/14154533

Academic Editor: Pouyan Talebizadeh Sardari

Received: 15 June 2021
Accepted: 21 July 2021
Published: 27 July 2021

1. Introduction

The challenge for heat exchanger designers is to achieve high heat transfer intensification, which is always associated with increased flow resistance in the channels. This phenomenon, increases the amount of energy consumed to pump the fluid, mainly due to the turbulisation of the flow. One of the most commonly used parameters to evaluate thermal channels is the thermal performance factor (TPF) or otherwise known as performance evaluation criteria (PEC) [1–3]. It compares the duct under test to a smooth round pipe, with the same thermal conditions and the same pumping power. If the PEC is greater than unity, then such a channel is considered to be more thermally efficient than a smooth pipe (the goal is to achieve the highest possible values), and if it is less than unity, the thermal efficiency of such a channel after modification is worse than a smooth pipe.

Many authors apply this method of evaluation in examining heat channels with turbulising inserts, porous material, or with internally ribbed walls [4–7]. These heat transfer intensification mechanisms rely on the fluid flow disturbance in the channel's entire cross-section (inserts) or only at the wall (micro-ribs). It can be said that all methods of convective heat transfer intensification are aimed at reducing the thickness of the laminar boundary layer or even breaking it locally, which, as is known, is the main resistance in convective heat transport.

The article presents the results of testing channels with an insert that does not turbulise the flow like most elements of this type but uses an additional heat transfer mechanism, i.e., thermal radiation. Due to the necessary radiation transmittance of the working fluid, this method can only be used for transparent gases.

The insert presented in this article is made in the form of a thin, smooth pipe and placed parallel and concentrically to the flow channel walls (Figure 1). In addition to the standard convective heat transport method between the pipe wall and the fluid, the phenomenon of thermal radiation between the wall, and the insert is also used as a heat transfer intensification mechanism. During the fully developed fluid flow, the insert placed in the centre of the tube does not cause fluid mixing between the boundary layer and the turbulent core. The temperature of the insert differs from that of the wall and depends on its diameter and flow conditions. Due to the temperature difference, heat is transferred by radiation between the wall and the insert surface. Assuming that the working gas is completely transparent and does not absorb radiation, the total radiant heat flux is transferred to the insert, which consequently becomes an additional heat transfer surface, giving up heat energy to the fluid on both sides by convection.

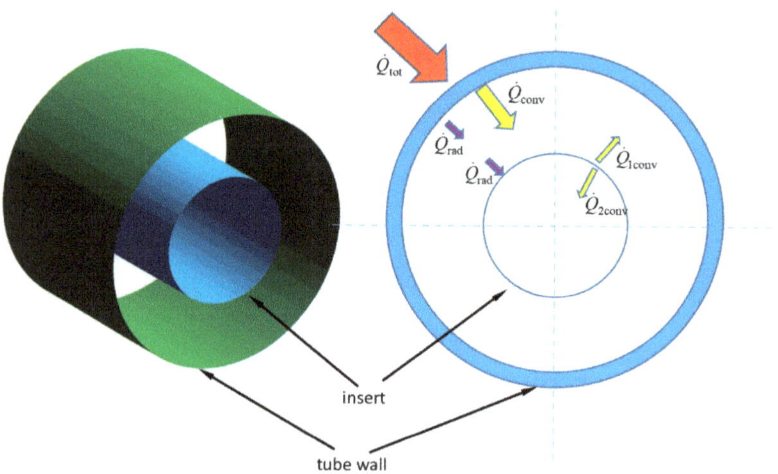

Figure 1. A fragment of a pipe with an insert and a diagram of heat fluxes.

Alijani and Hamidi, in their works [8,9], presented research of inserts, in which the mode of operation was most similar to that presented in this article. They experimentally studied inserts in the form of a solid rod inserted into a pipe [8], while in the second case [9], the insert had a honeycomb cross-section. Both inserts intensified the heat transfer using the effect of thermal radiation. The authors investigated rods with constant dimensions and at different temperatures of the pipe wall. The highest thermal efficiency values (up to 4–5) were obtained for the honeycomb insert, for the highest wall temperatures about 400 °C, and the lowest Reynolds numbers 5000–7000. In turn, Zhang et al. [10] investigated a fully developed channel flow with a porous core in the centre. The numerical analysis confirmed the effectiveness of the porous core in increasing the intensity of heat transfer. The proposed system has been particularly effective for a gaseous medium with low emittance and low fluid velocities. It was also observed that increasing the porous core diameter increases the heat transfer intensity, but the flow resistance also increases.

As previously mentioned, this article is a continuation of the previous work under the same title [11], and the presented results are based on the data from that paper. Numerical simulations were performed using the ANSYS-CFX computer code. It was chosen as a computational tool both because of its flexibility and reliability and also because of the extensive experience acquired over many years by our research groups, as applied to various thermal and flow problems [12–14].

This work's originality is based on the use of an additional heat transport mechanism, which is radiation from the pipe wall to the absorption insert. The presented results were obtained for a relatively small temperature difference $\Delta T = 100\ ^\circ C$, as for radiation processes. The concept of this type of insert may be of great importance, especially for high-temperature combustion and heat recovery processes, where thermal radiation is the dominant method of heat transfer.

In general, the main purpose of this work is to investigate the relationship between the geometric dimension of the insert, i.e., its diameter, in relation to the heat efficiency of the pipe. The heat transfer enhancement with respect to the pipe without insert, considering the criterion of the same pumping power shows the PEC coefficient.

2. Geometric Model of the Insert

According to the scheme in Figure 1, the total heat flux $\dot{Q}_{tot}$ delivered to the channel's outer wall is partly transferred by convection $\dot{Q}_{conv}$ to the fluid, and the rest of this flux $\dot{Q}_{rad}$ is transferred to the insert by radiation. Thus, the balance equation of such a system has this form:

$$\dot{Q}_{tot} = \dot{Q}_{conv} + \dot{Q}_{rad} = \dot{Q}_{conv} + \left(\dot{Q}_{1conv} + \dot{Q}_{2conv}\right) \quad (1)$$

During the fully developed channel flow, a temperature difference exists between the wall surface and the insert, causing the radiant heat flux $\dot{Q}_{rad}$ to be transferred to the insert. Due to the fact that the insert does not cause turbulising and mixing of the fluid, its temperature is a function of its radius. The radiative heat flux is absorbed by the insert, causing its temperature to rise slightly above the flowing gas's boundary layer's local temperature. Simultaneously, the insert is washed by the working gas (which flows both outside and inside the insert) and gives off the heat on both sides in a convective manner (heat fluxes $\dot{Q}_{1conv}$ and $\dot{Q}_{2conv}$, Figure 1). Such an insert can thus actually be treated as an additional heat transfer surface inside the duct.

The thickness of such an insert should be as small as possible and should be made of a material that conducts heat well. You can then ignore the wall thickness and assume that its temperature is the same on both sides. In order to obtain the highest possible radiative heat flux, the emissivity of both radiating surfaces: the pipe wall and the outer insert, should also be as high as possible.

The practical application of such an insert may be particularly appropriate in high-temperature heat exchangers. Due to a significant temperature difference, the radiation fraction in heat transfer is significant compared to convection, e.g., heat recovery from exhaust gases in gas boilers, cars, etc. Nevertheless, the use of such an insert even at a small temperature difference of 100 °C, as presented in the paper, significantly increases the heat transfer efficiency in the heat exchanger channel.

3. Boundary Conditions of Simulation and Numerical Model

Numerical simulations were performed for eight insert diameters d_i. In all cases, the same temperature difference $\Delta T = 100\ ^\circ C$ was maintained between the wall and the average gas temperature in the pipe's entire volume. The heat transfer boundary condition of the 1st kind was applied on the channel's outer wall, i.e., a constant wall temperature equal to 100 °C. The list of used boundary conditions are introduced in Table 1.

For gas, on the other hand, the mean volume temperature was kept constant at 0 °C. The working gas was the air with physical properties dependent on temperature and complete transparency for radiation. In Table 2, the coefficients of polynomial equations for thermal conductivity and dynamic viscosity of the air were shown.

Table 1. Boundary conditions and the parameters values.

Boundary Condition	Description	Parameter	Value/Type
Fluid Domain	Air	Mean Temperature	273 K
		Reference Pressure	1 atm
		Turbulence Model	SST k-ω
Subdomain	Subdomain was set in domain of air. The pressure gradient determined according to the flow direction.	Pressure gradient	0.025–8 Pa/m
		Volumetric heat flux	$-Q_{vol} = \frac{Q_{tot} \cdot A_2}{V_{air}}$ W
Wall	Boundary condition set on the wall of pipe in the form of constant Temperature	Temperature	373 K
Translational Periodicity	Translational Periodicity set on the inlet and outlet areas of Fluid Domain.	-	-
Rotational Symmetry	Rotational Symmetry set on the both sides of Fluid Domain.	-	-

Table 2. Coefficients of polynomial equation $k(T)$, $\mu(T) = a_0 + a_1 \times T + a_2 \times T^2 + a_3 \times T^3$ for the air temperature range 173–373 K.

		a_0	a_1	a_2	a_3
thermal conductivity	$k(T)$	-1.8650×10^{-3}	1.1018×10^{-4}	-5.6729×10^{-8}	1.6728×10^{-11}
dynamic viscosity	$\mu(T)$	2.0251×10^{-6}	6.3507×10^{-8}	-3.0959×10^{-11}	7.8708×10^{-15}

In order to simplify the notation and the possibility of reference to other geometrical dimensions, the insert diameters d_i are presented in a dimensionless form $D = d_p/d_i$ (Figure 2), in relation to the pipe's diameter d_p = 200 mm. The values of d_i and D, along with their exact dimensions, are given in Table 3.

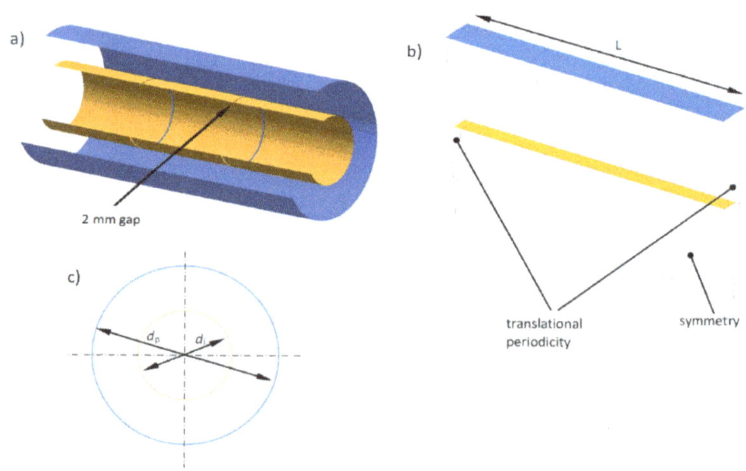

Figure 2. Repeatable insert segment: (**a**) 3D view of a pipe fragment with an insert; (**b**) computational domain; (**c**) channel cross-section diagram.

Table 3. Dimensions of the tested insert and the corresponding dimensionless diameters.

d_i [mm]	40	60	80	100	120	140	160	180
D [-]	0.2	0.3	0.4	0.5	0.6	0.7	0.8	0.9

Due to the axial nature of the flow and the lack of a rotational velocity component, it was assumed that each tested insert could be considered as a two-dimensional case. Therefore, in the simulations, as the computational domain, the geometry in the shape of a longitudinal section of a cylinder with an aperture angle of 10° and length L = 152 mm was used (Figure 2b). An axial symmetry— on the pipe section's side faces as the domain's side boundary conditions—was set.

The appropriate method of carrying out numerical simulations allowed us to obtain a fully developed flow in this relatively short domain. In order to achieve such an effect, a translational periodicity (as boundary conditions) at the inlet and outlet of the domain was set, and the fluid flow by a pressure gradient was forced. Thus, the computational domain has been reduced to a repetitive, periodic, and axisymmetric geometry representative of the entire channel. By reducing the domain size and the number of computational mesh nodes (up to 2D), it was possible to significantly shorten the computation time while maintaining the high mesh quality. Such a research method has also been presented in [15–17], while other, less important details, generally related to such a method of numerical modelling, are presented in [18–20]. In order to ensure the same static pressure on both sides of the insert, small gaps in the continuity of the insert with a 2 mm length were made (Figure 2a), representing about 1.5% of the total length of the insert. These gaps, due to their minimal size, did not disturb the flow.

For the mentioned boundary condition applied to the pipe wall (temperature of 100°), the heat flux supplied to the gas changes for each insert geometry and flow rate. In order to obtain a periodic thermal layer and preserve the energy balance in the domain, the heat flux supplied to the domain was compensated by applying a negative volumetric energy source $\dot{Q}_{vol}$, i.e., subtracting from the entire flow domain the same amount of heat energy that was supplied by the wall surface (Figure 3).

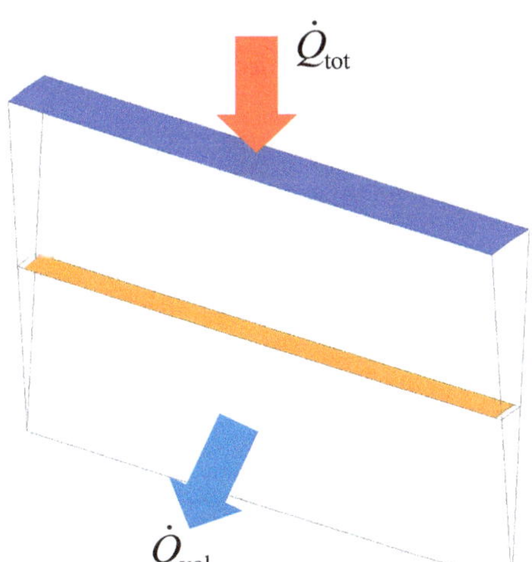

Figure 3. Scheme of the computational domain with supplied and withdrawn heat flux.

3.1. Governing Equations

Numerical calculations of the frictional resistance and the heat transfer were performed using the ANSYS-CFX code. During calculations are solved the basic equations of conservation of mass (2), momentum (3) and energy (4), which have the form [21]:

$$\frac{\partial \rho}{\partial t} + \nabla \cdot (\rho \overline{U}) = 0 \tag{2}$$

$$\frac{\partial(\rho\overline{U})}{\partial t} + \nabla\cdot(\rho\overline{U}\times\overline{U}) = -\nabla p + \nabla\cdot\mu_e(\nabla\overline{U} + (\nabla\overline{U})^T - \frac{2}{3}\delta\nabla\cdot\overline{U}) + (\rho - \rho_{ref})g \quad (3)$$

$$\frac{\partial(i_{tot}\rho)}{\partial t} = \frac{\partial\rho}{\partial t} + \nabla\cdot(\rho\overline{U}i_{tot}) = \nabla\cdot(k\nabla T) + \nabla\cdot(\overline{U}\tau_w) + S_E \quad (4)$$

where the term $\nabla\cdot(\overline{U}\tau_w)$ represents the work of viscous forces, S_E is a term of an energy sources and the total enthalpy i_{tot} is expressed as:$i + \frac{1}{2}\overline{U}^{\,2}$. For the turbulence model used, the governing equations are presented in detail in [17].

3.2. Turbulence Model

The turbulence model SST k-ω (Shear Stres Transport) in all numerical simulations was used due to the assumed turbulent fluid flow. It is one of the most frequently used models in CFD applications due to the much better mapping of flow-thermal phenomena in calculations than the standard k-ε model [17,21]. The SST model's main advantage is the ability to take into account a viscous boundary sublayer by applying the k-ω model near the wall and using the standard k-ε model in the turbulent core region. A special function (so-called blending function) implemented in the SST model is responsible for selecting an appropriate calculation model.

3.3. Radiation Model

Several radiation models are available in the ANYS-CFX calculation program, such as Rosseland, P1, Discrete Transfer and Monte Carlo. In a situation where the thermal radiation energy is transferred between two surfaces, and the medium is transparent to radiation with wavelengths in which most of the heat transfer takes place, the Monte Carlo model is the only one that applies—and this model was used in the simulations [21]. The remaining details of the radiation model validation are described in Part 1 of this article [11].

3.4. Grid Independence

The actual simulations were preceded by performing verification calculations for several geometries at different mesh densities. Table 4 gives an example of the grid under test for one of the insert diameter $D = 0.4$ and Re = 18650. It was observed that the deviation between the grid elements of 25,440 and 49,820 is only 0.9% for friction factor and 0.22% for Nu number. Thus, for further calculations, a structural and hexagonal mesh (Figure 4) with 25,440 elements was chosen and such quality, at which its further densification gives results differing by less than 1.5%. Due to the fact that for such geometries, the velocity field structure is not very complicated and similar to the flow in a smooth pipe; therefore, the computational grid is also simple. Only in the near-wall laminar sublayer areas, i.e., at the channel wall and on both sides of the insert, the mesh was additionally densified to obtain the appropriate y^+ value, which for the used SST k-ω turbulence model should not exceed 2 [2,16,17,21,22].

Table 4. Grid independent test for Re = 18650 and insert diameter $D = 0.4$.

No. of Elements	f	dev. %	Nu	dev. %
6140	0.0461	-	73.23	-
11,960	0.0508	9.38	78.05	6.18
18,230	0.0543	6.32	80.13	2.60
25,440	0.0551	1.45	80.62	0.61
49,820	0.0556	0.9	80.80	0.22

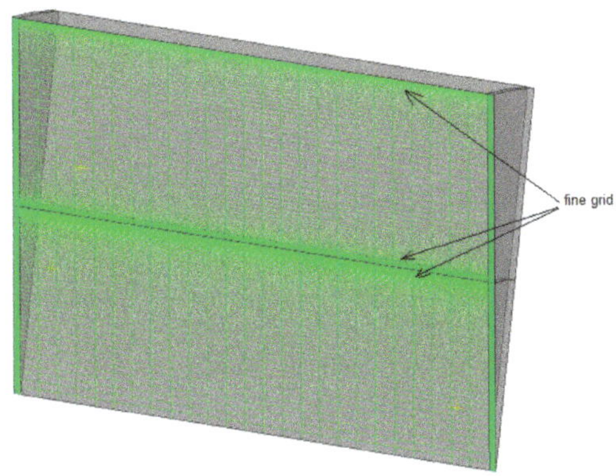

Figure 4. Computational mesh with densification areas at the walls.

Nevertheless, validation for numerical results was done, but for slightly different flow channel shapes and presented in papers [16,19,23].

4. Data Processing

To calculate the Nu number from numerical simulations, the following dependency was used:

$$Nu = \frac{\dot{Q}_{tot} \cdot d_p}{k \cdot (T_w - T_b) \cdot A_2} \tag{5}$$

while the Nu number for a smooth pipe without an insert was calculated from the Dittus-Boelter correlation:

$$Nu_s = 0.023 \cdot Re^{0.8} \cdot Pr^{0.4} \tag{6}$$

The Re number is a function of the average velocity u and the pipe diameter d_p:

$$\mathrm{Re} = \frac{u \cdot d_p}{\nu} \tag{7}$$

For the numerical calculations, the friction factor from the Darcy–Weisbach equation was calculated:

$$f = \frac{2 \cdot d_p}{\rho \cdot u^2} \cdot \frac{dp}{dx} \tag{8}$$

The quantity forcing the fluid flow was the pressure gradient $\mathrm{d}p/\mathrm{d}x$, while the average velocity in the pipe u resulted from numerical calculations.

To calculate the friction factor for a smooth pipe, the Blassius formula was used:

$$f_s = 0.3164 \cdot Re^{-0.25} \tag{9}$$

According to the same pumping power criterion, the PEC coefficient was calculated from the relationship:

$$PEC = \frac{Nu/Nu_s}{(f/f_s)^{\frac{1}{3}}} \tag{10}$$

Both the Nu number and the friction factor f were calculated on the basis of results obtained from the numerical simulation, using Equations (2) and (5) respectively. In Equation (2), the constant values were: wall temperature T_w, pipe diameter d_p and the related wall area A_2, while the resulting values were: total heat flux on the wall Q_{tot} and bulk temperature T_b. Whereby, T_b was calculated as the volume average, and Q_{tot} as the area average. Similarly, when calculating the friction factor from Equation (5), the constant values were pipe diameter d_p and pressure gradient dp/dx, which was the parameter forcing the fluid flow. The resulting value from the calculations was the mean velocity u in the channel cross-section.

5. Results and Discussion

This section may be divided by subheadings. It should provide a concise and precise description

5.1. Heat Fluxes Balance

The insert's operation, consisting of intensifying the pipe's heat transfer, is possible due to the wall's temperature difference and the insert. As previously mentioned, the insert's temperature depends on its diameter and the gas flow velocity and is a key parameter on which the radiation heat flux absorbed by the insert depends. Figure 5 shows the inserts' temperature as a function of the Re number, together with the temperature difference ΔT marked schematically.

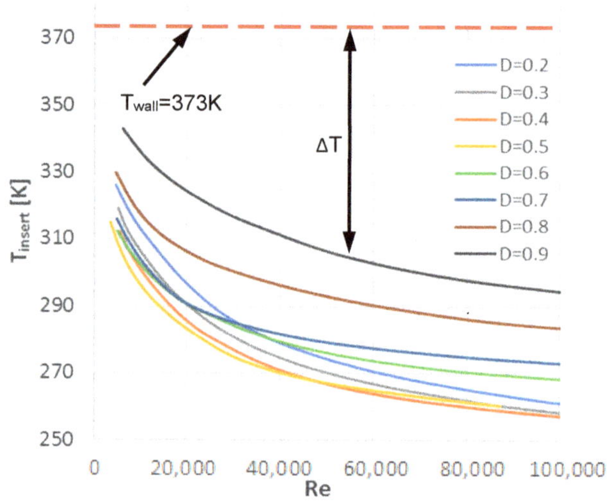

Figure 5. The tested inserts temperatures as a function of the Re number.

The highest insert temperatures occur for small Re numbers and decrease with increasing the Re numbers. As can be seen in Figure 5, in the tested Re number range, the insert temperature can change even by about 40–50 °C. As it results from the Stefan–Boltzmann law for grey bodies, the net radiative heat flux exchanged between the surfaces is a highly nonlinear function and mainly depends on the difference of the fourth powers of the surface temperatures (11).

$$Q_{rad} = \frac{\sigma \cdot A_1 \cdot (T_1{}^4 - T_2{}^4)}{\frac{1}{\varepsilon_1} + \frac{A_1}{A_2} \cdot \left(\frac{1}{\varepsilon_2} - 1\right)} = \varepsilon_{12} \cdot \sigma \cdot A_1 \cdot \left(T_1{}^4 - T_2{}^4\right) \quad (11)$$

The term of Equation (11), which takes into account the ratio of the heat transfer surfaces (pipe and insert) and their emissivity, is called the interchange factor (12).

$$\varepsilon_{12} = \frac{1}{\frac{1}{\varepsilon_1} + \frac{A_1}{A_2} \cdot \left(\frac{1}{\varepsilon_2} - 1\right)} \quad (12)$$

The greater the temperature difference between the surfaces, the greater radiative heat flux transferred between them. In the case of the tested inserts, the amount of this flux is also influenced by the heat transfer area ratio, which is different for each tested geometry due to the insert's changing diameter.

The total heat flux transferred to the fluid from the pipe wall consists of the radiation and convection parts (13), according to the diagram shown in Figure 1.

$$\dot{Q}_{tot} = \dot{Q}_{conv} + \dot{Q}_{rad} \tag{13}$$

In Figure 6, the fractions of the radiative (a) and convection flux (b) in relation to the total heat flux on the pipe wall are shown. As can be seen from the graphs, the greatest amount of heat energy is transferred by radiation to the inserts with the largest diameters, while the highest convective heat fluxes are observed for the smallest inserts. For all studied geometries, the tendency is that with the Re number increase, the fraction of the radiative heat flux decreases and the convective flux increases.

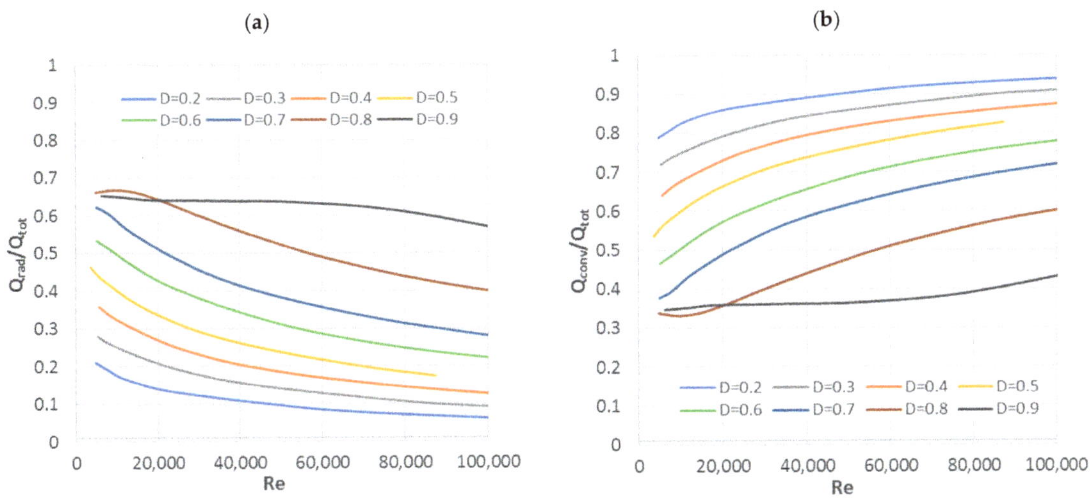

Figure 6. Fractions of radiative (**a**) and convective (**b**) heat fluxes in the total heat flux at the pipe wall.

The insert fully absorbs the radiative heat flux radiated by the pipe wall. As a result of absorbing certain thermal energy, the insert changes its temperature above the local gas temperature and gives off the heat on both sides in a convective manner (14), Figure 1.

$$\dot{Q}_{rad} = \dot{Q}_{1conv} + \dot{Q}_{2conv} \tag{14}$$

The heat flux $\dot{Q}_{1conv}$ is assumed by the flowing fluid outside the insert while $\dot{Q}_{2conv}$ is transferred to the fluid flowing inside the insert. Figure 7 shows the fractions of these heat fluxes in relation to the radiation flux absorbed by the insert. The presented graphs show the negative fractions and fractions greater than 1. Negative fractions, Figure 7a mean that above a certain Re number, the convective heat flux changes direction and the insert are additionally heated by gas from the outside. For the internal convective flux (associated with the external flux), the fractions greater than 1 were observed, Figure 7b. This means that thermal energy is transferred by convection from the insert's outer space to its inner space. If such a phenomenon occurs, it can be said that the insert only works one-sidedly, which significantly reduces the intensification of heat transfer. In Figure 7, it can be seen that the two inserts with dimensionless diameters $D = 0.8$ and $D = 0.9$, almost in the entire range of Re numbers, work this way. However, in the remaining inserts, such a phenomenon occurs only for larger Re numbers.

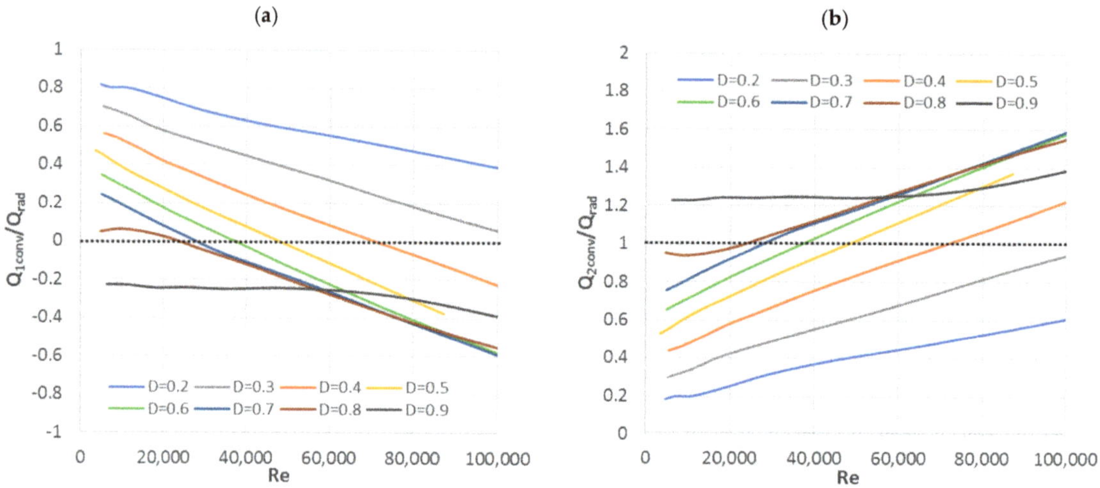

Figure 7. Fractions of convective heat fluxes on the insert surface in relation to the heat flux delivered to the insert by radiation Q_{rad}. (**a**) Q_{1conv}—heat flux on the outer surface of the insert, (**b**) Q_{2conv}—heat flux on the inner surface of the insert.

Figure 8 shows the pipe cross-section temperature profiles for a selected insert with a diameter of $D = 0.4$ and a few Re numbers. These examples show how the insert temperature and the gas temperature in the pipe change as a function of the radius, depending on the Re number. As can be seen, up to a Re number of about 40,000, on both sides of the insert, the gas temperature is lower than that of the insert, which means that it gives two-sided heat to the fluid by convection. For the highest number of Re = 86,500, the gas temperature in the annular section is higher than the inserts, so in this case, the insert is convectionally heated by the gas. This phenomenon causes the insert to return heat to the gas only through its inner surface, significantly reducing its thermal efficiency.

5.2. PEC Coefficient

In Part 1 of this article [11], the method of calculating the friction factor, Nu number and the analysis of the obtained results are presented. The thermal efficiency (PEC) determines how much the heat transfer's intensity will increase after the insertion in relation to the smooth pipe while maintaining the same pumping power criterion. This coefficient is calculated using the Formula (10). PEC values above unity are favourable and indicate the predominant effect of heat transfer intensification than flow resistance, while values below unity indicate greater flow resistance in relation to the benefit obtained from intensifying heat transfer of the tested geometry. Therefore, to calculate PEC, it is necessary to know the ratio f/f_s and Nu/Nu_s as a function of the Re number.

In Figure 9a the ratio f/f_s as a function of the Re number is presented, showing how many times the friction factor has increased for a given insert diameter in relation to the pipe without the insert. As you can see, for all geometries, the greatest increase is for small Re numbers, while from Re about 20,000, the characteristics are more or less horizontal, which means that the trend is similar to that for a smooth pipe. In turn, in Figure 9b, which shows the ratio f/f_s as a function of the dimensionless insert diameter D, one can observe this diameter's influence on a few selected Re numbers' friction factor. There is a clear maximum for all characteristics, ranging from 0.5–0.7D, which means the highest flow resistances for these insert diameters.

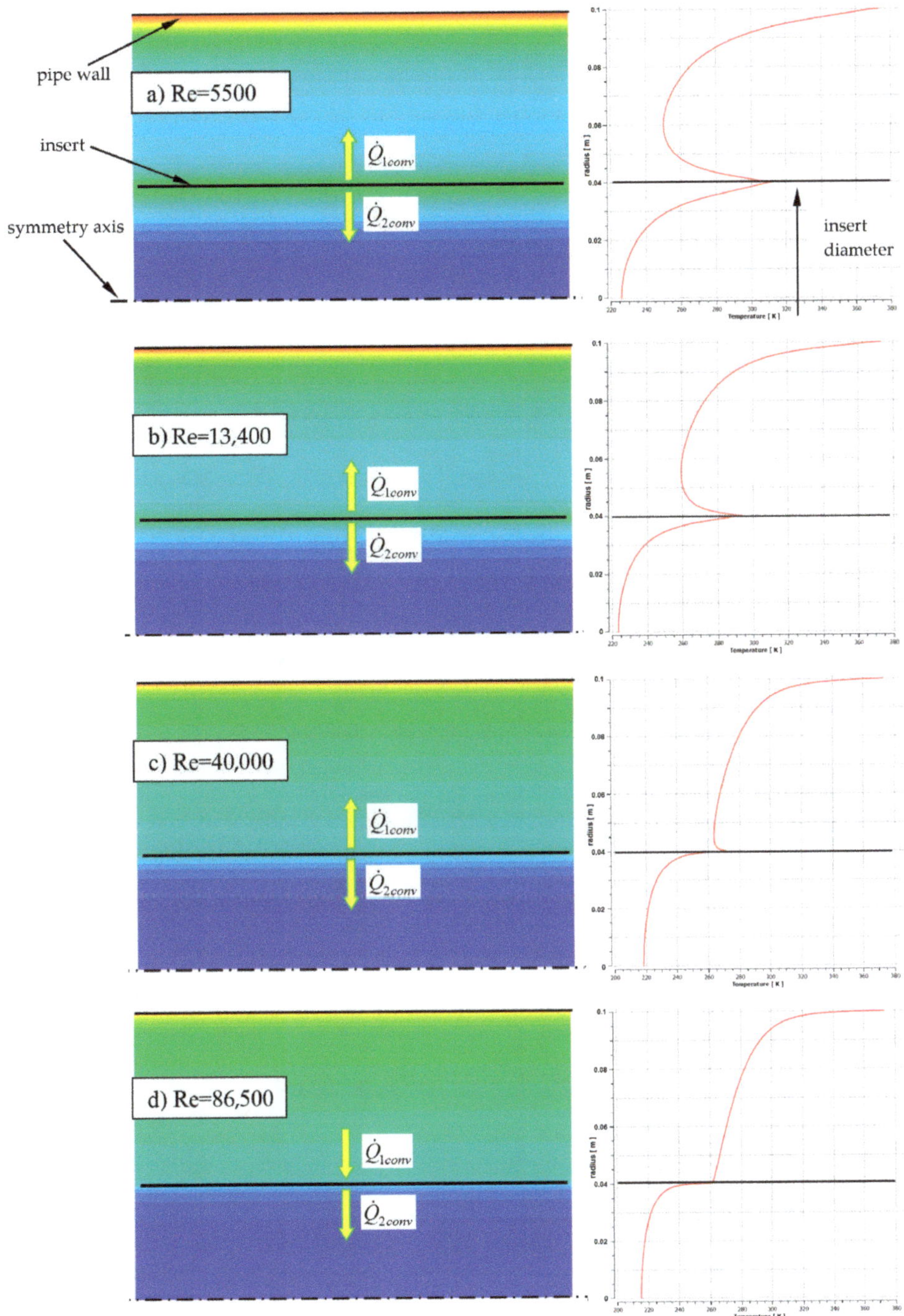

Figure 8. Temperature fields and profiles in the pipe cross-section for an exemplary insert with a diameter of $D = 0.4$ and several Re numbers.

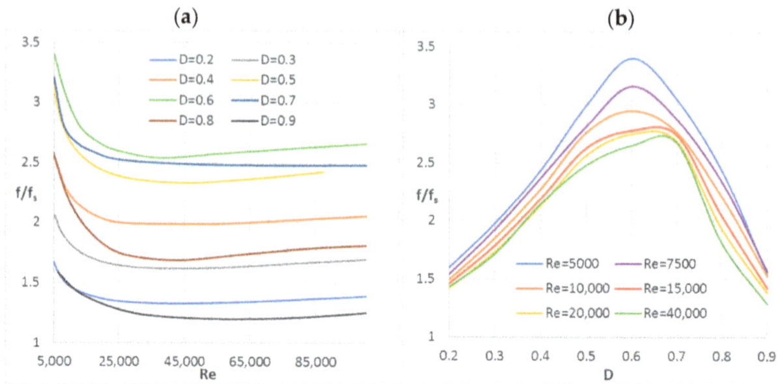

Figure 9. (**a**) the ratio $f/f_s(Re)$ for dimensionless diameters D; (**b**) the ratio $f/f_s(D)$ for several Re numbers.

The Nu number, calculated according to (1), characterises the heat transfer intensity on the channel flow wall, in this case, a round pipe. The insert divides the channel cross-section into two parts in the performed tests: annular (outside the insert) and tubular (inside the insert). This raises the question of calculating the temperature T_b, which can be related to (a) the entire volume of the pipe, (b) only the volume of the annular part that actually contacts with the channel wall. Both approaches have their justification.

ad. (a) T_b *calculated as the average volumetric temperature of the whole channel*. This method of calculation is most often used by most researchers, both for empty channels and those equipped with inserts or other turbulators, e.g., [22–32]. The heat flux is transferred to the fluid in contact with the channel wall, and its average temperature is a result of the heat transfer conditions and the medium's velocity on both sides of the insert. A certain disadvantage of this approach with the insert under consideration is the lack of fluids mixing from the annular and tubular spaces, as is the case in reality. For this reason, there are different velocity and temperature fields in these areas, and the temperature gradient in the boundary layer of the channel is completely different from what it would be with a classic insert.

ad. (b) T_b *calculated as the volumetric average temperature from the annular part only*. This approach is justified because the heat flux is directly transferred to the fluid in contact with the channel wall. Of course, its average temperature is also influenced by the amount of heat supplied to the insert by radiation and returned by the insert to the fluid by convection in the annular and tubular portions.

The author suggests that both ways of computing the Nu number can be considered, despite the fact that they give different results. Part 1 of this article [11] presents and analyses the Nu (Re) number characteristics calculated for the methods (a) and (b). A specific feature is that Nu numbers are smaller than for a smooth tube—mainly for the largest insert diameters and large Re numbers. With standard turbulising inserts, such a phenomenon should not occur because any insert placed in the pipe disturbs the flow, intensifying the convective heat flux simultaneously increasing the flow resistance.

In the case of the tested inserts, the appearance of Nu numbers lower than the reference level, which is the smooth pipe, can be explained by the creation of additional space between the pipe wall and the insert (annular section). The larger the insert diameter, the smaller the annular cross-section becomes, and at the same time, the average gas temperature in it increases, and its velocity decreases. This makes this space-gap a kind of insulation layer between the pipe wall and the main stream of gas flowing through it. With small diameters, the insert is washed on both sides by the fluid more or less evenly. On the other hand, with large diameters, the flowing gas's mass flow and its velocity are much smaller in the annular section than inside the insert. Therefore, the convective heat transfer is significantly reduced there, and the resistance from heat conduction of the gas inside the gap increases.

Figure 10a,b show the ratio of the Nu number of the tested inserts to the Nu number for a smooth pipe for the two above-mentioned calculation methods. Taking into account the average temperature Tb of the entire pipe volume (case a), it can be seen that for inserts with diameters $D = 0.9$, 0.8, 0.7 and 0.6, the values of the Nu/Nus characteristics are lower than unity. On the other hand, taking the average temperature Tb for the calculations only from the annular cross-section (case b), the function lines are arranged in an "ideal" order, i.e., as the insert's diameter increases, the heat transfer intensity also increases.

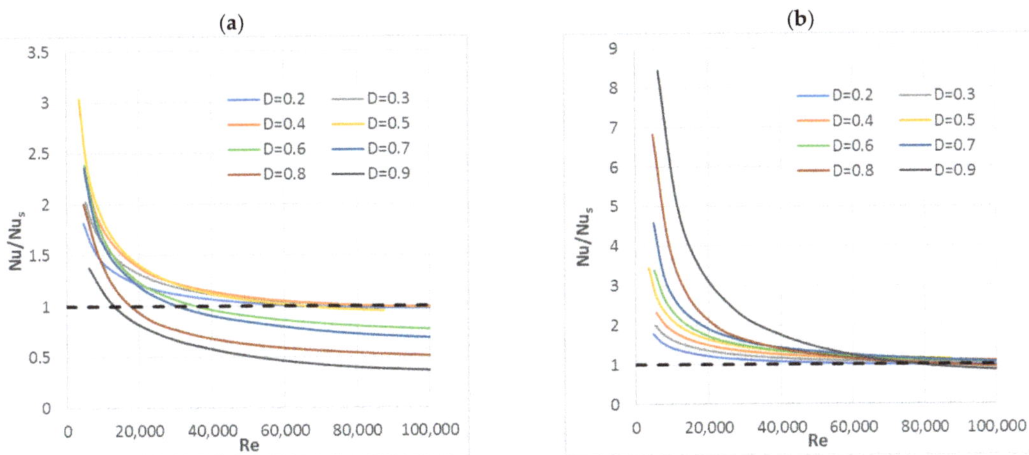

Figure 10. The ratio of the Nu number obtained from the simulation to the Nu number for a smooth pipe $Nu/Nu_S(Re)$, calculated for the mean volumetric temperature: (**a**) from the entire pipe, (**b**) from the annular section.

The thermal efficiency of the studied geometries, based on the two methods of calculating the Nu number, is shown in Figure 11. PEC coefficient calculated by method (a) reaches values much lower than that calculated by method (b). As you can see, the max PEC value calculated by method (a) is approx. 2 for the insert diameter $D = 0.6$, while the value calculated by method (b) reaches a value over 8 for $D = 0.9$.

When comparing Figures 10 and 11, a fairly large similarity can be seen in the position of the function lines. The PEC coefficients calculated in two ways quite well reflect the behaviour of the trend of Nu numbers' characteristics. This means that the decisive parameter influencing such channels' thermal efficiency is the Nu number and not the friction factor f.

In the large Re number range, i.e., above 30,000, all channels calculated by method (a) have a PEC below unity. For the largest diameters of the inserts $D = 0.8$–0.9, the PEC is greater than one only for small numbers of Re in the range up to approx. 10,000. The highest PEC values are observed for inserts with diameters $D = 0.5$ and $D = 0.4$ with the smallest Re numbers. Therefore, it appears that the effective range of use for such inserts is the Re numbers less than 15,000.

According to the insert dimension, the thermal performance curves calculated by method (b) are arranged in a very regular way. The lowest PEC values are observed for the smallest insert diameter $D = 0.2$. With the increase of the insert diameter, the PEC increases regularly, and the highest values are reached for $D = 0.9$, with the maximum reaching over 8. As for a heat-flow channel with a single-phase flow, these are very high values. For this reason, one should approach these results and the associated method of calculating the Nu number with great care. Therefore, as mentioned before, the author recommends instead (a) calculating the Nu number based on the average gas temperature in the pipe's entire volume.

(a) (b)

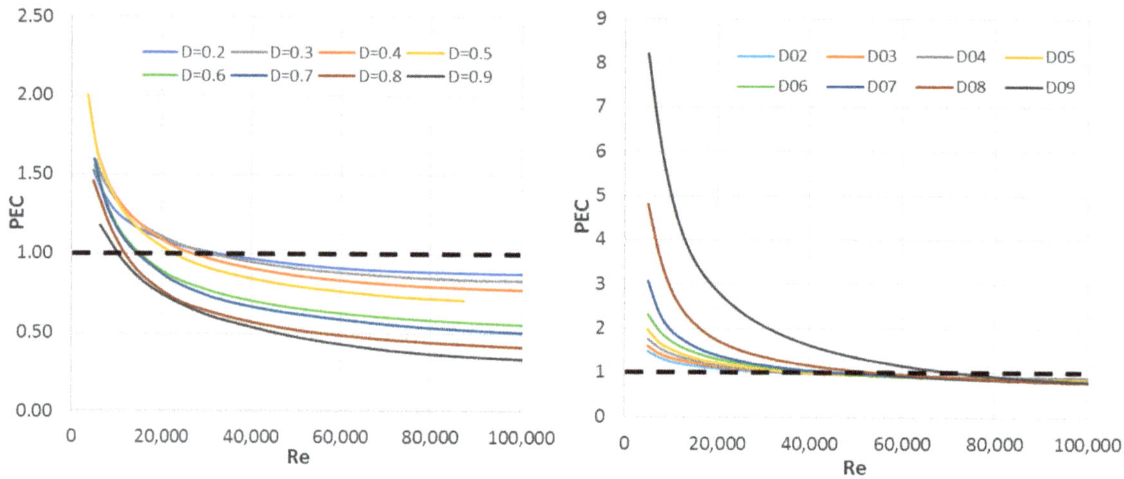

Figure 11. PEC(Re) characteristics calculated by two methods: (**a**) for the average temperature of the entire pipe section; (**b**) for the average temperature of the annular section.

Figure 12 shows the PEC characteristics for several Re numbers as a function of the dimensionless insert diameter D. For method (a), we can see the maxima of the function for diameters $D = 0.4$ and $D = 0.5$ at small Re numbers <10,000, while for Re numbers in the range 10,000–20,000 maximum shifts towards diameters $D = 0.3$ and $D = 0.4$. This means that for the smallest Re numbers, an insert with a diameter of $D = 0.5$ achieves the highest PEC, and with slightly larger Re numbers, up to about 20,000, an insert $D = 0.3$. With method (b), with increasing diameter D, the PEC characteristics trend is also upward, and the highest values are observed for the largest insert $D = 0.9$ and the smallest Re numbers.

(a) (b)

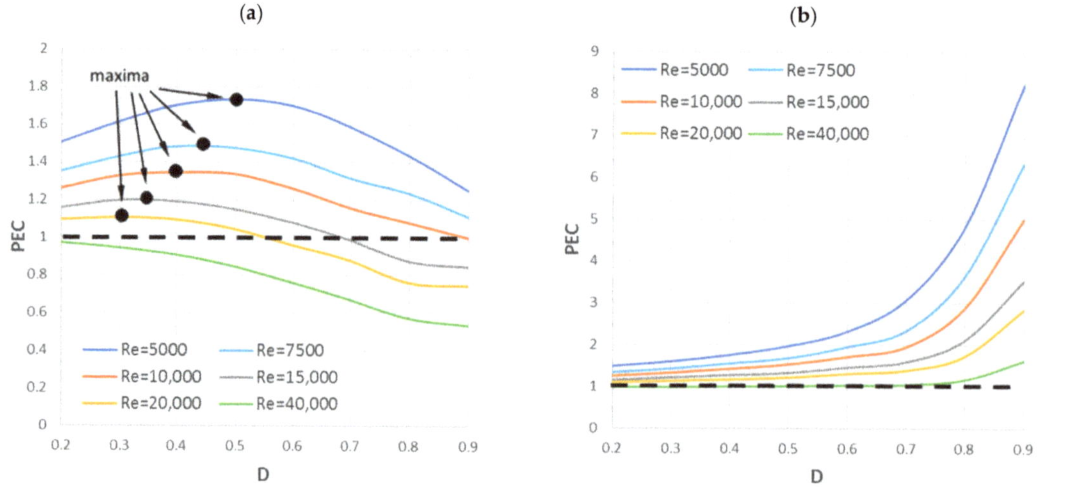

Figure 12. PEC(D) characteristics calculated by two methods: (**a**) for the average temperature of the entire pipe section; (**b**) for the average temperature of the annular section.

6. Summary and Conclusions

This work, which is Part 2 of the article, presents the results of numerical investigations for various sizes of heat transfer intensifying inserts, using the phenomenon of thermal radiation absorption. On the basis of the obtained results, heat flux balances were shown and described, and their characteristics were plotted. The diagrams in Figure 6 show the heat fluxes for the pipe wall, and Figure 7 illustrate the heat fluxes for the tested inserts. Figure 5 also shows the temperature insert characteristics as a function of the Re number, and Figure 8 shows the fields and temperature profiles in the pipe cross-section.

In order to calculate the thermal efficiency of the inserts, the relationships of the friction factor and Nu number in relation to a smooth pipe—$f/f_s(Re)$ and $Nu/Nu_s(Re)$ were presented. Based on the Formula (7), the PEC coefficient was calculated, and in Figure 11, its characteristics for the tested geometries are present—for the two described methods of calculating the Nu number.

The analysis of the results allowed for the formulation of the following conclusions:

(1) The highest insert temperatures are observed for small Re numbers, and they decrease with increasing Re numbers. As it is known from the Stefan–Boltzmann law for grey bodies (11), the greater the temperature difference among the surfaces, the greater the radiative heat flux transferred between them. In the case of the tested inserts, the magnitude of this flux is also influenced by the ratio of the heat transfer surfaces A_1/A_2, which is different in each geometry due to the changing insert diameter. According to Figure 5, the largest temperature differences ΔT and the greatest radiative heat fluxes are noted for the largest Re numbers.

(2) The fractions of these fluxes in relation to the total flux are shown in Figure 6. As can be seen, the largest radiative heat fluxes are absorbed by the inserts with the largest diameters, and the largest convective heat fluxes occur for the smallest insert's diameters. In general, for all studied inserts, as the Re number increases, the radiative heat flux decreases and the convective flux increases.

(3) The fractions of convective heat fluxes on both sides of the insert, in relation to the radiation flux absorbed by it, are shown in Figure 7. The presented graphs show negative values and values greater than 1. Negative fractions in Figure 7a mean that the convective heat flux has changed direction, and now the insert from the outer side is additionally heated by the flowing gas. Related to it is a convective heat flux directed to the inside of the insert, for which fractions greater than one were observed, Figure 7b. This notation means that thermal energy is transferred by convection from the insert's outer space to its interior. It can therefore be said that for such cases, the insert only acts one-sidedly, which generally significantly reduces the overall intensification of heat transfer. In Figure 8 can be seen that at high Re numbers, the gas's temperature in the annular cross-section is higher than the inserts, which causes its convection heating. With dimensionless diameters $D = 0.8$ and $D = 0.9$, two inserts work like this throughout almost the entire range of Re numbers.

(4) The thermal efficiency (PEC), depending on the calculation method, differs significantly from each other. As shown in Figure 11, the max PEC value calculated by method (a) is approx. 2 for the insert diameter $D = 0.5$, while the value calculated by method (b) reaches a value over 8 for $D = 0.9$. It can be noticed that the characteristics calculated by method (b) are arranged in a very orderly manner—their value increases with the insert diameter. The lowest PEC values are observed for the smallest insert diameter $D = 0.2$, and the highest values for $D = 0.9$, with the maximum reaching over 8.

Figure 12 shows the PEC characteristics for several Re numbers as a function of the dimensionless insert diameter D. Thus, for method (a), the maxima of the function are observed for diameters $D = 0.4$ and $D = 0.5$, and small Re numbers up to 10,000, while for Re numbers in the range 10,000–20,000 the maxima move towards the diameters $D = 0.3$ and $D = 0.4$. This means that an insert with a diameter of $D = 0.5$ has the highest PEC for the smallest Re numbers, while for slightly larger Re numbers reaching approx. 20,000, an insert with a diameter $D = 0.3$ has the highest PEC values. On the other hand, for manner (b), with increasing diameter D, the trend of the function on the graph is also upward, and the highest PEC values, as previously mentioned, are noticed for the largest insert $D = 0.9$.

Considering the criterion of thermal efficiency, the most optimal dimensions of the inserts are diameters $D = 0.4$ and $D = 0.5$, and for the range of small numbers, Re < 15,000.

(5) In numerical tests, the same emissivity of the pipe's inner surface and the insert equal to $\varepsilon = 0.9$ was used because these values better correspond to actual conditions. As it results from the Stefan–Boltzmann law and Equation (11), increasing the emissivity of such surfaces will increase the radiative heat flux and the thermal efficiency of such a tube.

The presented results were obtained for a relatively small temperature difference $\Delta T = 100$ °C, as for radiation processes. There is a need for further investigation of such inserts for larger temperature differences in order to determine their thermal efficiency and thermal-flow characteristics. Another interesting research issue is the consideration in calculations of the radiation properties of the working gases, i.e., water vapour and carbon dioxide, which are only partially permeable to radiation. This means that in addition to the convective heat transfer, these gases also heat up in their entire volume, absorbing part of the radiation.

It seems that the concept of such an insert may be of great importance, especially for high-temperature heat-flow processes, where radiation is the dominant method of heat transfer. Therefore, at large temperature differences, one can expect much higher thermal efficiencies of such channels.

Funding: This research received no external funding.

Institutional Review Board Statement: Not applicable.

Informed Consent Statement: Not applicable.

Data Availability Statement: Not applicable.

Conflicts of Interest: Author declare no conflict of interest.

Nomenclature

A	heat transfer area (m^2)
d_i	insert diameter (m)
d_p	pipe diameter (m)
D	diameter ratio, (d_i/d_p)
dp/dx	pressure gradient (Pa/m)
f	friction factor
k	thermal conductivity (W/mK)
L	domain length (m)
Nu	Nusselt number
$\dot{Q}_{tot}$	total heat flux (W/m^2)
$\dot{Q}_{rad}$	radiative heat flux (W/m^2)
$\dot{Q}_{conv}$	convective heat flux (W/m^2)
PEC	Performance Evaluation Criteria
Pr	Prandtl number
Re	Reynolds number
T_b	bulk temperature (K)
T_w	wall temperature (K)
u	average velocity (m/s)
$\overline{U}$	vector of velocity (m/s)
v	kinematic viscosity (m^2/s)
ΔT	temperature difference (K)
ε	emissivity

ε_{12} interchange factor
ρ density (kg/m^3)
μ dynamic viscosity (Pa·m)
τ_w wall shear stress (Pa)
indexes:
s smooth tube
1 for smaller heat transfer area
2 for bigger heat transfer area

References

1. Bas, H.; Ozceyhan, V. Heat transfer enhancement in a tube with twisted tape inserts placed separately from the tube wall. *Exp. Therm. Fluid Sci.* **2012**, *41*, 51–58. [CrossRef]
2. Jasiński, P.B. Numerical study of thermal-hydraulic characteristics in a circular tube with ball turbulators. Part 3: Thermal performance analysis. *Int. J. Heat Mass Transf.* **2017**, *107*, 1138–1147. [CrossRef]
3. Bitama, W.; Demaghb, Y.; Hachicha, A.A.; Benmoussa, H.; Kabar, Y. Numerical investigation of a novel sinusoidal tube receiver for parabolic trough technology. *Appl. Energy* **2018**, *218*, 494–510. [CrossRef]
4. Ozceyhan, V.; Gunes, S.; Buyukalaca, O.; Altuntop, N. Heat transfer enhancement in a tube using circular cross sectional rings separated from wall. *Appl. Energy* **2008**, *85*, 988–1001. [CrossRef]
5. Tu, W.; Wang, Y.; Tang, Y. A numerical study on thermal-hydraulic characteristics of turbulent flow through a circular tube fitted with pipe inserts. *Appl. Therm. Eng.* **2016**, *101*, 413–421. [CrossRef]
6. Wang, W.; Zhang, Y.; Lee, K.-S.; Li, B. Optimal design of a double pipe heat exchanger based on the outward helically corrugated tube. *Int. J. Heat Mass Transf.* **2019**, *135*, 706–716. [CrossRef]
7. Zheng, Z.-J.; Li, M.; He, Y.-L. Thermal analysis of solar central receiver tube with porous inserts and non-uniform heat flux. *Appl. Energy* **2017**, *185*, 1152–1161. [CrossRef]
8. Alijani, M.R.; Hamidi, A.A. Convection and radiation heat transfer in a tube with core rod insert at high temperature. *Heat Mass Transf.* **2014**, *51*, 911–919. [CrossRef]
9. Alijani, M.R.; Hamidi, A.A. Convection and radiation heat transfer in a tube with core rod and honeycomb network inserts at high temperature. *J. Mech. Sci. Technol.* **2013**, *27*, 3487–3493. [CrossRef]
10. Zhang, J.; Sutton, W.; Lai, F. Enhancement of heat transfer using porous convection-to-radiation converter for laminar flow in a circular duct. *Int. J. Heat Mass Transf.* **1996**, *40*, 39–48. [CrossRef]
11. Jasiński, P.B. Numerical study of heat transfer intensification in a circular tube using a thin, radiation-absorbing insert. Part 1: Thermo-hydraulic characteristics. *Energies* **2021**, in press.
12. Gutkowski, A. Numerical Analysis of Flame Behavior Near the Quenching Conditions During Passage from Wider to Narrower Tube Diameters. *Combust. Sci. Technol.* **2012**, *184*, 1616–1634. [CrossRef]
13. Kacprzak, K.; Liskiewicz, G.; Sobczak, K. Numerical investigation of conventional and modified Savonius wind turbines. *Renew. Energy* **2013**, *60*, 578–585. [CrossRef]
14. Fodemski, T.; Górecki, G.; Jasiński, P. Corrugated channels heat transfer efficiency Analysis based on velocity fields resulting from computer simulation and PIV Measurements. In Proceedings of the 8th International Conference on Heat Transfer, Fluid Mechanics and Thermodynamics, HEFAT 2011, Pointe Aux Piments, Mauritius, 11–13 July 2011.
15. Kryłłowicz, W.; Magiera, R.; Łagodziński, J.; Sobczak, K.; Liskiewicz, G. Aerodynamical and structural design of the diagonal blower and its numerical and experimental validation. *J. Vib. Eng. Technol.* **2014**, *2*, 459–468.
16. Jasiński, P.B. Numerical study of thermal-hydraulic characteristics in a circular tube with ball turbulators. Part 1: PIV experiment and pressure drop. *Int. J. Heat Mass Transf.* **2014**, *74*, 48–59. [CrossRef]
17. Jasiński, P.B. Numerical study of thermal-hydraulic characteristics in a circular tube with ball turbulators. Part 2: Heat transfer. *Int. J. Heat Mass Transf.* **2014**, *74*, 473–483. [CrossRef]
18. Woźniak, D.; Obidowski, D.; Papierski, A. CFD based optimization of a semi-open impeller centrifugal pump for circulatory support. *Int. J. Artif. Organs* **2015**, *38*, 380.
19. Jasiński, P. Numerical Study of Friction Factor and Heat Transfer Characteristics for Single-Phase Turbulent Flow in Tubes with Helical Micro-Fins. *Arch. Mech. Eng.* **2012**, *59*, 469–485. [CrossRef]
20. Jasiński, P. Numerical optimization of flow-heat ducts with helical micro-fins, using Entropy Generation Minimization (EGM) method. In Proceedings of the WSEAS International Conferences (HTE'11), Florence, Italy, 23–25 August 2011; pp. 47–54.
21. ANSYS-CFX Manual v. 18.1. Available online: http://www.ansys.com (accessed on 25 July 2021).
22. Gutkowski, A.N.; Łęcki, M.; Jasiński, P.; Jędrowiak, B. Flame behavior during propagation in small isothermal tubes characterized by different degrees of the end opening. *Combust. Sci. Technol.* **2018**, *191*, 711–725. [CrossRef]
23. Zheng, N.; Liu, P.; Wang, X. Numerical simulation and optimization of heat transfer enhancement in a heat exchanger tube fitted with vortex rod inserts. *Appl. Therm. Eng.* **2017**, *123*, 471–484. [CrossRef]
24. Huanga, S.; Chena, H.; Zhang, X. Experimental evaluation of thermal performance in a circular tube with Y-branch insert. *Int. Commun. Heat Mass Transf.* **2019**, *106*, 15–21. [CrossRef]

25. Jasiński, P.; Kowalczyk, M.; Romaniak, A.; Warwas, B.; Obidowski, D.; Gutkowski, A. Investigation of Thermal-Flow Characteristics of Pipes with Helical Micro-Fins of Variable Height. *Energies* **2021**, *14*, 2048. [CrossRef]
26. Hasanpour, A.; Farhadi, M.; Sedighi, K. A review study on twisted tape inserts on turbulent flow heat exchangers: The overall enhancement ratio criteria. *Int. Commun. Heat Mass Transf.* **2014**, *55*, 53–62. [CrossRef]
27. Chang, C.; Sciacovelli, A.; Wu, Z.; Li, X.; Li, Y.; Zhao, M.; Deng, J.; Wang, Z.; Ding, Y. Enhanced heat transfer in a parabolic trough solar receiver by inserting rods and using molten salt as heat transfer fluid. *Appl. Energy* **2018**, *220*, 337–350. [CrossRef]
28. Liu, G.; Yang, C.; Zhang, J.; Zong, H.; Xu, B.; Qian, J.-Y. Internal Flow Analysis of a Heat Transfer Enhanced Tube with a Segmented Twisted Tape Insert. *Energies* **2020**, *13*, 207. [CrossRef]
29. Wijayanta, A.T.; Pranowo; Mirmanto; Kristiawan, B.; Aziz, M. Internal Flow in an Enhanced Tube Having Square-cut Twisted Tape Insert. *Energies* **2019**, *12*, 306. [CrossRef]
30. Bellos, E.; Tzivanidis, C. Enhancing the Performance of Evacuated and Non-Evacuated Parabolic Trough Collectors Using Twisted Tape Inserts, Perforated Plate Inserts and Internally Finned Absorber. *Energies* **2018**, *11*, 1129. [CrossRef]
31. Ho, C.; Liu, Y.-C.; Yang, T.-F.; Ghalambaz, M.; Yan, W.-M. Convective heat transfer of nano-encapsulated phase change material suspension in a divergent minichannel heatsink. *Int. J. Heat Mass Transf.* **2021**, *165*, 120717. [CrossRef]
32. Ho, C.; Liu, Y.-C.; Ghalambaz, M.; Yan, W.-M. Forced convection heat transfer of Nano-Encapsulated Phase Change Material (NEPCM) suspension in a mini-channel heatsink. *Int. J. Heat Mass Transf.* **2020**, *155*, 119858. [CrossRef]

energies

MDPI

Article

Improvement of Mathematical Model for Sedimentation Process

Ivan Pavlenko [1], Marek Ochowiak [2,*], Praveen Agarwal [3], Radosław Olszewski [4], Bernard Michałek [4] and Andżelika Krupińska [2]

1 Department of Computational Mechanics Named after V. Martsynkovskyy, Sumy State University, 2, Rymskogo-Korsakova Str., 40007 Sumy, Ukraine; i.pavlenko@omdm.sumdu.edu.ua
2 Department of Chemical Engineering and Equipment, Poznan University of Technology, 5, M. Skłodowskiej-Curie Sq., 60-965 Poznan, Poland; andzelika.krupinska@put.poznan.pl
3 Department of Mathematics, Anand International College of Engineering, D-40, Shanti Path, Jawahar Nagar, Jaipur 303012, India; goyal.praveen2011@gmail.com
4 Faculty of Chemistry, Adam Mickiewicz University, 1, Wieniawskiego Str., 61-614 Poznan, Poland; radoslaw.olszewski@adob.com.pl (R.O.); bernard.michalek@adob.com.pl (B.M.)
* Correspondence: ochowiak@op.pl

Abstract: In this article, the fractional-order differential equation of particle sedimentation was obtained. It considers the Basset force's fractional origin and contains the Riemann–Liouville fractional integral rewritten as a Grunwald–Letnikov derivative. As a result, the general solution of the proposed fractional-order differential equation was found analytically. The belonging of this solution to the real range of values was strictly theoretically proven. The obtained solution was validated on a particular analytical case study. In addition, it was proven numerically with the approach based on the S-approximation method using the block-pulse operational matrix. The proposed mathematical model can be applied for modeling the processes of fine particles sedimentation in liquids, aerosol deposition in gas flows, and particle deposition in gas-dispersed systems.

Keywords: particle sedimentation; resistance force; fractional-order integro-differential equation; laplace transform; Mittag–Leffler function; block-pulse operational matrix

Citation: Pavlenko, I.; Ochowiak, M.; Agarwal, P.; Olszewski, R.; Michałek, B.; Krupińska, A. Improvement of Mathematical Model for Sedimentation Process. *Energies* **2021**, *14*, 4561. https://doi.org/10.3390/en14154561

Academic Editor: Pouyan Talebizadeh Sardari

Received: 22 June 2021
Accepted: 23 July 2021
Published: 28 July 2021

1. Introduction

Based on the experimental and numerical simulation results, available experience indicates the importance of considering the Basset force in studying the processes of deposition and sedimentation of small particles moving close to rigid boundaries [1]. In this regard, differential formulation of the Basset force for its numerical calculation is an urgent problem in the field of computational mechanics [2]. Recently, the solution of this problem has been associated with the fractional calculus application [3].

The fractional-order differential equations allows us to generalize existing approaches in the fields of mechanical and chemical engineering, particularly for modeling hydromechanical processes, such as the sedimentation of particles in a viscous fluid, deposition of aerosols in separation channels, pneumatic classification of fine particles, nutrient release from mineral fertilizers and migration of mineral components in soil, and gas-cleaning.

However, despite the existing numerical approaches for solving fractional-order differential equations, their analytical solutions' approaches remain an incompletely studied problem. Therefore, this article aims to develop analytical techniques for solving the fractional-order differential equation of particle sedimentation considering the fractional origin of the Basset force.

To achieve this goal, the following objectives were set: substantiation of fractional origin for the Basset force, obtaining the fractional-order equation of the sedimentation of particles, solving the obtained equation analytically, validating the obtained general solution analytically, and approximating the numerically obtained case studies by obtained analytical dependencies.

The problems of particle deposition and sedimentation in a heterogeneous dispersed system have not been thoroughly investigated. The first works devoted to solving small particle motion equations in a viscous fluid and a nonuniform flow were presented by Leal [4], Maxey, and Riley [5], respectively. Mainly, weak inertia and non-Newtonian effects on the dynamics of rigid particles in an unbounded fluid were considered. However, the proposed analytical approaches were not considered the fractional origin of the Basset force. In addition, Loussaief et al. [6] investigated a spherical particle's motion in a viscous fluid and applied the Thomas algorithm to determine the characteristics of motion for a spherical particle along a slip wall. However, the proposed methodology does not allow one to obtain the general solution of the particle sedimentation equation analytically.

Coimbra and Kobayashi [7] studied the small particle motion in a viscous medium in the rotating cylinder considering the Saffman fractional-order lift force. However, this force acts on particles in vortex flows. Moreover, the authors only presented a comparison with the results obtained by considering the drag force as a dominant one. Oppenheimer et al. [8] studied the coupled thermal and hydromechanical particle motion problem in a viscous fluid at low Reynolds numbers. As a result, a general analytical expression to determine the force and its torque on a particle was derived based on the Lorentz reciprocal theorem. However, this presented analytical solution does not consider the fractional-order Basset force.

Moreno-Casas and Bombardelli [9] presented a general numerical approach for the calculation of the Basset force. A method for approximating the Basset force was proposed by van Hinsberg et al. [10]. However, such an approach does not allow one to estimate the sedimentation velocity analytically and, therefore, the sedimentation time accurately.

A number of recent research works have aimed at applying fractional calculus in the field of engineering. Particularly, Chung [11] described a general approach for solving fractional Newton's mechanics problems based on fractional-order differential equations. As a result, an approach to solve fractional-order differential equations of Newton dynamics approximately has been proposed based on infinite power series. However, such an approach cannot be applied to obtain the analytical solution of the fractional-order particle sedimentation equation rigorously.

Agila [12] proposed applying fractional Euler–Lagrange equations for solving the problem of free damped oscillations. He et al. [13] analyzed the dynamic response of viscosimeters based on fractional-order differential equations. As a result, the memory-free Yuan–Agrawal's approach for numerical integrating fractional-order differential equations was developed and proven for the particular case studies. The presented approach discovered perspectives in measuring the viscosity of fluids.

Tomovski and Sandev [14] proposed an analytical treatment of the wave equation considering fractional friction and obtained the solution based on the Mittag–Leffler-type functions. Rossikhin and Shitikova [15] studied the dynamic behavior of nonlinear oscillatory systems described by fractional-order time derivatives. The proposed approach allowed them to obtain approximations of particular oscillatory modes for the fixed equilibrium position case study.

Therefore, the following research gaps in particle deposition and sedimentation in a heterogeneous dispersed system should be stated due to the critical review mentioned above. First, the fractional-order particles sedimentation equation considering the Basset force's fractional origin should be substantiated and solved analytically. Second, the general solution of the proposed equation should be found numerically. Finally, the obtained solution should be validated analytically for the available case study and proven numerically.

2. Materials and Methods

2.1. The Particle Motion Equation Considering the Basset Force

According to the equation of motion for a small rigid sphere in a nonuniform flow [16], the particle motion equation moving with a time-varying velocity in projection to the positive tangential direction has the following form:

$$m_p \frac{dv_p(t)}{dt} = F_g - F_A - F_d - F_f - F_B, \quad (1)$$

where m_p—mass of a particle (kg), v_p—particle velocity (m/s), t—time (s), $F_g = m_p g$—gravity force (N), g—acceleration gravity (m/s), $F_A = m_f g$–Archimedes' force, (N), m_f—added mass of a flow (kg), $F_d = 6\pi\mu a v_p(t)$—drag force (N), μ—dynamic viscosity (Pa·s), a—radius of a particle (m), $F_f = \frac{1}{2} m_f \frac{dv_p(t)}{dt}$—force of the flow added mass (N), and F_B—Basset force, determined as follows [17]:

$$F_B = 6\pi\mu a^2 \int_0^t \frac{\frac{dv_p(\tau)}{d\tau}}{\sqrt{\pi\nu(t-\tau)}} d\tau, \quad (2)$$

where ν—kinematic viscosity (m^2/s) and τ—time parameter (s).

The expressions mentioned above allow us to rewrite Equation (1) as follows:

$$\left(m_p + \frac{1}{2} m_f\right) \frac{dv_p(t)}{dt} + 6\sqrt{\pi\rho\mu} a^2 \int_0^t \frac{\frac{dv_p(\tau)}{d\tau}}{\sqrt{t-\tau}} d\tau + 6\pi\mu a v_p(t) = \left(m_p - m_f\right) g, \quad (3)$$

where ρ—fluid density (kg/m^3).

The introduction of the parameters

$$\alpha = \frac{3\pi\sqrt{\rho\mu} a^2}{m_p + \frac{1}{2} m_f}; \; n = \frac{6\pi\mu a}{m_p + \frac{1}{2} m_f}; \; v_{p\infty} = \frac{\left(m_p - m_f\right) g}{6\pi\mu a} \quad (4)$$

allows us to rewrite integro-differential Equation (3) in the following form:

$$\frac{dv_p(t)}{dt} + \frac{2\alpha}{\sqrt{\pi}} \int_0^t \frac{\frac{dv_p(\tau)}{d\tau}}{\sqrt{t-\tau}} d\tau + n v_p(t) = n v_{p\infty}, \quad (5)$$

where n—relaxation factor (s^{-1}), α—coefficient of the Basset force ($s^{-1/2}$), and $v_{p\infty} = \lim_{t\to\infty} v(t)$—stationary velocity (m/s).

2.2. The Fractional-Order Differential Equation of Particles Sedimentation

For further analytically solving Equation (5), the Riemann–Liouville integral and derivative [18] and Grünwald–Letnikov fractional derivative are considered [19]:

$$\begin{gathered} I_{t_0,t}^{\beta} v_p(t) = \frac{1}{\Gamma(\beta)} \int_{t_0}^t v_p(\tau)(t-\tau)^{\beta-1} d\tau; \\ D_{t_0,t}^{\beta} v_p(t) = \frac{1}{\Gamma([\beta]-\beta)} \frac{d^{[\beta]}}{dt^{[\beta]}} \int_{t_0}^t v_p(\tau)(t-\tau)^{\beta-[\beta]+1} d\tau; \\ D_{t_0,t}^{\beta} v_p(t) = \lim_{N\to\infty} \left(\frac{t-t_0}{N}\right)^{\beta} \sum_{i=0}^{N} \frac{\Gamma(\alpha+i)}{i!\Gamma(\alpha)} v_p\left[t - \frac{i}{N}(t-t_0)\right] = I_{t_0,t}^{-\beta} v_p(t), \end{gathered} \quad (6)$$

where β—fractional-order, t_0—initial time (s), and $\Gamma(\beta)$—gamma function [20]:

$$\Gamma(\beta) = \int_0^{\infty} e^{-x} x^{\beta-1} dx, \; Re(\beta) > 0 \quad (7)$$

Considering the connection between fractional derivatives [21]:

$$D^{\gamma}_{t_0,t}\left[\frac{d^r}{dt^r}v_p(t)\right] = \frac{d^r}{dt^r}\left[D^{\gamma}_{t_0,t}v_p(t)\right] - \frac{(t-t_0)^{-\gamma-r+i}}{\Gamma(-\gamma-r+i+1)}\left[\frac{d^i}{dt^i}v_p(t)\right]_{t=t_0}, \tag{8}$$

in the case of $\gamma = \frac{1}{2}$, $r = 1$ for initial time $t_0 = 0$, and considering $\Gamma\left(\frac{1}{2}\right) = \sqrt{\pi}$, the following equation can be written:

$$I^{\frac{1}{2}}_{0,t}\left[\frac{dv_p(t)}{dt}\right] = D^{\frac{1}{2}}_{0,t}\left[v_p(t)\right] - \frac{v_p(0)}{\sqrt{\pi t}}. \tag{9}$$

Comparing Equations (2) and (9) with the Riemann–Liouville integro-differential (6) for the value of $\beta = \frac{1}{2}$ and initial time $t_0 = 0$, as well as for the case of zero initial condition ($v_p(0) = 0$), allows us to determine the Basset force in terms of fractional calculus:

$$F_B = 6\pi a^2\sqrt{\mu\rho}I^{\frac{1}{2}}_{0,t}\left[\frac{dv_p(t)}{dt}\right] = 6\pi a^2\sqrt{\mu\rho}D^{\frac{1}{2}}_{0,t}\left[v_p(t)\right]. \tag{10}$$

Therefore, Equation (5) takes the following form:

$$\frac{dv_p(t)}{dt} + 2\alpha\frac{d^{\frac{1}{2}}v_p(t)}{dt^{\frac{1}{2}}} + nv_p(t) = nv_{p\infty}. \tag{11}$$

Notably, this formula is the fractional-order differential equation of particle sedimentation that considers the Basset force. This equation can be applied for more precise modeling of the processes of fine particles sedimentation in liquids, aerosol deposition in a gas flow, and particle deposition in gas-dispersed systems.

3. Results

3.1. The General Solution of the Fractional-Order Differential Equation of Particles Sedimentation

For solving Equation (11) analytically, the Laplace transform [22] is used for zero initial condition ($v_p(0) = 0$). In this case, the following equational operation can be obtained:

$$(s + 2\alpha\sqrt{s} + n)V_p(s) = \frac{nv_{p\infty}}{s}, \tag{12}$$

where s—complex frequency parameter (s^{-1}) and $V_p(s)$—Laplace transform of the particle velocity (m):

$$V_p(s) = \int_0^{\infty} v_p(t)e^{-st}dt, \tag{13}$$

which can be determined from the algebraic Equation (12):

$$V_p(s) = \frac{nv_{p\infty}}{s(s+2\alpha\sqrt{s}+n)} = nv_{p\infty}\left(\frac{a_1}{\sqrt{s}} + \frac{a_2}{s} + \frac{a_3}{\sqrt{s}-\theta_1} + \frac{a_4}{\sqrt{s}-\theta_2}\right), \tag{14}$$

where $\theta_{1,2} = -\alpha \pm \sqrt{\alpha^2 - n}$—a couple of roots of the square equation $\theta^2 + 2\alpha\theta + n = 0$.

Notably, $\theta = \sqrt{s}$. The unknown parameters a_1, a_2, a_3, and a_4 are obtained from the following matrix equation:

$$\begin{bmatrix} 1 & 0 & 1 & 1 \\ -(\theta_1+\theta_2) & 1 & -\theta_2 & -\theta_1 \\ \theta_1\theta_2 & -(\theta_1+\theta_2) & 0 & 0 \\ 0 & \theta_1\theta_2 & 0 & 0 \end{bmatrix}\begin{Bmatrix} a_1 \\ a_2 \\ a_3 \\ a_4 \end{Bmatrix} = \begin{Bmatrix} 0 \\ 0 \\ 0 \\ 1 \end{Bmatrix}. \tag{15}$$

Based on the inverse matrix method, the following dependencies can be obtained after identical transformation:

$$a_1 = -\frac{2\alpha}{n^2};\ a_2 = \frac{1}{n};\ a_{3,4} = \pm\frac{1}{2\left(\alpha \mp \sqrt{\alpha^2 - n}\right)}. \tag{16}$$

Introduction of the parameters $\theta = -\theta_1$ and $\theta' = -\theta_2$, after considering the equality $L^{-1}\left(\frac{1}{\sqrt{s}}\right) = \frac{1}{\sqrt{\pi t}}$ allows us to apply inverse Laplace transform [23] to Equation (14):

$$v_p(t) = n v_{p\infty}\left[\frac{a_1}{\sqrt{\pi t}} + a_2 H(t) + a_3 L^{-1}\left(\frac{1}{\sqrt{s}+\theta}\right) + a_4 L^{-1}\left(\frac{1}{\sqrt{s}+\theta'}\right)\right], \tag{17}$$

where *H(t)*—Heaviside step function.

Notably, the absence of singularity for this solution is proven below. For further consideration, Mittag–Leffler function [24] is used:

$$E_{\gamma,\beta}(z) = \sum_{i=0}^{\infty}\frac{z^i}{\Gamma(\gamma i + \beta)}. \tag{18}$$

It has the following peculiarity in terms of its Laplace transform [25]:

$$L\left[t^{\beta-1}E_{\gamma,\beta}(\theta t^{\gamma})\right] = \frac{s^{\gamma-\beta}}{s^{\gamma}-\theta}. \tag{19}$$

Consequently, the following inverse Laplace transform for the case of $\gamma = \frac{1}{2}$ and $\beta = 1$ can be written as follows:

$$L\left[E_{\frac{1}{2},1}\left(\theta\sqrt{t}\right)\right] = \frac{1}{\sqrt{s}\left(\sqrt{s}-\theta\right)} = \frac{1}{\theta}\left(\frac{1}{\sqrt{s}-\theta} - \frac{1}{\sqrt{s}}\right). \tag{20}$$

Therefore, the following inverse Laplace transform can be obtained:

$$L^{-1}\left(\frac{1}{\sqrt{s}-\theta}\right) = \frac{1}{\sqrt{\pi t}} + \theta E_{\frac{1}{2},1}\left(\theta\sqrt{t}\right). \tag{21}$$

In addition, the decomposition

$$\frac{1}{s-\theta^2} = \frac{1}{2\theta}\left(\frac{1}{\sqrt{s}-\theta} - \frac{1}{\sqrt{s}+\theta}\right) \tag{22}$$

with equality $L^{-1}\left(\frac{1}{s-\theta^2}\right) = e^{\theta^2 t}$ with Equation (21) allows us to obtain the following inverse Laplace transform:

$$L^{-1}\left(\frac{1}{\sqrt{s}+\theta}\right) = \frac{1}{\sqrt{\pi t}} + \theta E_{\frac{1}{2},1}\left(\theta\sqrt{t}\right) - 2\theta e^{\theta^2 t}. \tag{23}$$

Therefore, Equation (16) can be rewritten as follows:

$$v_p(t) = n v_{p\infty}\left[\frac{a_1 + a_3 + a_4}{\sqrt{\pi t}} + a_2 H(t) + a_3\theta E_{\frac{1}{2},1}\left(\theta\sqrt{t}\right) + a_4\theta' E_{\frac{1}{2},1}\left(\theta'\sqrt{t}\right) - 2\left(a_3\theta e^{\theta^2 t} + a_4\theta' e^{\theta'^2 t}\right)\right], \tag{24}$$

Notably, $a_1 + a_3 + a_4 \equiv 0$ due to Equation (16), and $H(t) = 1$ for $t > 0$. Therefore, Equation (24) can be rewritten in the following form:

$$v_p(t) = v_{p\infty}\left\{1 - n\left[2\left(a_3\theta e^{\theta^2 t} + a_4\theta' e^{\theta'^2 t}\right) - a_3\theta E_{\frac{1}{2},1}\left(\theta\sqrt{t}\right) - a_4\theta' E_{\frac{1}{2},1}\left(\theta'\sqrt{t}\right)\right]\right\}. \tag{25}$$

Remarkably, in practical applications, the inequality $\alpha^2 < n$, the obtained solution, seems complex because of the complexity of parameters $\theta_{1,2}$. Consequently, the belonging of this solution to the real range of values must be proven strictly theoretically. For this purpose, the complex parameters $\theta_{1,2} = -\alpha \pm j\sqrt{n-\alpha^2}$ are considered (j–imaginary unit: $j^2 = -1$). The substitution of these parameters to Equation (25), as well as the use of Newton's binomial theorem [26] and properties of hyperbolic functions [27], allows us to obtain the following expressions:

$$\begin{aligned} a_3\theta e^{\theta^2 t} + a_4\theta' e^{\theta'^2 t} &= \tfrac{1}{n}\sum_{i=0}^{\infty}\tfrac{t^{\frac{i}{2}}}{\Gamma\left(\frac{i}{2}+1\right)}\sum_{r=0}^{i}\begin{pmatrix} i \\ r \end{pmatrix}\alpha^{i-r}\left(n-\alpha^2\right)^{\frac{r}{2}}\left(\cos\tfrac{\pi i}{2} - \tfrac{\alpha}{\sqrt{n-\alpha^2}}\sin\tfrac{\pi i}{2}\right); \\ a_3\theta E_{\frac{1}{2},1}\left(\theta\sqrt{t}\right) + a_4\theta' E_{\frac{1}{2},1}\left(\theta'\sqrt{t}\right) &= \tfrac{1}{n}e^{-(n-2\alpha^2)t}\left(\cos 2\alpha\sqrt{n-\alpha^2}t - \tfrac{\alpha}{\sqrt{n-\alpha^2}}\sin 2\alpha\sqrt{n-\alpha^2}t\right). \end{aligned} \tag{26}$$

Therefore, the Velocity (25) belongs to the real range of values.

Finally, the general Equation (25) of the fractional-order differential Equation (11) of particles sedimentation that considers the Basset force takes the following form:

$$\begin{aligned} v_p(t) = v_{p\infty}\Big[&1 - 2e^{-(n-2\alpha^2)t}\left(\cos 2\alpha\sqrt{n-\alpha^2}t - \tfrac{\alpha}{\sqrt{n-\alpha^2}}\sin 2\alpha\sqrt{n-\alpha^2}t\right) + \\ &+ \sum_{i=0}^{\infty}\tfrac{t^{\frac{i}{2}}}{\Gamma\left(\frac{i}{2}+1\right)}\sum_{r=0}^{i}\begin{pmatrix} i \\ r \end{pmatrix}\alpha^{i-r}\left(n-\alpha^2\right)^{\frac{r}{2}}\left(\cos\tfrac{\pi i}{2} - \tfrac{\alpha}{\sqrt{n-\alpha^2}}\sin\tfrac{\pi i}{2}\right)\Big]. \end{aligned} \tag{27}$$

3.2. The Particular Case Study

For validating the obtained general Equation (27) of the fractional-order differential Equation (11), the particular case study for $\alpha = 0$ is considered. In this case, the solution takes the following form:

$$v_{p0}(t) = v_{p\infty}\left[1 - 2e^{-nt} + E'_{\frac{1}{2},1}\left(\sqrt{nt}\right)\right], \tag{28}$$

where the following modified Mittag–Leffler function is introduced:

$$E^{\langle c\rangle}_{\gamma,\beta}(z) = \sum_{i=0}^{\infty}\frac{z^i}{\Gamma(\gamma i+\beta)}\cos\frac{\pi i}{2}, \tag{29}$$

which differs from the traditional one (18) by the multiplier $\cos\frac{\pi i}{2}$.

Remarkably, the modified Mittag–Leffler function has the following peculiarity:

$$E^{\langle c\rangle}_{\frac{1}{2},1}(z) \equiv e^{-z^2}. \tag{30}$$

In this case, a particular Equation (28) takes the form

$$v_{p0}(t) = v_{p\infty}\left(1 - e^{-nt}\right), \tag{31}$$

which corresponds to the traditional one [28].

Thus, the general Equation (27) discovers new areas in studying the process of particle sedimentation in a fluid flow.

3.3. Analysis of Leading Orders of the General Solution

To analyze the leading orders of the general Equation (27), the dimensionless velocity $\overline{v}_p(\tau) = \frac{v_p(\tau)}{v_{p\infty}}$, the dimensionless time $\tau = nt$, and the dimensionless coefficient of the Basset force $\varepsilon = \frac{\alpha}{\sqrt{n}}$ are introduced. Accordingly, Equation (27) takes the following dimensionless form:

$$\begin{aligned} \overline{v}_p(\tau) = &1 - 2e^{-\tau}e^{2\varepsilon^2\tau}\left(\cos 2\varepsilon\sqrt{1-\varepsilon^2}\tau - \tfrac{\varepsilon}{\sqrt{1-\varepsilon^2}}\sin 2\varepsilon\sqrt{1-\varepsilon^2}\tau\right) + \\ &+ \sum_{i=0}^{\infty}\tfrac{\tau^{\frac{i}{2}}}{\Gamma\left(\frac{i}{2}+1\right)}\sum_{r=0}^{i}\begin{pmatrix} i \\ r \end{pmatrix}\varepsilon^{i-r}\left(1-\varepsilon^2\right)^{\frac{r}{2}}\left(\cos\tfrac{\pi i}{2} - \tfrac{\varepsilon}{\sqrt{1-\varepsilon^2}}\sin\tfrac{\pi i}{2}\right). \end{aligned} \tag{32}$$

For the first approximation, in case of relatively small values of ε << 1, after expansion into a Maclaurin series leaving the first-order terms ε only, Equation (27) considering identity Equation (30) and dependence Equation (31) can be rewritten approximately as follows:

$$\overline{v}_p(\tau) = 1 - e^{-\tau} - \varepsilon E^{\langle s\rangle}_{\frac{1}{2},1}(\sqrt{\tau}) = \overline{v}_{p0}(\tau) - \delta\overline{v}_p(\tau), \tag{33}$$

where the first-order variation has been introduced:

$$\delta\overline{v}_p(\tau) = \varepsilon E^{\langle s\rangle}_{\frac{1}{2},1}(\sqrt{\tau}). \tag{34}$$

It contains the modified Mittag–Leffler function

$$E^{\langle s\rangle}_{\gamma,\beta}(z) = \sum_{i=0}^{\infty} \frac{z^i}{\Gamma(\gamma i + \beta)} \sin\frac{\pi i}{2} \tag{35}$$

for values $\gamma = \frac{1}{2}$, $\beta = 1$ and argument $z = \sqrt{\tau}$.

Due to the abovementioned results, the leading terms of the Equation (27) are the first terms in the double sum for the case of $i = r = 0$, where $\varepsilon^{i-r} = \varepsilon^0 = 1$:

1. Multiplier before $\cos\left(2\alpha\sqrt{n-\alpha^2}t\right) = \cos\left(2\varepsilon\sqrt{1-\varepsilon^2}\tau\right)$, when the use of identity Equation (30) allows obtaining the leading component Equation (31). This component is responsible for the sedimentation velocity without considering the Basset force.
2. Multiplier before $\sin\left(2\alpha\sqrt{n-\alpha^2}t\right) = \sin\left(2\varepsilon\sqrt{1-\varepsilon^2}\tau\right)$. It allows us to obtain the velocity variation (34) proportional to the dimensionless parameter ε. This component allows us to reduce the general Equation (27) to the most simplified variation form (33) with respect to dimensionless parameters ε and τ.

Notably, due to Equation (33), an increase in the Basset force (the dimensionless parameter ε) leads to decreased sedimentation velocity.

For practical purposes, Simplification (33) can be applied for values of the dimensionless coefficient $\varepsilon \leq 0.23$ because the maximum relative deviation of the sedimentation velocity from the accurate analytical Solution (27) is less than 10%. Therefore, if the dimensionless coefficient ε is more than this critical value, Equation (27) or (32) should be given preference.

Figure 1 presents the dimensionless solution Equation (32) of the particle sedimentation Equation (11) for different dimensionless parameters ε in a wide range (from 0 to 1).

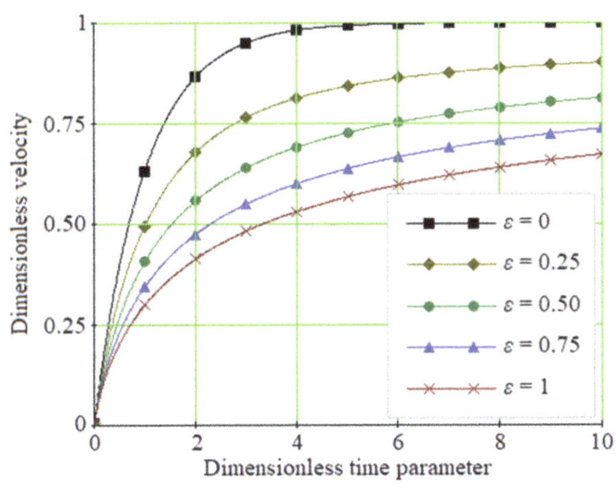

Figure 1. Comparative analysis of the dimensionless solutions for the fractional-order integro-differential equations of particle sedimentation.

Notably, all the solutions asymptotically approach the horizontal lines of the maximum sedimentation velocities.

3.4. Comparison with the Numerical Simulation Results

The next step for proving the obtained general solution Equation (27) is its comparison with the numerical simulation results.

The direct integration of the fractional-order differential Equation (11) can be realized with approach based on the S-approximation method [29] using the block-pulse operational matrix ($S(t)$) [30], the elements of which are as follows:

$$S_i(t) = \frac{1}{2}\{sign(t - i\Delta t) - sign[t - (i+1)\Delta t]\}, \tag{36}$$

where i—plot index and $\Delta t = t_{max}/N$—timestep (t_{max}—the maximum time range; N—number of plots for numerical integration).

In this case, the initial fractional-order differential Equation (11) is transformed to the following fractional-order integral equation:

$$v_p(t) + 2\alpha I_{0,t}^{\frac{1}{2}}[v_p(t)] + nI_{0,t}^{1}[v_p(t)] = nv_{p\infty}t, \tag{37}$$

which satisfies the zero initial condition, as $v_p(0) = 0$.

The numerical solution of this integral equation is based on the following operational analog [31]:

$$\left([E] + 2\alpha\left[P^{\langle 0.5\rangle}\right] + n\left[P^{\langle 1\rangle}\right]\right)\{Y\} = \{F\}, \tag{38}$$

where $\{Y\}$—operational column-vector of the solution, $[E]$—the identity matrix with dimensions of $N \times N$, and $[P^{<0.5>}]$, $[P^{<1>}]$—elements of the operational matrix $[P^{<\beta>}]$ ($\beta = 0.5$, and $\beta = 1$, respectively), the elements of which are determined as follows:

$$P_{i,j}^{\langle\beta\rangle} = p(\beta,\, i-j), \tag{39}$$

where $p(\beta, r)$—the following function ($|r| = 0, 1, \ldots, N\text{–}1$):

$$P_{i,j}^{\langle\beta\rangle} = \frac{\Delta t^{\beta}}{\Gamma(\beta+2)} \cdot \begin{cases} 1, \ i = j; \\ (r+1)^{\beta+1} - 2r^{\beta+1} + (r-1)^{\beta+1}, \ i > j \ , \\ 0, \ i < j. \end{cases} \tag{40}$$

Elements F_i ($i = 1, 2, \ldots, N$) of the operational column-vector of external impact $\{F\}$ are determined as follows:

$$F_i = \frac{1}{h}\int_{i\Delta t}^{(i+1)\Delta t} nv_{p\infty}t dt = \left(i + \frac{1}{2}\right)\Delta t \cdot nv_{p\infty}. \tag{41}$$

Using the inverse matrix method [32], the operational column-vector of the solution

$$\{Y\} = \left([E] + 2\alpha\left[P^{\langle 0.5\rangle}\right] + n\left[P^{\langle 1\rangle}\right]\right)^{-1}\{F\} \tag{42}$$

allows us to obtain the following numerical solution of the initial Equation (33):

$$v_p(t) = \sum_{i=0}^{N-1} Y_i S_i(t). \tag{43}$$

Figure 1 also presents the numerically obtained results of solving the particle sedimentation Equation (37). Approximation lines for all the data completely coincide with the analytically obtained function Equation (32). This fact additionally proves the analytical approach's adequacy for solving the fractional-order differential equation of particle sedimentation.

Remarkably, the Basset force significantly increases the dimensionless sedimentation time determined from the condition $v_p = (1-\delta)v_{p\infty}$, where δ—deviation of the sedimentation velocity from the stationary one $v_{p\infty}$, usually chosen as equal to $\delta = 0.05$.

Particularly, if the Basset force is not considered ($\varepsilon = 0$), then the dimensionless sedimentation time is equal to 3.0 (Figure 1). However, even considering the parameter $\varepsilon = 0.05$, this dimensionless time parameter increases by 43% and becomes equal to 4.3.

The fact mentioned above explains the use of significant correction factors in traditional calculations of apparatuses for the sedimentation and deposition of the particles. However, using the fractional-order differential equation of particle sedimentation avoids unjustified correction factors in modeling processes and designing related chemical technology equipment.

4. Discussion

The fractional-order differential Equation (11) and its analytical solution Equation (27) eliminate the gap in studying the particle sedimentations and deposition processes.

Notably, Wan [33] obtained the fractional-order differential equation's analytical solution for the free-falling process. However, the considered hydraulic drag as a linear combination of the velocity v_p and its square v_p^2 was extended to a more general case of velocity analog based on the fractional derivative of particle displacement with the order in a range from 1 to 2. However, this approach does not consider the Basset force at all. Moreover, consideration of this force makes it possible to use the fractional derivative of particle displacement with the order of 3/2 only.

Visitskii et al. [34] investigated spherical particles' sedimentation in a viscous fluid considering the Basset force. However, the solution was obtained numerically for particular case studies. In addition, numerical simulation results can be obtained using the operational matrix based on the Legendre polynomials proposed by Saadatmandi and Dehghan [35]. However, the resulting curves presented in Figure 1 remain unchanged in terms of their approximation by the obtained analytical solution Equation (27).

Sobral et al. [36] applied the "Maple" software for obtaining the particular solution of the sedimentation equation. However, this solution is quite complicated for the analysis due flaws, including complex variables that need to reduce hyperbolic functions by trigonometric ones with imaginary coefficients. Moreover, in the case of zero Basset force, the corresponding solution should be reduced to the solution of the relaxation equation.

All these disadvantages were eliminated using the Mittag–Leffler function and its particular properties. In addition, the validity of the proposed solution for the imaginary arguments was justified. Finally, the singularity of the fractional-order differ-integral sedimentation equation was proven analytically.

Notably, the proposed approach can be applied not only for particle sedimentation and deposition problems but also for solving nutrient release problems from mineral fertilizers and the pneumatic classification of fine particles. Particularly, Equation (11) and its general solution Equation (27) allow us to clarify the concentration distribution for nutrients during their washing out from the fertilized soil [37]. The migration process of mineral components in the soil [38] can also be studied more precisely using the proposed fractional-calculus approach. The proposed methodology also extends the understanding of the distribution of fine particles' concentration and the height of the rhomb-shaped pneumatic classificators [39].

The fractional-order sedimentation equation can be further expanded to solve the problem of dispersed particle's deposition in separation channels [40] and for further precise modeling of the gas emission process [41] and designing gas-cleaning equipment [42].

5. Conclusions

In this article, the fractional origin for the Basset force was substantiated. As a result of modifying the particle motion equation moving with a time-varying velocity, the fractional-order equation of particles' sedimentation was obtained. This equation differs from the traditional one by the semi-derivative of the particle velocity.

As a result of the integration of the proposed equation, its general solution was found analytically. This solution varies from the traditional one by the component based on the modified Mittag–Leffler function. Notably, for the particular case of zero Basset force, this equation reduces to the traditional one. Moreover, the belonging of this solution to the real range of values was theoretically proven.

In addition, the particle sedimentation's fractional-order differential equation was solved numerically using the analogous fractional-order integral equation. The numerical simulation was realized based on the S-approximation method using the block-pulse operational matrix. For normalizing the obtained solutions, the dimensionless particle velocity was introduced.

As a result, approximations of the numerically obtained case studies were presented graphically for a single variable dimensionless system parameter ranging from 0 to 1. The obtained analytical dependencies approximated the corresponding block-pulse curves. The adequacy of the proposed analytical approach for solving the fractional-order differential equation of particle sedimentation was proven analytically and numerically.

Notably, the Basset force significantly increases the sedimentation time. For example, if the Basset force is not considered, then the dimensional sedimentation time is equal to 3.0. However, when considering the Basset force with the dimensionless parameter equal to 0.05, this time increases by 43%.

Finally, the proposed mathematical model can be applied for modeling the processes of particle sedimentation in liquids, aerosol deposition in gas flows, and particle deposition in gas-dispersed systems as essential processes in the fields of mechanical and chemical engineering.

Author Contributions: Conceptualization and methodology—I.P.; software and validation—I.P., M.O., A.K. and R.O.; formal analysis—P.A.; investigation and resources—I.P., R.O. and B.M.; data curation—I.P. and M.O.; writing—original draft preparation—I.P.; writing—review and editing—M.O.; visualization—I.P. and A.K.; supervision—I.P. and M.O.; project administration—I.P. and P.A.; funding acquisition—M.O. All authors have read and agreed to the published version of the manuscript.

Funding: This research was funded by the Ministry of Education and Science of Poland: SBAD.

Institutional Review Board Statement: Not applicable.

Informed Consent Statement: Not applicable.

Data Availability Statement: The data presented in this study are available on request from the corresponding author.

Acknowledgments: The authors acknowledge the research project of Sumy State University (state reg. no. 0120U102036) ordered by the Ministry of Education and Science of Ukraine and the International Association for Technological Development and Innovations, as well as Poland Ministry of Education and Science. The research results allow fulfilling the objectives of the perspective development plan of Sumy State University within the scientific direction "Technical Sciences".

Conflicts of Interest: The authors declare no conflict of interest.

References

1. Bombardelli, F.A.; González, A.E.; Niño, Y.I. Computation of the particle Basset force with a fractional-derivative approach. *J. Hydraul. Eng.* **2008**, *134*, 1513. [CrossRef]
2. Parmar, M.; Annamalai, S.; Balachandar, S.; Prosperetti, A. Differential formulation of the viscous history force on a particle for efficient and accurate computation. *J. Fluid Mech.* **2018**, *844*, 970–993. [CrossRef]

3. Rossikhin, Y.A.; Shitikova, M.V. Application of fractional calculus for dynamic problems of solid mechanics: Novel trends and recent results. *Appl. Mech. Rev.* **2009**, *63*, 010801. [CrossRef]
4. Leal, L.G. Particle motions in a viscous fluid. *Annu. Rev. Fluid Mech.* **1980**, *12*, 435–476. [CrossRef]
5. Maxey, M.R.; Riley, J.J. Equation of motion for a small rigid sphere in a nonuniform flow. *Phys. Fluids* **1983**, *26*, 883–889. [CrossRef]
6. Loussaief, H.; Pasol, L.; Feuillebois, F. Motion of a spherical particle in a viscous fluid along a slip wall. *Q. J. Mech. Appl. Math.* **2015**, *68*, 115–144. [CrossRef]
7. Coimbra, C.; Kobayashi, M. On the viscous motion of a small particle in a rotating cylinder. *J. Fluid Mech.* **2002**, *469*, 257–286. [CrossRef]
8. Oppenheimer, N.; Navardi, S.; Stone, H.A. Motion of a hot particle in viscous fluids. *Phys. Rev. Fluids* **2016**, *1*, 014001. [CrossRef]
9. Moreno-Casas, P.A.; Bombardelli, F.A. Computation of the Basset force: Recent advances and environmental flow applications. *Environ. Fluid Mech.* **2016**, *16*, 193–208. [CrossRef]
10. Van Hinsberg, M.A.T.; Ten Thije Boonkkamp, J.H.M.; Clercx, H.J.H. An efficient, second order method for the approximation of the Basset history force. *J. Comput. Phys.* **2011**, *230*, 1465–1478. [CrossRef]
11. Chung, W.S. Fractional Newton mechanics with conformable fractional derivative. *J. Comput. Appl. Math.* **2015**, *290*, 150–158. [CrossRef]
12. Agila, A.; Baleanu, D.; Eid, R.; Irfanoglu, B. A freely damped oscillating fractional dynamic system modeled by fractional Euler–Lagrange equations. *J. Vib. Control* **2018**, *24*, 1228–1238. [CrossRef]
13. He, X.; Liu, Q.; Huang, X.; Chen, Y. Dynamic response analysis of the fractional-order system of MEMS viscometer. *CMES-Comput. Model. Eng. Sci.* **2015**, *108*, 159–169. [CrossRef]
14. Tomovski, Ž.; Sandev, T. Effects of a fractional friction with power-law memory kernel on string vibrations. *Comput. Math. Appl.* **2011**, *62*, 1554–1561. [CrossRef]
15. Rossikhin, Y.A.; Shitikova, M.V. New approach for the analysis of damped vibrations of fractional oscillators. *Shock Vib.* **2009**, *16*, 387676. [CrossRef]
16. Hu, S.; Nadim, A. Three models for rectilinear particle motion with the basset history force. *Electron. J. Differ. Equ.* **2015**, *2015*, 1–19.
17. Pavlenko, I.; Liaposhchenko, O.; Pitel, J.; Sklabinskyi, V. Parameter identification of the Basset force acting on particles in fluid flow induced by the oscillating wall. *J. Appl. Math. Comput. Mech.* **2019**, *18*, 53–63. [CrossRef]
18. Ahmad, B.; Alghanmi, M.; Alsaedi, A. Existence results for a nonlinear coupled system involving both Caputo and Riemann-Liouville generalized fractional derivatives and coupled integral boundary conditions. *Rocky Mt. J. Math.* **2021**, *50*, 1901–1922. [CrossRef]
19. Brzeziński, D.W. Fractional order derivative and integral computation with a small number of discrete input values using Grünwald-Letnikov formula. *Int. J. Comput. Methods* **2020**, *17*, 1940006. [CrossRef]
20. Qi, F.; Guo, B.-N. From inequalities involving exponential functions and sums to logarithmically complete monotonicity of ratios of gamma functions. *J. Math. Anal. Appl.* **2021**, *493*, 124478. [CrossRef]
21. Azennar, R.; Mentagui, D. On some properties of the conformable fractional derivative. *Moroc. J. Pure Appl. Anal.* **2020**, *6*, 210–217. [CrossRef]
22. Liu, L.; Xue, D.; Zhang, S. Closed-loop time response analysis of irrational fractional-order systems with numerical Laplace transform technique. *Appl. Math. Comput.* **2019**, *350*, 133–152. [CrossRef]
23. Yüce, A.; Tan, N. Inverse Laplace transforms of the fractional order transfer functions. In Proceedings of the 2019 11th International Conference on Electrical and Electronics Engineering (ELECO), Bursa, Turkey, 28–30 November 2019; IEEE: Piscataway, NJ, USA, 2019; pp. 775–779. [CrossRef]
24. Du, F.; Jia, B. Finite time stability of fractional delay difference systems: A discrete delayed Mittag-Leffler matrix function approach. *Chaos Solitons Fractals* **2020**, *141*, 110430. [CrossRef]
25. Jumarie, G. Laplace's transform of fractional order via the Mittag-Leffler function and modified Riemann-Liouville derivative. *Appl. Math. Lett.* **2009**, *22*, 1659–1664. [CrossRef]
26. Abbas, S. Proving the extended binomial theorem using ordinary differential equations. *Math. Mag.* **2020**, *93*, 33–35. [CrossRef]
27. Tokibetov, J.; Abduakhitova, G.; Assadi, A. Representations of the solutions of the first-order elliptic and hyperbolic systems via harmonic and wave functions respectively. *Complex Var. Elliptic Equ.* **2020**, *65*, 1565–1574. [CrossRef]
28. Lytvynenko, A.; Yukhymenko, M.; Pavlenko, I.; Pitel, J.; Mizakova, J.; Lytvynenko, O.; Ostroha, R.; Bocko, J. Ensuring the reliability of pneumatic classification process for granular material in a rhomb-shaped apparatus. *Appl. Sci.* **2019**, *9*, 1604. [CrossRef]
29. Vasylyev, V.V.; Simak, L.O.; Zelenkov, O.A.; Voronova, O.S.; Rybnikova, G.M. *Analysis and Mathematical Modeling of Dynamical Systems Based on Non-Classical Operational Calculus*; National Aviation Institute: Kyiv, Ukraine, 2006.
30. Li, Y.; Sun, N. Numerical solution of fractional differential equations using the generalized block pulse operational matrix. *Comput. Math. Appl.* **2011**, *62*, 1046–1054. [CrossRef]
31. Vasilyev, V.V.; Simak, L.A. *Fractional Calculus and Approximating Approaches in Modeling Dynamic Systems*; National Academy of Sciences of Ukraine: Kyiv, Ukraine, 2008.
32. Khan, A.; Jahanzaib, L.S.; Trikha, P. Fractional inverse matrix projective combination synchronization with application in secure communication. In Proceedings of the International Conference on Artificial Intelligence and Applications, Maharaja Surajmal

Institute of Technology, New Delhi, India, 6–7 February 2020; Bansal, P., Tushir, M., Balas, V., Srivastava, R., Eds.; Advances in Intelligent Systems and Computing; Springer: Singapore, 2021; Volume 1164, pp. 93–101. [CrossRef]

33. Wan, Y.; French, R.M. Fractional calculus of hydraulic drag in the free falling process. In *Conference Proceedings of the Society for Experimental Mechanics Series*; Proulx, T., Ed.; Experimental and Applied Mechanics; Springer: New York, NY, USA, 2011; Volume 6, pp. 529–539. [CrossRef]
34. Visitskii, Y.V.; Petrov, A.G.; Shunderyuk, M.M. The motion of a particle in a viscous fluid under gravity, vibration and Basset's force. *J. Appl. Math. Mech.* **2009**, *73*, 548–557. [CrossRef]
35. Saadatmandia, A.; Dehghanb, M. A new operational matrix for solving fractional-order differential equations. *Comput. Math. Appl.* **2010**, *59*, 1326–1336. [CrossRef]
36. Sobral, Y.D.; Oliveira, T.F.; Cunha, F.R. On the unsteady forces during the motion of a sedimenting particle. *Powder Technol.* **2007**, *178*, 129–141. [CrossRef]
37. Vakal, V.; Pavlenko, I.; Vakal, S.; Hurets, L.; Ochowiak, M. Mathematical modeling of nutrient release from capsulated fertilizers. *Period. Polytech. Chem. Eng.* **2020**, *64*, 562–568. [CrossRef]
38. Chernysh, Y.Y.; Plyatsuk, L.D.; Yakhnenko, O.M.; Trunova, I.O. Modelling of the vertical migration process of phosphogypsum components in the soil profile. *J. Eng. Sci.* **2017**, *4*, G6–G11. [CrossRef]
39. Lytvynenko, A.; Pavlenko, I.; Yukhymenko, M.; Ostroha, R.; Pitel, J. Hydrodynamics of two-phase upflow in a pneumatic classifier with the variable cross-section. In *Advances in Design, Simulation and Manufacturing III. DSMIE 2020*; Ivanov, V., Pavlenko, I., Liaposhchenko, O., Machado, J., Edl, M., Eds.; Lecture Notes in Mechanical Engineering; Springer: Cham, Switzerland, 2020; pp. 216–227. [CrossRef]
40. Sklabinskyi, V.; Liaposhchenko, O.; Pavlenko, I.; Lytvynenko, O.; Demianenko, M. Modelling of liquid's distribution and migration in the fibrous filter layer in the process of inertial-filtering separation. In *Advances in Design, Simulation and Manufacturing. DSMIE 2018*; Ivanov, V., Rong, Y., Trojanowska, J., Venus, J., Liaposhchenko, O., Zajac, J., Pavlenko, I., Edl, M., Perakovic, D., Eds.; Lecture Notes in Mechanical Engineering; Springer: Cham, Switzerland, 2019; pp. 489–497. [CrossRef]
41. Moiseev, V.; Manoilo, E.; Hrubnik, A.; Vasyliev, M.; Davydov, D. Cleaning and disposal of gas emissions from the production of calcinated soda ash. *J. Eng. Sci.* **2017**, *4*, B1–B6. [CrossRef]
42. Plyatsuk, L.D.; Ablieieva, I.Y.; Vaskin, R.A.; Yeskendirov, M.; Hurets, L.L. Mathematical modeling of gas-cleaning equipment with a highly developed phase contact surface. *J. Eng. Sci.* **2018**, *5*, F19–F24. [CrossRef]

energies

MDPI

Article

Numerical Investigation of Microchannel Heat Sink with Trefoil Shape Ribs

Sadiq Ali [1], Faraz Ahmad [2,*], Kareem Akhtar [1,*], Numan Habib [3], Muhammad Aamir [4], Khaled Giasin [5], Ana Vafadar [4] and Danil Yurievich Pimenov [6]

1 Department of Mechanical Engineering, University of Engineering & Technology, Peshawar 25120, Pakistan; Sadiq.ali@uetpeshawar.edu.pk
2 Department of Mechanical Engineering, Aerospace and Aviation Campus, Air University Islamabad, Kamra 43570, Pakistan
3 Department of Mechanical Engineering, CECOS University of Information Technology and Emerging Sciences, Peshawar 25000, Pakistan; numanhabib@cecos.edu.pk
4 School of Engineering, Edith Cowan University, Joondalup, WA 6027, Australia; m.aamir@ecu.edu.au (M.A.); a.vafadarshamasbi@ecu.edu.au (A.V.)
5 School of Mechanical and Design Engineering, University of Portsmouth, Portsmouth PO1 3DJ, UK; khaled.giasin@port.ac.uk
6 Department of Automated Mechanical Engineering, South Ural State University, Lenin Prosp. 76, 454080 Chelyabinsk, Russia; danil_u@rambler.ru
* Correspondence: faraz.ahmad0460@gmail.com (F.A.); kareemakhtar@uetpeshawar.edu.pk (K.A.)

Citation: Ali, S.; Ahmad, F.; Akhtar, K.; Habib, N.; Aamir, M.; Giasin, K.; Vafadar, A.; Pimenov, D.Y. Numerical Investigation of Microchannel Heat Sink with Trefoil Shape Ribs. *Energies* **2021**, *14*, 6764. https://doi.org/10.3390/en14206764

Academic Editors: Dimitris Drikakis and Pouyan Talebizadeh Sardari

Received: 9 September 2021
Accepted: 14 October 2021
Published: 17 October 2021

Abstract: The present study investigates the thermo-hydraulic characteristics of a microchannel sink with novel trefoil Shaped ribs. The motivation for this form of rib shape is taken from the design of lung alveoli that exchange oxygen and carbon dioxide. This study has been conducted numerically by using a code from the commercially available Fluent software. The trefoil shaped ribs were mounted on the centerline of different walls of the microchannel in three different configurations. These consisted of base wall trefoil ribs (MC-BWTR), sidewall trefoil ribs (MC-SWTR), all wall trefoil ribs (MC-AWTR) and smooth channel (MC-SC) having no ribs on its wall. The streamline distance between the ribs was kept constant at 0.4 mm, and the results were compared by using pressure drop (Δp), Nusselt number (Nu), thermal resistance (R_{th}) and thermal enhancement factor (η). The results indicated that the addition of trefoil ribs to any wall improved heat transfer characteristics at the expense of an increase in the friction factor. The trends of the pressure drop and heat transfer coefficient were the same, which indicated higher values for MC-AWTR followed by MC-SWTR and a lower value for MC-BWTR. In order to compare the thermal and hydraulic performance of all the configurations simultaneously, the overall performance was quantified in terms of the thermal enhancement factor, which was higher than one in each case, except for MC-AWTR, in $100 < Re < 200$ regimes. The thermal enhancement factor in the ribbed channel was the highest for MC-SWTR followed by MC-BWTR, and it was the lowest for MC-AWTR. Moreover, the thermal enhancement factor increases with the Reynolds number (Re) for each case. This confirms that the increment in the Nusselt number with velocity is more significant than the pressure drop. The highest thermal enhancement factor of 1.6 was attained for MC-SWTR at Re = 1000, and the lowest value of 0.87 was achieved for MC-AWTR at Re = 100.

Keywords: microchannel heat sink; trefoil ribs; thermal enhancement; thermal resistance

1. Introduction

The increasing demand for digitalization has shifted the trend of the modern world towards the miniaturization of electronic equipment where hundreds of thousands of transistors are installed on a single silicon chip. Due to the reduction in the sizes of these equipment, the power density increased rapidly and can be as high as up to 1 W/mm^2. However, great care is needed to maintain such equipment for its safety and improved life cycle. Therefore, modern techniques including cooling using liquid such as a micro-jet, heat pipe, spray impingement and microchannel cooling are introduced to fulfill the requirement of higher heat dissipation [1]. Over time, the cooling techniques have been improved from cooling by force convection and air to more advanced techniques.

The microchannel cooling system was first coined in 1981 by Tuckerman and Pease [2]. Microchannel cooling is a passive technique for dissipating heat from electronic devices without occupying a larger space than other dissipating setups. A microchannel heat sink (MCHS) with smooth geometry is used to dissipate heat from different devices. The usage and capacity of microelectronic equipment increased drastically; as a result, an increased amount of heat is produced inside this equipment. To dissipate this large amount of heat, a smooth MCHS will be unable for that much dissipation. Therefore, the thermal performance of smooth MCHS needs improvement. The thermal performance of MCHS can be improved by performing some modifications to the smooth one and then some thermal performance parameters such as thermal enhancement factor (η), friction factor (f) and Nusselt number (Nu) will be compared with a smooth channel after modification (adding some extended surfaces or cavities) to the channel. In most cases, the better thermal performance of MCHS is associated with irreversibility in the flow field, which is computed in terms of friction factor (f), entropy generation (Sgen) and thermal resistance (R_{th}). The main objective of many researchers is to increase the heat dissipation rate (Nusselt number (Nu)) with minimal pressure loss [3–6].

Many researchers worked on the effect of extended surfaces (fins or ribs) on the performance improvement of MCHS. For example, Rehman et al. [7] added dimple shaped ribs to the smooth channel for enhancement in the overall performance of the smooth channel. In their study, different cases were used, which included base wall ribs, sidewall ribs and all walls' ribs. Moreover, aligned, staggered and mixed configurations of each case were investigated numerically. Their results concluded that the case in which all walls were ribbed had better overall performance for each configuration. The worst performance was noted for the case in which ribs were added to the sidewall. Similarly, numerical studies and comparisons of heat transfer, friction factor and thermal enhancement factor of six different micro-channels were carried out by Zhu et al. [8], and they intended to report the best one. The study compared smooth rectangular micro-channel, side grooved rectangular micro-channel, side grooved with diamond shape ribs, side grooved with rectangular shape ribs, side grooved with elliptic shape ribs and side grooved with forwarding triangular shape ribs micro-channel. The results concluded that the combination of grooves and ribs has better performance than simple ribbed micro-channel. The side grooved with rectangular shape ribs micro-channel has the best overall performance at Re = 500, relative rib width = 0.25 mm and minimum rib-groove distance = 0.1 mm.

Xiao et al. [9] worked on the thermal improvement of micro-channel by producing turbulence in a flow field with the help of V-shape ribs. A significant enhancement in the overall performance of micro-channel has been noted from this numerical study. Similarly, Hassani et al. [10] conducted a numerical and experimental study to investigate the enhancement in the overall performance of MCHS by adding pins and chamber shape cavity to it. In their study, four different cases were used, which included smooth micro-channel, smooth micro-channel with pins, smooth micro-channel with chamber shape cavity and smooth micro-channel with pins and chamber shape cavity. Their results concluded that the addition of pins improved heat transfer by 44.8% and chamber shape cavity by 3.7%, while the simultaneous addition of pins and cavity improved heat transfer by 37.5%. The highest enhancement factor was noted for the pin located at the middle of the cavity. Furthermore, Ahmad et al. [11] used cylindrical shape ribs and cavities to numerically investigate the performance enhancement in MCHS with different cases. In this study, a base wall with cylindrical shape ribs, a sidewall with cylindrical shape ribs and all walls with cylindrical shape ribs were compared. Moreover, the effects of rib spacing for all wall ribs cases and comparisons of all wall ribs, all wall cavities and the combination of both were also taken into account in this study. It was concluded that the case of all wall ribs performed best among all cases, based on entropy generation, thermal enhancement factor, thermal resistance and transport efficiency.

Yang et al. [12] proposed three different pin fin cross sections attached to micro-channel heat sinks to improve heat dissipation capacity. These cross-sections include rhombus, hydrofoil and sinusoidal shapes. Their experimental and numerical results concluded that the performance of the sinusoidal cross section performed best among the three cases compared. Similarly, Naqiuddin et al. [13] studied the effects of structure parameters on rectangular micro-channel and found that for any structure, the performance in terms of heat transfer increases at the cost of higher pressure drop. The average temperature of the channel was reduced by 33 K with an increase of 1000 kPa pressure drop. The optimum performance was noted for the coolant's flow rate of less than 200 mL/min. Similarly, Wang et al. [14] carried a numerical study of rectangular micro-channel with sidewall truncated ribs. Three different ribs parameters, including truncation gap height, width and ribs arrangement, were used to study the improvement in the hydro-thermal behavior of MCHS in flow regimes of $100 < Re < 1000$. Their study confirmed that the results of truncated ribs are far better than their counterpart traditional ribs; the parallel arrangement results in a higher overall heat transfer coefficient.

Chuan et al. [15] used porous ribs instead of solid ribs with a straight channel to assess the performance of the micro-channel. They examined that, with porous ribs, the increment in pressure drop is lower than compared to solid ribs of exact dimensions. Moreover, the same porous rib increased the thermal resistance at the same time. Similarly, Wang et al. [16] proposed a multi-objective genetic optimization technique for the optimal design of MCHS. In their study, double-layered MCHS with semi-porous ribs was considered. It was concluded that the decrement in thermal resistance is linked with an increment in pumping power. For an optimum design, pumping power is reduced by 16.40% and thermal resistance by 14.06% compared to the reference design. Furthermore, Ahmad et al. [17] proposed four different novel shapes of ribs attached to the sidewall of rectangular microchannel heat sinks, including trapezoidal, rectangular, elliptical and hydrofoil shapes for the improvement of the hydro-thermal performance of MCHS. The results revealed that each case improved overall heat transfer performance. Moreover, elliptical ribs have better thermal performance; however, the hydrofoil file has better performance in terms of friction factor and pressure loss.

Keeping in view the above literature, it is very obvious that the addition of each type of ribs enhances heat transfer with an increase in pumping power. Considerable research has been performed on traditional shapes of ribs; however, no significant work was found on unique shape ribs. In this study, bio-inspired rib shape (trefoil) is used where ribs are added in three different wall configurations consisting of ribs added to the base wall (BWTR), Sidewall (SWTR) and (AWTR). The addition of trefoil shape ribs is expected to significantly improve the overall performance of MCHS. The novelty of this work is due to the unique bio-inspired shape of ribs, and the motivation is basically taken from the lungs of humans; the lung's internal structure consists of trefoils such as shape alveoli which are involved in the exchange of oxygen and carbon dioxide.

The present study aims to investigate the performance of MCHS numerically with novel trefoil ribs using ANSYS Fluent. This study aims to analyze the performance of MCHS by employing a novel shaped rib at different walls. The proposed models of novel ribs include trefoil shape ribs. The use of trefoil rib is the special novelty of this study, and it has not been investigated in the past literature. In most of the previous papers, the ribs were mounted on the sidewalls of MCHS in the transverse configuration. However, in this study, the ribs were mounted in the centerline configuration and mounted on different walls, i.e., base wall, sidewall and all walls. This study assesses the performance of MCHS with trefoil shape ribs by using numerical simulation in a laminar flow regime with Reynolds number (Re) in the range of 10–1000. Moreover, the performance of the base wall with trefoil ribs (BWTR), sidewall with trefoil ribs (SWTR) and all walls with trefoil ribs (AWTR) was compared with each other and with smooth micro-channel based on thermal resistance, pressure drop, friction factor ratio, heat transfer coefficient, Nusselt number ratio and thermal enhancement factor.

2. Materials and Methods

In the present study, the micro-channel is made of copper, and water is used as a coolant. Numerical simulations of the microchannel (3D) using CFD techniques (RANS Simulations) are analyzed. During numerical simulations, the smooth micro-channel and micro-channel with trefoil shape ribs attached to the centerline of base, side and to all walls with a streamline distance of 0.4 mm were modelled. The model was meshed and imported to CFD Fluent in order to calculate Δp, h, f/f_o, Nu/Nu_0 , R_{th} and η with the addition of trefoil shape ribs. The details of the material used and methodology adopted for the present study are given in the following section respectfully.

2.1. Geometrical Shape of MCHS

In the present study, a rectangular microchannel is used because of its superior thermal performance among circular, hexagonal, rectangular and triangular cross-sections [18]. Ribs in the trefoil shape are attached to the base wall, sidewall and all walls with a stream-wise distance (pitch) of 0.4 mm; the computational domain of each case and geometric details of the trefoil rib are shown in Figure 1. Moreover, the thermophysical properties of both the fluid (water) and solid (copper) domains are kept constant and are given in Table 1.

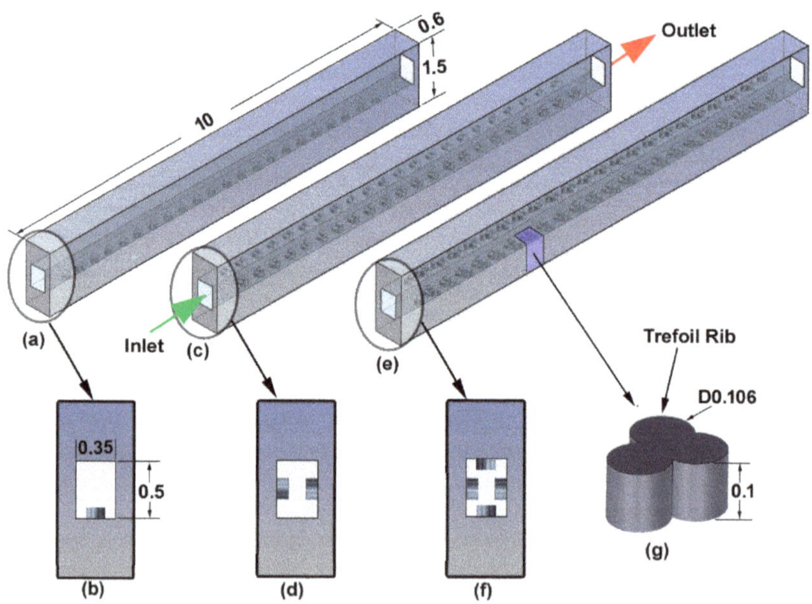

Figure 1. Computational domain of (**a**,**b**) MC-BWTR, (**c**,**d**) MC-SWTR, (**e**,**f**) MC-AWTR and (**g**) geometric details of trefoil rib.

Table 1. Thermophysical properties of materials.

	ρ (kg/m^3)	Cp (/kg/K)	k (W/m/K)	μ (kg/m/s)
Copper	8978	381	387.6	
Water	998.2	4812	0.6	0.001003

2.2. Governing Equations

The fluid is a Newtonian fluid with constant thermophysical properties, and the flow is laminar, steady-state and incompressible. Moreover, the effects of radiation and body forces are negligible and can be neglected. With all these assumptions, the governing equations for solid as well as a fluid domain can be presented as follows:

Continuity equation for fluid:

$$\frac{\partial u}{\partial x}+\frac{\partial v}{\partial y}+\frac{\partial w}{\partial z}=0 \quad (1)$$

where u, v and w are the velocity component of fluid in the x, y and z directions, respectfully. The movement equations in x-direction, y-direction and z-direction are written as stated below:

$$u\frac{\partial u}{\partial x}+v\frac{\partial u}{\partial y}+w\frac{\partial u}{\partial z}=-\frac{1}{\rho_f}\frac{\partial p}{\partial x}+\frac{\mu_f}{\rho_f}\left(\frac{\partial^2 u}{\partial x^2}+\frac{\partial^2 u}{\partial y^2}+\frac{\partial^2 u}{\partial z^2}\right) \quad (2)$$

$$u\frac{\partial v}{\partial x}+v\frac{\partial v}{\partial y}+w\frac{\partial v}{\partial z}=-\frac{1}{\rho_f}\frac{\partial p}{\partial y}+\frac{\mu_f}{\rho_f}\left(\frac{\partial^2 v}{\partial x^2}+\frac{\partial^2 v}{\partial y^2}+\frac{\partial^2 v}{\partial z^2}\right) \quad (3)$$

$$u\frac{\partial w}{\partial x}+v\frac{\partial w}{\partial y}+w\frac{\partial w}{\partial z}=-\frac{1}{\rho_f}\frac{\partial p}{\partial z}+\frac{\mu_f}{\rho_f}\left(\frac{\partial^2 w}{\partial x^2}+\frac{\partial^2 w}{\partial y^2}+\frac{\partial^2 w}{\partial z^2}\right) \quad (4)$$

where ρ_f represents the density, and μ_f is the dynamic viscosity of the working fluid.

The energy equation for fluid and solid domains can be written as follows:

$$u\frac{\partial T_f}{\partial x}+v\frac{\partial T_f}{\partial y}+w\frac{\partial T_f}{\partial z}=\frac{k_f}{\rho_f C_{Pf}}\left(\frac{\partial^2 T_f}{\partial x^2}+\frac{\partial^2 T_f}{\partial y^2}+\frac{\partial^2 T_f}{\partial z^2}\right) \quad (5)$$

$$k_s\left(\frac{\partial^2 T_s}{\partial x^2}+\frac{\partial^2 T_s}{\partial y^2}+\frac{\partial^2 T_s}{\partial z^2}\right)=0 \quad (6)$$

where T_f is the fluid temperature, k_f is the thermal conductivity and C_{Pf} is the specific heat of the working fluid. T_s is the temperature, and k_s is the thermal conductivity of the solid domain.

The following boundary conditions were used to solve the above equations.

At the inlet of the channel where x = 0 mm, the velocity-inlet boundary conditions with a fluid inlet temperature of 293.15 K are assumed to be as follows.

$$u=u_{in} \quad (7)$$

$$T_f=T_{in} \quad (8)$$

The different values for u_{in} are calculated by using an Re number in the range of 100 to 1000.

At the outlet of the channel, where x = 10 mm, we have the following.

$$p=p_{out}=1\ atm \quad (9)$$

At the bottom surface, the uniform heat flux is given by the following.

$$q=100\frac{w}{cm^2} \quad (10)$$

At inner wall of the channel, no slip boundary conditions are applied as follows.

$$u=v=w=0 \quad (11)$$

$$-k_s\frac{\partial T_s}{\partial n} = -k_f\frac{\partial T_f}{\partial n} \tag{12}$$

At all the remaining walls, the adiabatic conditions are used as follows.

$$\frac{\partial T_s}{\partial y} = \frac{\partial T_s}{\partial z} = 0 \tag{13}$$

2.3. Numerical Method and Mesh Independence

ANSYS Fluent is utilized for the numerical solution in the present study. The second-order difference schemes for discretization of diffusion terms and second-order upwind schemes for convection term are used in the present case. The convergence criteria of 10^{-6} are used for all variables. Different mesh sizes for the MC-BWT case for Re = 500 are solved, and the relative error in Δp and Nu is calculated to check the reliability of the solution and summarized in Table 2:

$$e\% = \frac{P_2 - P_1}{P_1} * 100 \tag{14}$$

where P_1 is the value of any parameter with the smallest mesh size, and P_2 is the value with other mesh sizes.

Table 2. Mesh independence test.

Mesh No	No of Elements	Pressure Drop (Pa)	Error (%)	Nu	Error (%)
1	568,956	4342.478	0.58	11.68255	0.83
2	691,194	4350.475	0.34	11.65281	0.58
3	811,646	4360.415	0.17	11.63914	0.46
4	963,976	4366.139	0.038	11.65263	0.058
5	1,231,560	4367.813		11.58557	

As shown in Table 2, for Mesh No. 3, the relative error in pressure was less than 0.2%, and the relative error in Nu was less than 0.5%; thus, this mesh, with 811,646 number of elements, was used for analysis in the present study.

2.4. Data Reduction

In MCHS, the η is used for performance measurement of different micro-channels; it is a function of both ratios of Nu and f and is given by the following equation [19].

$$\eta = \frac{(Nu/Nu_0)}{(Nu/Nu_0)^{\frac{1}{3}}} \tag{15}$$

In this equation, f is the Darcy friction factor [20], and Nu is the Nusselt number given by the following:

$$f = \frac{2D_h\Delta p}{L_{ch}\rho_f u_{m^2}} \tag{16}$$

$$Nu = \frac{h_{avg}D_h}{k_f} \tag{17}$$

where D_h is the hydraulic diameter, Δp is the pressure drop, L_{ch} is the length of the channel and h_{avg} is the average heat transfer coefficient and can be calculated by following relationships:

$$D_h = \frac{2H_{ch}W_{ch}}{H_{ch} + W_{ch}} \tag{18}$$

$$h_{avg} = \frac{q_w A_{base}}{2(W_{ch} + H_{ch})L_{ch}\Delta T} \tag{19}$$

where H_{ch} is the height, W_{ch} is the width of the channel, q_w is the heat flux and A_{base} is the area of the base wall. ΔT is the temperature difference between channel wall temperature (T_w) and fluid mean temperature (T_f), and they are given by the following.

$$\Delta T = T_w - T_f \tag{20}$$

$$T_w = \frac{\int T_{w-x,y} dydx}{\int dydx} \tag{21}$$

$$T_b = \frac{\int T_{b-x,y} dydx}{\int dydx} \tag{22}$$

$$T_f = \frac{\int T_{f-i,x} \rho_{f-i,x} \left| \vec{v}.d\vec{A} \right| dx}{\int \rho_{f-i,x} \left| \vec{v}.d\vec{A} \right| dx} \tag{23}$$

The overall thermal resistance is given as follows.

$$R_{tot} = R_{cond} + R_{conv} + R_{cap} \tag{24}$$

$$R_{tot} = \frac{T_b - T_{ch}}{q_w A_b} + \frac{T_{ch} - T_f}{q_w A_b} + \frac{T_f - T_{in}}{q_w A_b} = \frac{T_b - T_{in}}{q_w A_b} \tag{25}$$

3. Results

The accuracy of numerical schemes used for the present study is first validated with available experimental work in literature. The validated numerical modal is then used for calculating the Δp, h, f/f_o, Nu/Nu_0, R_{th} and η for MC-BWTR, MC-SWTR and MC-AWTR. The calculated values for ribbed channels are compared with the reference case (MC-SC), which are discussed in detail in the following sections.

3.1. Model Validation

As shown in Figure 2, the Nu and f values were calculated and compared to the experimental work of Wang et al. [21] with the uniform inlet velocity, Tin = 293 K, which is calculated from Re and a heat flux of 100 W/cm^2 with the same conditions as used in the experimental study [21]. The obtained results were close to the experimental results available in the literature with less than 5% deviation; hence, the present model can be used for calculating the thermal-hydraulic perimeters of MCHS.

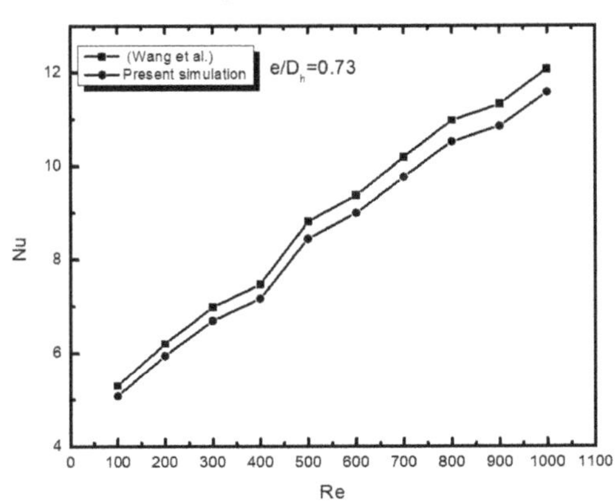

(a)

Figure 2. *Cont.*

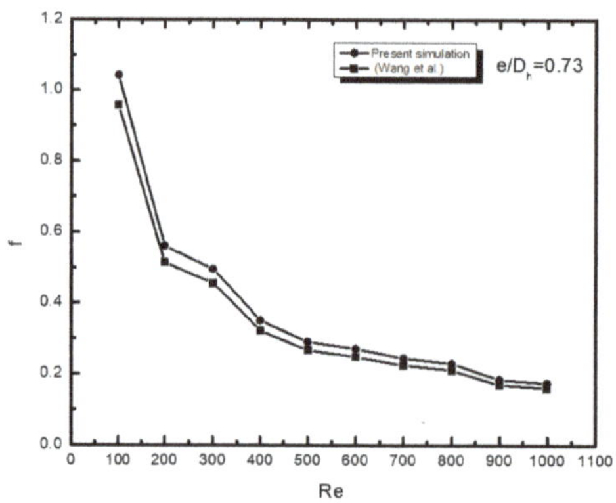

(b)

Figure 2. Validation of (**a**) Nusselt number and (**b**) friction factor with Wang et al. [21].

3.2. Characteristics of Temperature and Pressure Distribution

The temperature distribution for MCHS with trefoil shaped ribs was added to the base wall (BWTR), sidewall (SWTR) and all wall (AWTR) cases together with the smooth channel, which is shown in Figure 3. This was performed in order to study the temperature distribution (cooling) with the addition of trefoil shape ribs. MC-AWTR case showed better cooling in terms of temperature drop followed by MC-SWTR and MC-BWTR, respectively.

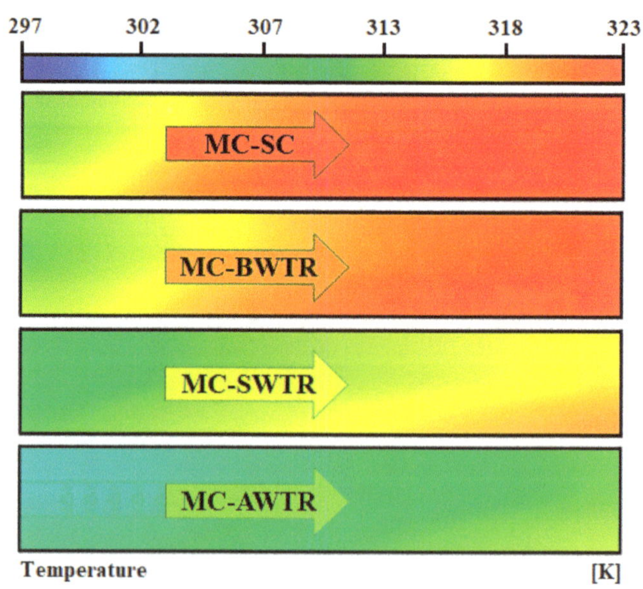

Figure 3. Temperature distribution of MCHS with trefoil shaped ribs attached to different walls.

The increase in thermal performance was at the cost of an increase in power required to pump the same amount of fluid because the ribs blocked the flow and increased the pressure requirement. The pressure distribution was the determining factor for pumping power; the pressure distributions of all ribbed cases along with smooth channel are presented in Figure 4. The pressure drop in the smooth channel was due to the friction of the wall, while for the ribbed channel, it was the combined effect of wall roughness and resistance offered to the flow by ribs. The pressure drop of MC-SC is minimum, and the pressure drop for MC-AWTR is highest because of the presence of a higher number of ribs.

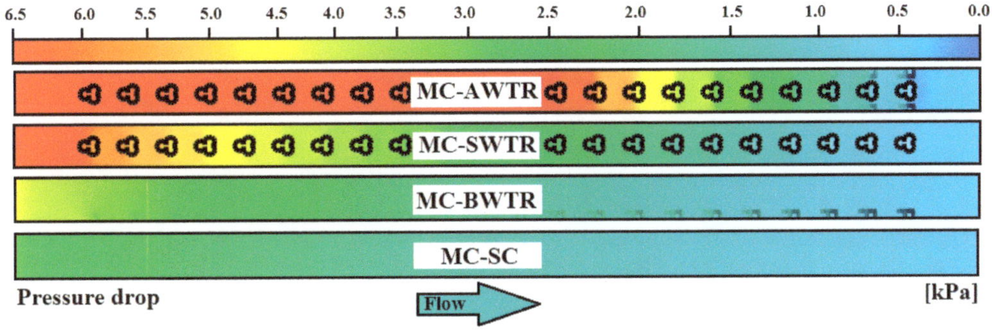

Figure 4. Pressure distribution of MCHS with trefoil shape ribs attached to different walls.

The temperature distribution (cooling diagram) recommended the MC-AWTR case, while the pressure distribution recommended MC-BWTR. Other performance parameters can be analyzed for final conclusions.

3.3. Performance Comparison

Performance measurement of all cases was carried out for Re = 100–1000 with a step of 100, and the results are presented in contrasting graphs to indicate the improvements made by the trefoil shaped ribs. In addition to the performance improvements in the different cases and the reference case, MC-SCs were compared by plotting Δp, h, f/f_o, Nu/Nu_0, R_{th} and η.

The Δp of the MCHS with trefoil shaped ribs and a straight channel are displayed in Figure 5. The lowest Δp was attained in the MC-SC case, which was only produced by the wall's roughness. The Δp in the ribbed channel cases, MC-BWTR, MC-SWTR and MC-AWTR, was a combination of Δp by the wall's roughness and Δp produced by the blockage offered by ribs. The Δp for the MC-BWTR case was minimal compared with the other ribbed cases, and this was due to the ribs being attached to only one wall. However, in the MC-SWTR and MC-AWTR cases, the ribs were attached to either two or four walls, respectively. In the MC-SWTR case, the ribs were attached to two walls; therefore, the number of ribs, in this case, was higher than for MC-BWTR. As such, the Δp was more than the MC-SC and MC-BWTR cases. A maximum Δp was noted for the MC-AWTR case because each wall contained ribs and blocked the flow, thereby causing the Δp to rise. At Re < 250, the Δp for MC-SC, MC-BWTR and MC-SWTR was almost the same, whereas the MC-AWTR case displayed higher Δp. Moreover, in the range of 100 < Re < 600, the MC-BWTR Δp was almost equal to MC-SC compared to Δp in the other cases, which was significantly higher. The pressure drop increases with an increase in Re for all cases; however, this increase is minimal in the low Re range, while the increase is maximal in the higher Re range, i.e., Re > 500. Moreover, the pressure drop behaviour against Re of MC-AWTR is stiffer when compared to other ribbed cases such as MC-SWTR and MC-BWTR, which is flattened and indicates that the increase in Δp for MC-AWTR case is higher than MC-SWTR and MC-BWTR.

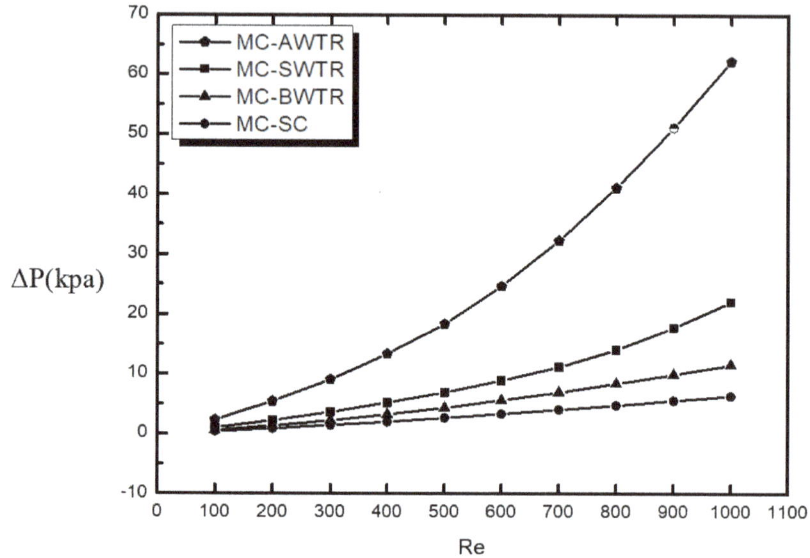

Figure 5. Comparison of pressure drop for various wall configurations MCHS.

Figure 6 shows the comparison of h for the various MCHSs together with the MC-SC. It is clear that the h value in all ribbed cases was significantly higher than for the MC-SC. The addition of trefoil shaped ribs caused better mixing and produced higher h values as a result. The heat transfer coefficient for the MC-AWTR case was highest, and it was the lowest for MC-BWTR. This could be because the number of ribs in MC-AWTR is greater than that of MC-BWTR, which provides more heat transfer area and better mixing of fluid. Moreover, the MC-SWTR case had better performance in terms of heat transfer than the MC-SC and MC-BWTR cases. The increase in h at Re < 200 from MC-SC to the ribbed channel was smaller; however, at Re > 200, the increase was significantly higher because the turbulence was higher at Re, which caused high heat transfer. Furthermore, h increases with an increase in Re. The highest increase is noted for each case compared to the straight channel (MC-SC) at Re = 1000, which is 3, 2.4 and 1.7 times for MC-AWTR, MC-SWTR and MC-BWTR, respectively.

The friction caused by the addition of ribs to the base wall, sidewalls and all walls resulted in an increase in friction factor. The reltive increase in friction factor with trefoil shaped ribs was calculated in terms of friction factor ratio (f/f_o) and plotted against Re, as shown in Figure 7. It is clear that the f/f_o for MC-BWTR and MC-SWTR was considerably smaller than the ratio for MC-AWTR. The f/f_o at Re = 1000 for MC-AWTR and MC-SWTR is 1.9 and 5.39 times higher when compared to MC-BWTR, respectively, because the friction factor is the measure of the resistance to flow, which is at the maximum in the case of the low flow area of MC-AWTR. The f/f_o ratio rose with the increase in Re, while an increase in the SC-BWTR case was almost negligible, as indicated by the flatness of the curve. This was because very few turbulence enhancers were present at the base wall. The increase in the f/f_o for MC-SWTR with an increase in the Re was insignificant up to Re < 700 regimes; however, the increase was significant above Re > 700 regimes for the MC-AWTR case. The f/f_o rose with the increase in Re number throughout due to the presence of ribs at all walls.

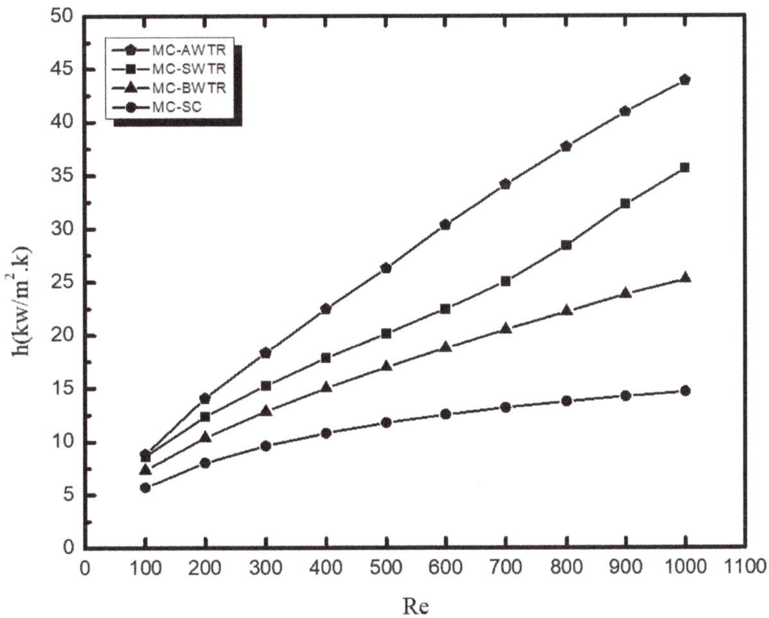

Figure 6. Comparison of heat transfer coefficient for different wall configurations of MCHS.

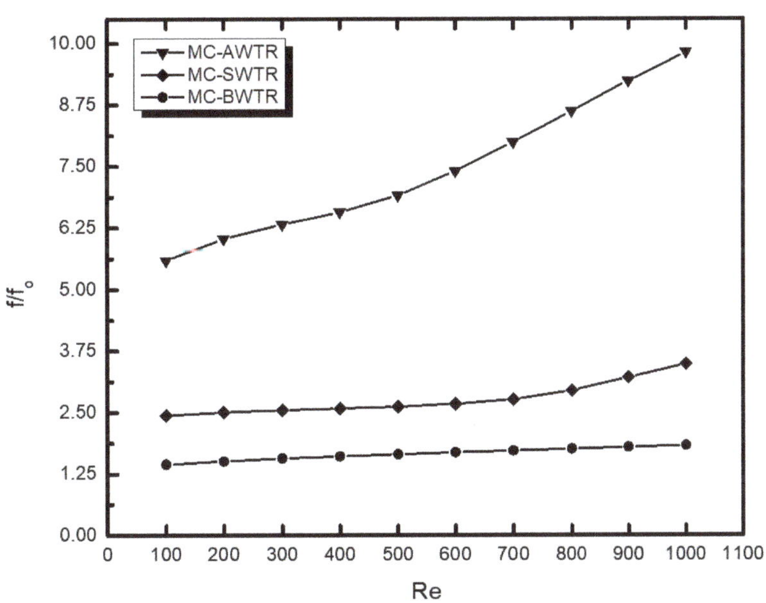

Figure 7. Comparison of friction factor ratio for different wall configurations MCHS.

Figure 8 shows the comparison of the Nu for various MCHSs along with the MC-SC. The Nu is a dimensionless number used to indicate the capability of heat transfer by a channel. The higher the Nu number, the more effective heat transfer will be. Figure 8 indicates that the minimum value of Nu was observed for MC-SC, while the maximum value was noted for MC-AWTR. Conversely, MC-BWTR had a smaller Nu than MC-SWTR and MC-AWTR. The average percentage increase in Nu is 142%, 90% and 51% for MC-AWTR, MC-SWTR and MC-BWTR, respectively, when compared to MC-SC. The Nu was directly proportional to the effective heat transfer area, which was maximum for MC-AWTR and minimum for MC-SC. The effective heat transfer area for MC-SWTR was higher than MC-BWTR, and this explained why MC-SWTR performed better in terms of Nu than MC-BWTR. Moreover, Nu inclined linearly with Re for all cases except MC-SWTR after Re > 600. This was because the ribs were attached to both sides in MC-SWTR, which produced turbulence compared to other cases at higher velocity.

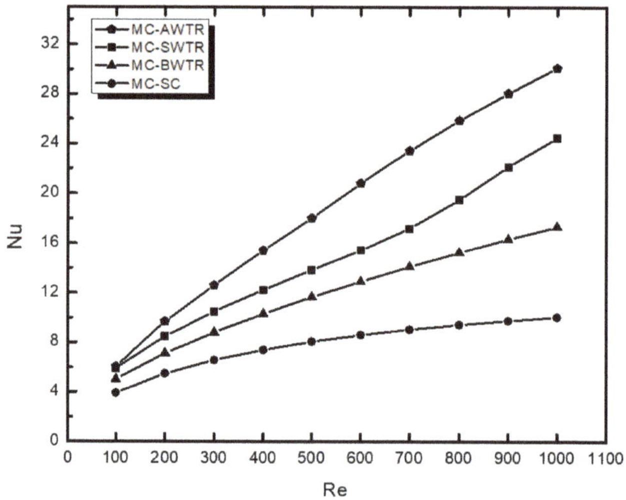

Figure 8. Comparison of Nusselt number for different wall configurations MCHS.

In order to compare the effectiveness of the heat transfer among ribbed channels, the Nu/Nu_0 of the ribbed channel and the MC-SC was plotted against the Re, and the results are presented in Figure 9. It is evident that MC-AWTR performed exceptionally well in comparison to the MC-SC in terms of Nu; the performance of MC-AWTR was three times superior to MC-SC at Re = 1000. MC-BWTR had a slight increase in Nu compared to MC-SC: the maximum value of Nu/Nu_0 is 1.5 at Re = 1000, half of the MC-AWTR at the same Re. MC-SWTR performance in terms of Nu ratio was better than MC-BWTR and slightly lower than MC-AWTR; therefore, this case's performance was 2.5 times higher than the MC-SC at Re = 1000. The value of Nu/Nu_0 increased smoothly with Reynolds number except for MC-SWTR in Re > 600. The increase in the value of Nu/Nu_0 was higher with Re and MC-BWTR in Re < 200, where the curve was nearly flat, representing a minor increase in Nu/Nu_0 with Re.

MCHS performed better thermally with the addition of ribs to any wall; however, the addition of ribs declined the hydrodynamic performance by increasing pressure drop, which was not required. As a result, there should be a trade-off between thermal improvement and pressure drop, which could be calculated in terms of overall η. The η for trefoil shaped ribs of different wall configuration MCHS was plotted against the Re number and is presented in Figure 10. It was clear that MC-SWTR had a higher value of η among all cases for any Re, although its heat transfer was less than MC-AWTR; however, its Δp was quite low, making the η high. The η for MC-AWTR was lowest among all cases; the increase in Nusselt number for this configuration is associated with a higher Δp, making the value of η the lowest. The value of η for MC-BWTR was higher than MC-AWTR and lower than MC-SWTR. Figure 10 shows that the value of η for all cases was higher than one except for MC-AWTR in Re < 200 in which the increase in Nu did not compensate for the increase in Δp. Moreover, the value of η increased in increment in Re for each case, which confirms that the increment in Nu was more significant than the increment in Δp. The highest η value of 1.6 was attained for MC-SWTR at Re = 1000, whereas the lowest value of 0.87 was achieved for MC-AWTR at Re = 100. The thermal enhancement for MC-BWTR and MC-AWTR was almost the same in 700 < Re < 900. The increase in η with Re for MC-BWTR was nearly uniform throughout. By contrast, MC-SWTR and MC-AWTR were non-uniform. For MC-AWTR, the η increase was more significant at Re < 700 and less effective at Re > 700 because the increase in Δp was very high than compared to Nu. For MC-SWTR in Re > 700 regimes, the increase in η was more rapid with Re, indicating that the increment in Nu at high Re was more significant than the increment in Δp for sidewall trefoil ribs.

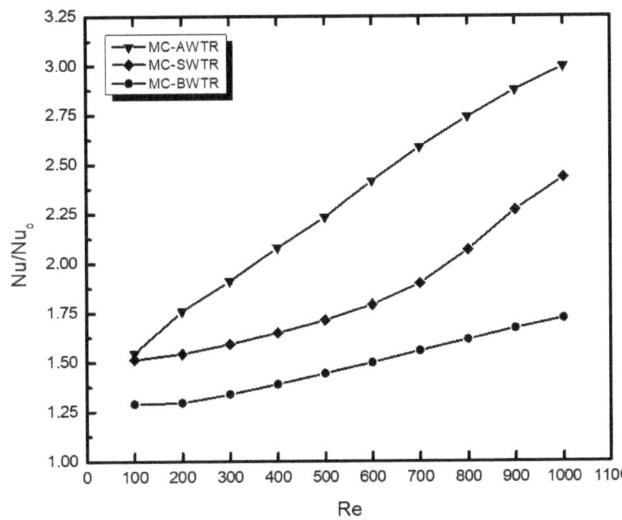

Figure 9. Comparison of Nusselt number ratio for different wall configurations MCHS.

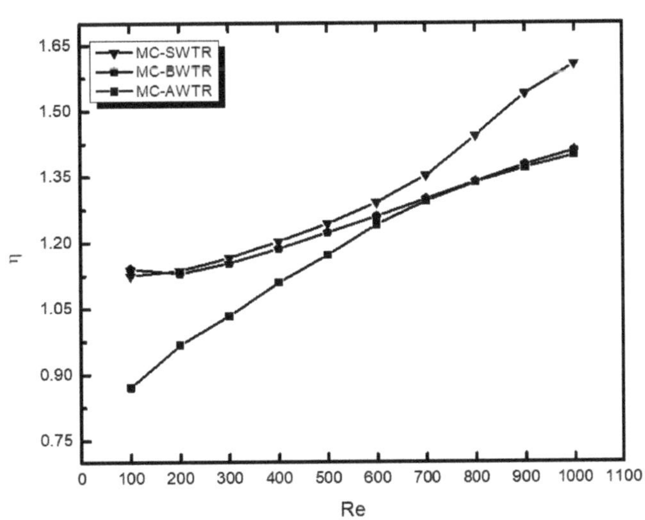

Figure 10. Comparison of thermal enhancement factor for different wall configurations MCHS.

Figure 11 shows the total R_{th} for various ribbed channel configurations against Re in a range of 100–1000. Total R_{th} demonstrates the capability of the micro-channel to transfer heat with given pumping power. The higher the value R_{th} is, the lower the heat transfer will be with respect to pumping power. MC-SC had a higher value than any other ribbed case confirming the evidence based on the η: The performance of each ribbed case was better than the MC-SC. MC-AWTR had minimum R_{th} compared to MC-BWTR, which had maximum values among ribbed channels; however, MC-SWTR had the highest η, and its R_{th} was more elevated than MC-AWTR due to the low number of ribs that caused lower heat transfer for the same power in MC-SWTR.

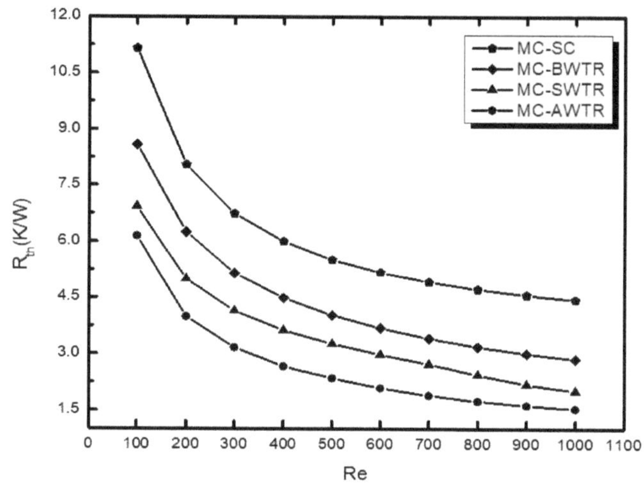

Figure 11. Comparison of total thermal resistance for different wall configurations MCHS.

4. Conclusions

In this study, the bio-inspired rib shape (trefoil) is used where ribs are added in three different wall configurations consisting of ribs added to the base wall (BWTR), Sidewall (SWTR) and (AWTR). The addition of trefoil shaped ribs significantly improved the overall performance of MCHS. The novelty of this work is due to the unique bio-inspired shape of the rib, which is basically taken from lungs of human, and the lung's internal structure consists of trefoils-like shaped alveoli that are actually involved in the exchange of oxygen and carbon dioxide.

The present study utilizes the Fluent code to investigate the heat transfer and flow field of MCHS with trefoil shaped ribs. The trefoil shaped ribs in this study were mounted to the center of the base wall, sidewall and all walls with a streamline distance of 0.4 mm. The performance improvement has been calculated in terms of Δp, h, f, Nu, R_{th} and η for Reynolds number in the range of Re = 100 to Re = 1000. Based on the results discussed in the present study, the following significant points are concluded as follows:

- The addition of trefoil ribs to any wall improved heat transfer characteristic (h and Nu) of MC-SC; the improvement was greater in MC-AWTR followed by MC-SWTR and minimum for MC-BWTR.
- The addition of ribs to any wall has increased the friction factor and, hence, the pumping power. This increase is minimum for MC-BWTR and maximum for MC-AWTR, whereas MC-SWTR is moderate. The addition of ribs improves heat transfer at the expense of an increase in the friction factor (pumping power), which is quantified in terms of η. This factor was higher than one in each case, except for MC-AWTR in $100 < Re < 200$. The η for the ribbed channel was highest for MC-SWTR followed by MC-BWTR and lowest for MC-AWTR.
- Although the heat transfer capability of MC-AWTR is superior to others, its overall performance is inferior, thereby confirming that the heat transfer coefficient is not the only criteria for modification of MC-SC. On the other hand, the overall heat transfer

enhancement of MC-SWTR was higher than any other case throughout the study, and its value was significantly high in Re > 700.
- Moreover, η increases with Re for each case, which confirms that the increment in Nu with Re is more significant than the increment in Δp. The highest η value of 1.6 is attained for MC-SWTR at Re = 1000. By contrast, the lowest value of 0.87 was achieved for MC-AWTR at Re = 100.

Author Contributions: Conceptualization, F.A., S.A. and K.A.; methodology, K.A., S.A, F.A. and N.H.; validation, K.A., S.A, F.A., N.H., M.A., K.G., A.V. and D.Y.P.; investigation, S.A., K.A., F.A. and N.H.; writing—original draft preparation, S.A. and F.A.; writing—review and editing, M.A., K.G, A.V. and D.Y.P. All authors have read and agreed to the published version of the manuscript.

Funding: This research received no external funding.

Institutional Review Board Statement: Not applicable.

Informed Consent Statement: Not applicable.

Data Availability Statement: The data presented in this study are available upon request.

Acknowledgments: The authors acknowledge the support of UET Peshawar and Air University, Islamabad, for providing the computational facilities to conduct this research study.

Conflicts of Interest: The authors declare no conflict of interest.

Nomenclature

MCHS	Microchannel heat sink
MC-AWTR	All wall trefoil ribbed microchannel
MC-AWSR	Side wall trefoil ribbed microchannel
MC-AWBR	Base wall trefoil ribbed microchannel
Re	Reynolds number
Nu	Nusselt number
f	Friction factor
L	Length of the MCHS (mm)
Hch	Height of the channel (mm)
Wch	Width of the channel (mm)
Dh	Hydraulic diameter (mm)
ΔT	Temperature difference (K)
Δp	Pressure drop (Pa)
qw	Wall flux (W/cm^2)
Cp	Specific heat capacity at constant pressure (J/kg K)
μ	Dynamic viscosity (kg/ms)
k	Thermal conductivity (W/mK)
ρ	Density (kg/m^3)
η	Thermal enhancement factor

References

1. Xu, M.; Lu, H.; Gong, L.; Chai, J.C.; Duan, X. Parametric numerical study of the flow and heat transfer in microchannel with dimples. *Int. Commun. Heat Mass Transf.* **2016**, *76*, 348–357. [CrossRef]
2. Tuckerman, D.B.; Pease, R.F.W. High-performance heat sinking for VLSI. *IEEE Electron Device Lett.* **1981**, *2*, 126–129. [CrossRef]
3. Ahmad, F.; Cheema, T.A.; Ur Rehman, M.M.; Abbas, A.; Woo Park, C. Thermal enhancement of microchannel heat sink using rib surface refinements. *Numer. Heat Transf. Part A Appl.* **2019**, *76*, 851–870. [CrossRef]
4. Liu, X.; Yu, J. Numerical study on performances of mini-channel heat sinks with non-uniform inlets. *Appl. Therm. Eng.* **2016**, *93*, 856–864. [CrossRef]
5. Rehman, M.M.U.; Cheema, T.A.; Ahmad, F.; Khan, M.; Abbas, A. Thermodynamic Assessment of Microchannel Heat Sinks with Novel Sidewall Ribs. *J. Thermophys. Heat Transf.* **2020**, *34*, 243–254. [CrossRef]
6. Ali, S.; Ahmad, F.; Hassan, M.; Rehman, Z.; Wadood, A.; Ahmad, K.; Park, H. Thermo-fluid Characteristics of Microchannel Heat Sink with Multi-configuration NACA 2412 Hydrofoil Ribs. *IEEE Access* **2021**, *9*, 128407–128416. [CrossRef]

7. Rehman, M.M.U.; Cheema, T.A.; Ahmad, F.; Abbas, A.; Malik, M.S. Numerical investigation of heat transfer enhancement and fluid flow characteristics in a microchannel heat sink with different wall/design configurations of protrusions/dimples. *Heat Mass Transf.* **2020**, *56*, 239–255. [CrossRef]
8. Zhu, Q.; Chang, K.; Chen, J.; Zhang, X.; Xia, H.; Zhang, H.; Wang, H.; Li, H.; Jin, Y. Characteristics of heat transfer and fluid flow in microchannel heat sinks with rectangular grooves and different shaped ribs. *Alex. Eng. J.* **2020**, *59*, 4593–4609. [CrossRef]
9. Xiao, H.; Liu, Z.; Liu, W. Turbulent heat transfer enhancement in the mini-channel by enhancing the original flow pattern with v-ribs. *Int. J. Heat Mass Transf.* **2020**, *163*, 120378. [CrossRef]
10. Hassani, S.M.; Khoshvaght-Aliabadi, M.; Mazloumi, S.H.; Rehman, S.; Alimoradi, A. Improving thermal performance of microchannels by combining rectangular pin with chamber. *Appl. Therm. Eng.* **2021**, *186*, 116373. [CrossRef]
11. Ahmad, F.; Cheema, T.A.; Mohib Ur Rehman, M.; Ilyas, M.; Park, C.W. Thermodynamic analysis of microchannel heat sink with cylindrical ribs and cavities. *J. Heat Transf.* **2020**, *142*, 092503. [CrossRef]
12. Yang, D.; Jin, Z.; Wang, Y.; Ding, G.; Wang, G. Heat removal capacity of laminar coolant flow in a micro channel heat sink with different pin fins. *Int. J. Heat Mass Transf.* **2017**, *113*, 366–372. [CrossRef]
13. Naqiuddin, N.H.; Saw, L.H.; Yew, M.C.; Yusof, F.; Ng, T.C.; Yew, M.K. Overview of micro-channel design for high heat flux application. *Renew. Sustain. Energy Rev.* **2018**, *82*, 901–914. [CrossRef]
14. Wang, G.; Chen, T.; Tian, M.; Ding, G. Fluid and heat transfer characteristics of microchannel heat sink with truncated rib on sidewall. *Int. J. Heat Mass Transf.* **2020**, *148*, 119142. [CrossRef]
15. Chuan, L.; Wang, X.D.; Wang, T.H.; Yan, W.M. Fluid flow and heat transfer in microchannel heat sink based on porous fin design concept. *Int. Commun. Heat Mass Transf.* **2015**, *65*, 52–57. [CrossRef]
16. Wang, T.H.; Wu, H.C.; Meng, J.H.; Yan, W.M. Optimization of a double-layered microchannel heat sink with semi-porous-ribs by multi-objective genetic algorithm. *Int. J. Heat Mass Transf.* **2020**, *149*, 119217. [CrossRef]
17. Ahmad, F.; Cheema, T.A.; Khan, A.; Mohib-Ur-Rehman, M.; Yildizhan, H. Hydrothermal Investigation of the Performance of Microchannel Heat Sink with Ribs Employed on Side Walls. *J. Non-Equilib. Thermodyn.* **2021**. [CrossRef]
18. Moradikazerouni, A.; Afrand, M.; Alsarraf, J.; Mahian, O.; Wongwises, S.; Tran, M.D. Comparison of the effect of five different entrance channel shapes of a micro-channel heat sink in forced convection with application to cooling a supercomputer circuit board. *Appl. Therm. Eng.* **2019**, *150*, 1078–1089. [CrossRef]
19. Karwa, R.; Sharma, C.; Karwa, N. Performance evaluation criterion at equal pumping power for enhanced performance heat transfer surfaces. *J. Sol. Energy* **2013**, *37*, 23–32. [CrossRef]
20. Steinke, M.E.; Kandlikar, S.G. Single-phase liquid friction factors in microchannels. *Int. J. Therm. Sci.* **2006**, *45*, 1073–1083. [CrossRef]
21. Wang, G.; Niu, D.; Xie, F.; Wang, Y.; Zhao, X.; Ding, G. Experimental and numerical investigation of a microchannel heat sink (MCHS) with micro-scale ribs and grooves for chip cooling. *Appl. Therm. Eng.* **2015**, *85*, 61–70. [CrossRef]

energies

MDPI

Article

Solidification Enhancement in a Triple-Tube Latent Heat Energy Storage System Using Twisted Fins

Xinguo Sun [1,*], Jasim M. Mahdi [2], Hayder I. Mohammed [3], Hasan Sh. Majdi [4], Wang Zixiong [5] and Pouyan Talebizadehsardari [6,*]

1 Jiangsu Smart Factory Engineering Research Centre, College of Management and Engineering, Huaiyin Institute of Technology, Huai'an 223003, China
2 Department of Energy Engineering, University of Baghdad, Baghdad 10071, Iraq; jasim@siu.edu
3 Department of Physics, College of Education, University of Garmian, Kurdistan, Kalar 46021, Iraq; hayder.i.mohammad@garmian.edu.krd
4 Department of Chemical Engineering and Petroleum Industries, Al-Mustaqbal University College, Babylon 51001, Iraq; hasanshker1@gmail.com
5 China Water Resources Pearl River Planning Surveying & Designing Co, Ltd., Guangzhou 510610, China; xiongwz2020@126.com
6 Centre for Sustainable Energy Use in Food Chains, Institute of Energy Futures, Brunel University London, Kingston Lane, Uxbridge UB8 3PH, UK
* Correspondence: sunxinguo2021@163.com (X.S.); pouyan.talebizadehsardari@brunel.ac.uk (P.T.)

Abstract: This work evaluates the influence of combining twisted fins in a triple-tube heat exchanger utilised for latent heat thermal energy storage (LHTES) in three-dimensional numerical simulation and comparing the outcome with the cases of the straight fins and no fins. The phase change material (PCM) is in the annulus between the inner and the outer tube, these tubes include a cold fluid that flows in the counter current path, to solidify the PCM and release the heat storage energy. The performance of the unit was assessed based on the liquid fraction and temperature profiles as well as solidification and the energy storage rate. This study aims to find suitable and efficient fins number and the optimum values of the Re and the inlet temperature of the heat transfer fluid. The outcomes stated the benefits of using twisted fins related to those cases of straight fins and the no-fins. The impact of multi-twisted fins was also considered to detect their influences on the solidification process. The outcomes reveal that the operation of four twisted fins decreased the solidification time by 12.7% and 22.9% compared with four straight fins and the no-fins cases, respectively. Four twisted fins improved the discharging rate by 12.4% and 22.8% compared with the cases of four straight fins and no-fins, respectively. Besides, by reducing the fins' number from six to four and two, the solidification time reduces by 11.9% and 25.6%, respectively. The current work shows the impacts of innovative designs of fins in the LHTES to produce novel inventions for commercialisation, besides saving the power grid.

Keywords: triple-tube heat exchanger; twisted fin array; phase change material; thermal energy storage; solidification

Citation: Sun, X.; Mahdi, J.M.; Mohammed, H.I.; Majdi, H.S.; Zixiong, W.; Talebizadehsardari, P. Solidification Enhancement in a Triple-Tube Latent Heat Energy Storage System Using Twisted Fins. *Energies* **2021**, *14*, 7179. https://doi.org/10.3390/en14217179

Academic Editor: Luisa F. Cabeza

Received: 16 September 2021
Accepted: 28 October 2021
Published: 1 November 2021

1. Introduction

The electrical and power generation equipment's application generally faces a considerable heat flux [1–3]. The operation of these machines may be thermally affected if the applicable thermal control measure is absent, causing a failure in the operation process [4–6]. Latent heat thermal energy storage (LHTES) could consider as a passive heat control measure for related thermal applications [7,8]. The phase change material (PCM) based in the LHTES can hold the temperature at a constant value by releasing or absorbing heat during the phase change process [9–11]. The simple structure, high and constant performance, and no extra power spending, make the LHTES appropriate for space application [12,13]. Still, a major concern for the LHTES system is PCM's minimal thermal efficiency, which decreases the phase change rate [14–17]. Researchers developed several techniques to improve the heat transfer rate of such systems, including the expansion of the heat transfer surface area [18–20], adding micro or nano-sized particles [21–24], using cascade layer PCM [25], encapsulation techniques [26,27], changing the location of the heat transfer fluid (HTF) channel [28–30], fins combinations [31–33], conductive foams [34–36], and using magnetic fields [37,38].

A lot of studies related to enhancing the thermal performance in the solidification process have been achieved [39]. Tao [40] was the first researcher to investigate the solidification process in cylindrical geometry. He developed a numerical model to predict interface moving issues during the phase change process. Gortych et al. [41] experimentally and numerically analysed the discharge process of the PCM located in a horizontal annular channel. They assumed a constant wall temperature, which is not a real condition, and they detected a moderate range of the natural convection coefficient. Abdollahzadeh and Esmaeilpour [42] investigated the thermal energy storage (TES) with a wavy wall, and nanofluid used as HTF. They found that the configuration of the system and the nanofluid have great influences on the thermal performance of the system. Shahsavar et al. [43] examined the effect of the wavy channel combined with the metal foam on the latent heat system (LHS). They found that the system configuration and the porous medium have a considerable effect on the thermal performance of the solidification process due to increasing the heat transfer surface area and enhancing the average thermal conductivity of the system. Choi and Kim [44] evaluated the circular fins for the discharging improvement in the LHS. Their work stated that the fins enhance the heat transfer coefficient by 3 times over the case without fins. Wang et al. [45] numerically studied the solidification process in a 2-D zigzag shape heat exchanger. Unlike the inlet velocity, they found that the average velocity of the HTF has a noticeable influence on the thermal performance. Sardari et al. [46] studied the modification of the LHS using a zigzag configuration. They confirm that the unit with the zigzag angle of 60° accelerated the storage time by 1/3 times over the time of the case with a 30° zigzag angle.

Applying double and triple pipe as the heat exchanger has been widely used in the TES to steady the effects of various parameters such as fins, temperature, and velocity of HTF [47,48]. Shokouhmand and Kamkari [49] numerical evaluated the charging process of the PCM in the double pipe heat exchanger. They stated that the phase change process is strongly affected by the fins placed in the inner tube. Bazai et al. [50] numerically studied an elliptical tube implanted in an annulus channel, they investigated the effects of various aspect ratios and the angular position of the inner ellipse diameters during the charging process. They found that the maximum charging rate accelerated by 61% and the system performance improved by 26%. Shahsavar et al. [51] numerically studied the phase change processes in a wavy double-pipe LHTES unit. They found that the essential time to charge and discharge the PCM decreases by 29% and 58%, respectively, utilising wavy tubes compared with the straights. In a separate study, Shahsavar et al. [52] assessed the efficiency of the phase change process in a wavy double-channel TES system. Increasing the inlet temperature, average velocity, and wave amplitude increases the performance of the system. Xu et al. [53,54] examined a horizontal double-pipe TES combined with a porous medium and optimised the position of the porous injected in the system. They found that the system with a partially filled foam at the base part has the same effect as the system with totally filled by the foam with 80% enhancement of the melting rate. Researchers also applied triple pipe in the TES heat exchanger. Ghalambaz et al. [55] studied the impact of the fins array in a triple-tube LHTES during the melting process. They found that the charging rate for the case of utilising four straight fins was 8.3% lower than that compared with the fins-less case. In two separate works, Mahdi et al. [56,57] studied the performance of the charging rate of the PCM in a triplex tube system. Li et al. [58] studied the effect of the metal foam and nanoparticles on the PCM in a triplex tube LHS. The main outcome of their work was that increasing the loading of nanoparticles or decreasing the porosity of the porous medium accelerates the phase change rate of the PCM.

Fins are considered the best technique to solve the issue of the low thermal conductivity of the PCM and improve the general heat transfer performance in the LHS. Longitudinal, annular, pin, triangular, radial, array, and tree-like fins are the shapes studied by the researcher [59]. Mat et al. [60] utilised a longitudinal fin in the LHS, and they detected a 58% reduction in the phase change time at the constant HTF velocity and 86% under a constant inlet temperature of the HTF. Darzi et al. [61] numerically analysed the solidification process for the PCM in the TES combined with radial fins. They stated that utilising fins increase the solidification process due to increasing the surface area of the heat transfer, and this effect diminishes during the melting process due to annihilation of the natural convection. Pizzolato et al. [62] detected an increase in the melting and solidification rates by 37% and 15%, respectively, when high conductive fins were implanted in a small size TES. Yıldız et al. [63] studied the effect of the fins dimensions and structure (using tree-shaped fins) on the phase change rate, and they found that the rectangular shape has a stronger influence on the system. Yu et al. [64] studied the performance of the LHS using tree-shaped fins, they found that using such fins decreases the melting time by 27% and increases the heat storage rate by 45% than the conventional fins. Rathod and Banerjee [65] stated that the fins improve the TES with both charging and discharging processes. The main enhancements were found as 11%, 12%, and 15% with utilising internal, internal with external, and external triangle fins over the case with longitudinal fins. The rectangular fins supply an improvement rate of 15% over the triangle fins when the evaluation of the fins' configuration is achieved by Abdulateef et al. [66]. Shahsavar et al. [67] examined the influence of the fins locations in the vertical pipe LHS for the charging and discharging systems. The time of the melting and solidification reduces by 41% and 10% by using a uniform fins array compared with the non-uniform array.

Twisted fins have been recently used to improve the heat transfer characteristics in heat exchangers [68]. Providing a higher heat transfer area in the length unit of the heat exchanger is the main advantage of twisted-fins implementation in heat exchangers. Moreover, they generate a swirling flow in the liquid phase, leading to an enhancement in flow mixing and thermal boundary layer disturbance which in turn increases the heat transfer [69]. There are limited studies in the literature on the use of twisted-fins array in latent heat storage systems. Ghalambaz et al. [9,55] studied the twisted-fin array as an advanced form for increasing the phase change rate of the PCM in the shell-and-tube unit during the melting process. They stated that in a double-tube heat exchanger after optimising the geometrical parameters of the fin [9], the use of five twisted fins array improves the melting rate by 42% and the storage rate by 63% compared with the case with straight longitudinal fins using similar geometrical parameters. In a triple-tube heat exchanger, they showed that the use of four twisted fins reduced the melting time and melting rate by 18% and 25%, respectively, compared with the cases of using the same number of straight fins and no-fins considering a similar PCM mass.

In this study, three-dimensional numerical modelling of the PCM solidification process is simulated in a triplex tube LHS combined with twisted fins. The use of twisted fins along the inner perimeter of the annulus hosting the PCM in the triple-tube heat exchanger during the solidification is considered a new contribution to the existing literature. Fins are inserted into the PCM in the centre of the tube and located in a staggered alignment. According to the above comprehensive review, there is no published study regarding twisted-fins application in the triple-tube heat exchanger for intensifying the PCM solidification. It should be noted that the authors studied a similar geometry in their previous study [55] during the melting mode of the PCM and in this study, the solidification process is investigated as the process of heat transfer is different during the solidification compared with the melting due to natural convection especially for vertical geometries. Different simulations were run via ANSYS FLUENT 17.0 (Canonsburg, PA, USA) to evaluate the effectiveness of the innovative design of the fins compared to the base cases of straight fins and no-fins during the solidification. The main purpose was to find suitable and efficient fins number and the best values of the mass flow rate and the inlet temperature of the HTF.

Liquid and temperature contour plots and solidification rates are analysed scientifically to evaluate the discharge process. The results of this work provide guidelines for the novel structure of latent heat storage units.

2. Problem Description

A triple-tube LHS system with twisted fins (Figure 1a) was investigated during the solidification process compared with no-fin and straight fin cases, shown in Figure 1. The system was positioned vertically, and the PCM region was located in the middle tube. Hot water with the uniform inlet temperature of 10, 15, and 20 °C and Reynolds numbers of 500, 1000, and 1500 passed through the inner and outer tubes using RT35 as the PCM. Note that the Reynolds number is changed by the variation of inlet velocity of the HTF. The velocity of the HTF for the Reynolds number of 1000 is 0.055 m/s. The adiabatic outer tube was chosen to neglect heat loss from the system to the environment. A pressure outlet was applied for the outlet, and uniform inlet temperature and velocity were employed for the inlet. For the wall surfaces, the no-slip boundary condition was applied. Note that because of the advantages of counter-current flow directions for the working fluid to have a higher melting rate, this method was employed in this study [58]. The inner, middle, and outer diameters of the system were 20, 42, and 64 mm, respectively. The thickness of the inner and middle tubes was considered 2 mm, considering copper for the material of the fins and inner tube. To compare the effect of twisted fins addition with straight fins and no-fin cases, four copper fins were added to the system which the fins were externally and internally attached to the inner and the middle tubes, respectively, as shown in Figure 1. Then, a different number of 2, 4, and 6 fins with twisted configurations were also investigated. The fin pitch for the case of twisted fins was 3 cm. The initial temperature of the PCM is considered 50 °C. It is worth noting that this work was achieved with analysis of the numerical results only, and no experiment works were included.

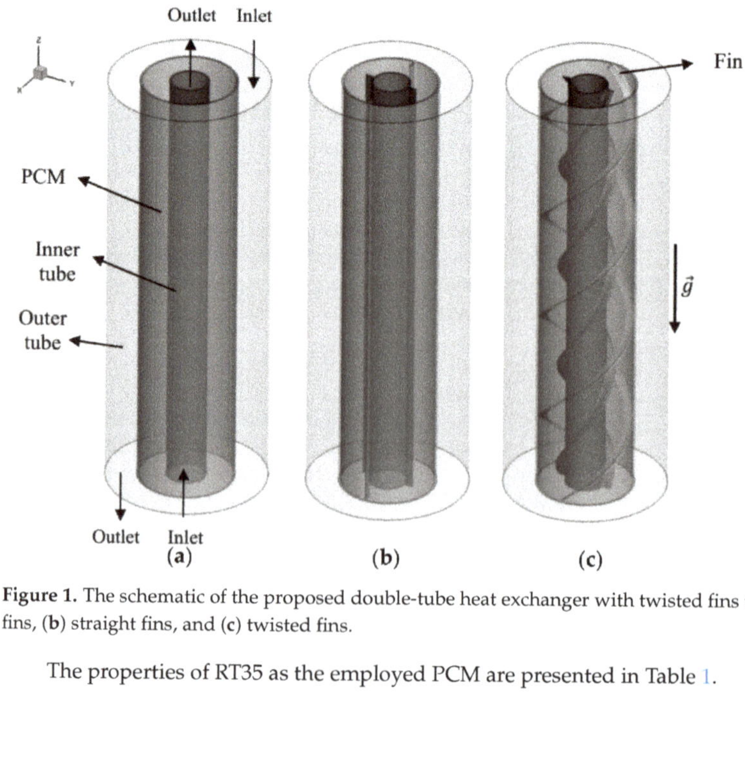

Figure 1. The schematic of the proposed double-tube heat exchanger with twisted fins using: (**a**) no fins, (**b**) straight fins, and (**c**) twisted fins.

The properties of RT35 as the employed PCM are presented in Table 1.

Table 1. Thermo-physical properties of RT35 [70].

Property	RT35
Specific heat (kJ/kg K)	2
Viscosity (Pa s)	0.023
Heat of fusion (kJ/kg)	170
Liquidus temperature (°C)	35
Density (kg/m^3)	815
Thermal conductivity (W/m K)	0.2
Solidus temperature (°C)	29
Thermal expansion coefficient (1/K)	0.0006

3. Mathematical Modeling

To calculate the phase change process numerically, the enthalpy–porosity approach was employed where, in each cell, the porosity and the liquid fraction were considered equal [71]. The Newtonian free convection flow of melted PCM was generated because of the buoyancy forces, which were transient and placed in the laminar flow regime because of the range of fluid velocity in the domain. The Boussinesq approximation was also employed in the momentum equation because of the small temperature gradient. Thus, the governing equations were derived based on these assumptions and are as follows neglecting Viscous dissipation [72]:

$$\frac{\partial \rho}{\partial t} + \nabla . \rho \vec{V} = 0 \tag{1}$$

$$\rho \frac{\partial \vec{V}}{\partial t} + \rho \left(\vec{V} . \nabla \right) \vec{V} = -\nabla P + \mu \left(\nabla^2 \vec{V} \right) - A_m \frac{(1-\lambda)^2}{\lambda^3 + 0.001} \vec{V} - \rho_{ref} \beta \left(T - T_{ref} \right) \vec{g} \tag{2}$$

$$\frac{\rho C_p \partial T}{\partial t} + \nabla \left(\rho C_p \vec{V} T \right) = -\left[\frac{\partial \rho \lambda L_f}{\partial t} + \nabla \left(\rho \vec{V} \lambda L_f \right) \right] + \nabla (k \nabla T) \tag{3}$$

where $\vec{V}$, T, λ, and P are the velocity vector, temperature, liquid volume fraction, and pressure, respectively; while t is time. T_{ref} and ρ_{ref} are the reference temperature and density. The third term on the right-hand side of Equation (2) represents the momentum sink for the phase change in the mushy zone [56]. The symbols L_f, ρ, C_p, k, μ, A_m, and β are the latent heat of fusion, density, specific heat capacity, thermal conductivity, dynamic viscosity, mushy, and volume expansion coefficient, respectively.

It is worth mentioning that the volume expansion of the PCM changing from the solid-state to the liquid-state was neglected, and the mushy zone constant was considered 10^{-5} based on the validation process and literature [71]. To simulate the flow of the water in the inner tube, the governing equations were the same as the above equations, ignoring the additional source of body forces and phase change. The liquid fraction, λ, is introduced as per Equation (4) [73]:

$$\lambda = \frac{\Delta H}{L_f} = \left\{ \begin{array}{ccc} 0 & if & T \leq T_S \\ \frac{(T-T_S)}{(T_L-T_S)} & if & T_S < T < T_L \\ 1 & if & T \geq T_L \end{array} \right\}, \tag{4}$$

where the subscripts S and L denote the solidus and liquidus states of PCM, and ΔH is the enthalpy variation during the phase change. The solidification or discharging rate $\dot{Q}$ is introduced as per Equation (5) [53]:

$$\dot{Q} = \frac{Q}{t_m} = \frac{m\left(\int_S C_p dT + L_f + \int_L C_p dT \right)}{t_m}, \tag{5}$$

where t_m is the melting time and m is the mass of PCM. The total enthalpy (H) is achieved as per Equation (6):

$$H = \Delta H + h, \tag{6}$$

where,

$$h = \int_{T_{ref}}^{T} C_p dT + h_{ref}. \tag{7}$$

A detailed description of the mathematical model can be found in the author's previous work [71].

4. Numerical Process

ANSYS computational fluid dynamic software (FLUENT) was employed to solve the problem using the SIMPLE algorithm for the pressure–velocity coupling scheme. The QUICK scheme was used to discretise the terms of the derivatives in the momentum and energy equations, while the PRESTO scheme was used for the continuity equation. For different equations governed, 10^{-6} was used as the convergence criteria. The grid independence analysis was performed before the main simulations considering different mesh and time-step sizes to determine the results independent from the grid number and the time step size. The grid independence test was performed using different cell numbers of 2,302,000, 2,357,000, and 2,451,000 for the case with six twisted fins shown in Figure 1c. The melting time was considered as the criteria to find the mesh independent from the number of cells. The melting time for the system with different cell numbers of 2,302,000, 2,357,000, and 2,451,000 are 2004, 2083, and 2098 s, respectively. The results showed that for the case of twisted f0.33inned triple-tube with six fins, 2,357,000 cells were enough to have independent results from the number of grids tested. The difference between the melting times for the cases with 2,357,000 and 2,451,000 cells is less than 0.3%. Different time step sizes of 0.1, 0.2, and 0.4 s are studied to find the results independent from the size of the time step. The results showed similar melting times for different sizes of time step, and therefore, the size of the time step was also selected equal to 0.2 s. The configuration of the final mesh is shown in Figure 2.

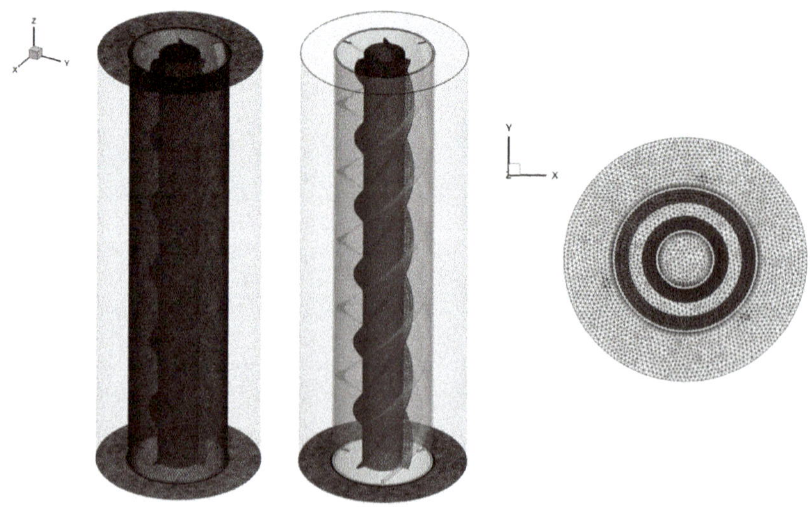

Figure 2. The configuration of the mesh after grid independence analysis.

The numerical model is verified using the experimental and numerical results of Mat et al. [60], where the effect of fins attached to both outer and inner surfaces of the tubes in the PCM zone (RT58) in a double-tube LHSHE unit was studied. In this study, constant wall temperature was implemented for the walls of the heat exchanger. As seen from Figure 3, the presented results are in line with the experimental data as well as numerical results for the temperature and numerical data for the melt fraction of Mat et al. [60]. It should be noted that the study of Mat et el. has been used in various studies in the literature to validate different codes.

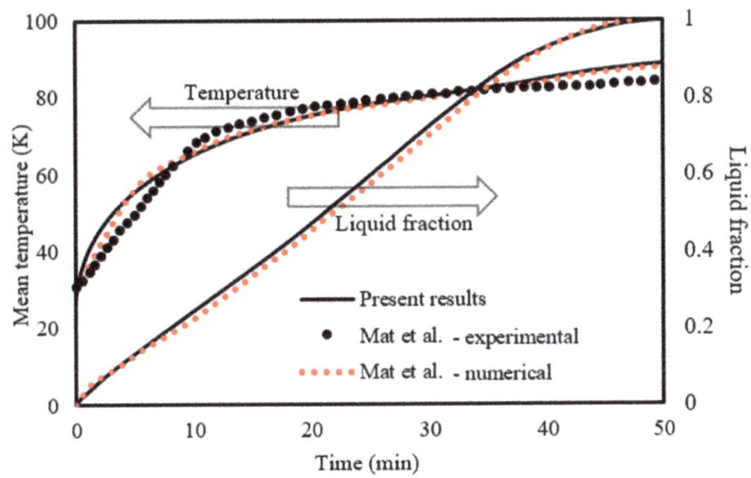

Figure 3. Verification of the numerical model.

5. Results and Discussion

Several simulation tests have been conducted in order to assess the potential of twisted fins on intensifying the solidification rate of paraffin (RT35) in the vertical TES triple-tube system. Three cases with no fins, straight fins, and twisted fins, are dealt with in this study. There are two, four, and six fins involved in the case of twisted fins. The total mass of PCM is fixed at 0.335 kg, which is equal to the mass of the PCM in the case without any fins, in all of these cases to enable making meaningful performance comparisons. It should be noted that including a denser material of the enhancer (fins) does help faster rates of solidification but also impact the mass/volume of PCM being occupied in the TES unit. Therefore, the storage capacity of the system is negatively affected. To reveal the impact of twisted fins on the system's thermal response, the present results were studied in terms of the liquid-fraction contours, isothermal contour distribution, and temporal fluid-fraction profiles. In any of these cases, it was supposed that the scenario to achieve the total solidification started when the PCM at an initial temperature (T_{int} = 305 K) was above the PCM liquidus temperature (T_s = 302 K) while the HTF (water) circulating at a lower temperature (T_{HTF} = 323 K). This supports the formation of a solidifying layer next to the thermally-active walls so that the PCM molecules near the cooling walls initiate the solidifying phase earlier than the other PCM parts. Over time the layers grow progressively to intrude the entire PCM domain when additional amounts of heat are removed by the heat-transfer fluid (HTF) flowing outside. The presence of twisted or straight fins serves as an extra promoter for better heat communication between the HTF and PCM so that faster heat removal rates from the PCM are achieved, as seen in the next sections. It would be worthy to mention that achieving a faster time response for heat removal in actual TES application is a critical part to consider when designing a TES system. It indicates the TES system's ability to achieve a continuous and stable operation of the energy recovery [74]. If the system fails to timely respond to the energy discharging duties on the PCM side, this causes delays in attaining the cyclic solidification within the time limit, and consequently, a failure of the system's design becomes more probable.

5.1. Evolution of the Temperature Field

Isotherm evolution for the three cases of no fins, straight fins, and twisted fins considered in this study over the three-time spans of solidification 600, 1200, and 1800 s are shown in Figure 4. During the initial span ($t \leq 600$ s), the isotherms of low temperatures around 290 K (the blue zones) seem to be the dominant group over the entire cavity in all examined cases. This is attributed to the fact that a major part of PCM is still liquid during this duration, and so, not much influence of solidification can be seen. Therefore, the isotherms seem similar between and close to the fin ligaments. During the subsequent span ($600 < t \leq 1200$ s), the isotherms of high temperatures above 305 K (the red zones) slightly start to shrink throughout the entire domain, particularly in the cases with fins. By this time, the cooling effect of the heat-transfer fluid (HTF) on the PCM becomes more effective so that the solidifying layer increases in size to occupy the whole domain. Moreover, the existence of fins further supports the heat communication between the PCM and the HTF, as the existence of fins allows for better heat removal from the PCM domain. Comparing the cases with fins to the base case of no fins suggests a larger shrink in the layer of isotherms with high temperatures (the red zone), particularly when moving downward. This implies that there is little influencing role of convection compared to the role of conduction in the case of fins, particularly in the upper portion of the domain. During the final span ($t \geq 1800$ s), the isotherms seem to be more uniform and consistent in shape than in previous durations, particularly in the lower portion of the domain. This would be due to the dominating role of heat conduction in this region, which helps earlier completion of the solidification process. Indeed, the existence of twisted fins further supports the role of heat conduction due to its curving structure, which limits the role of natural convection, and the relatively larger heat transfer area, which assists the overall heat transfer process.

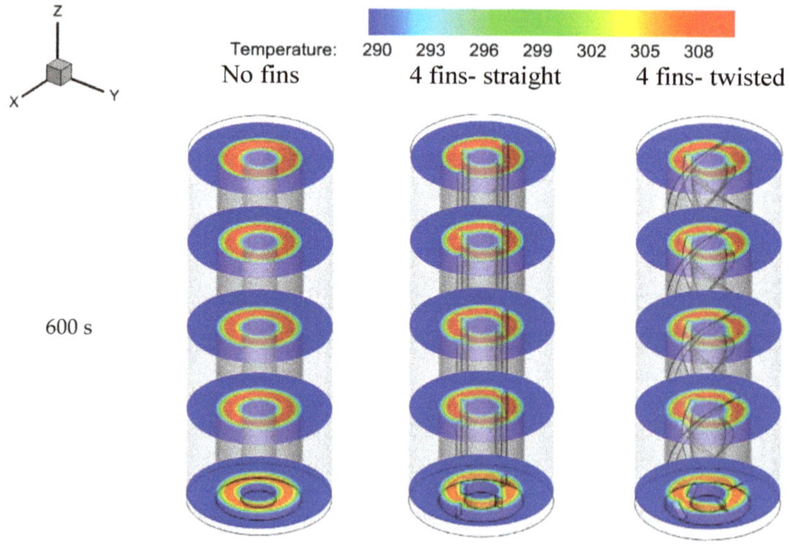

Figure 4. *Cont.*

1200 s

1800 s

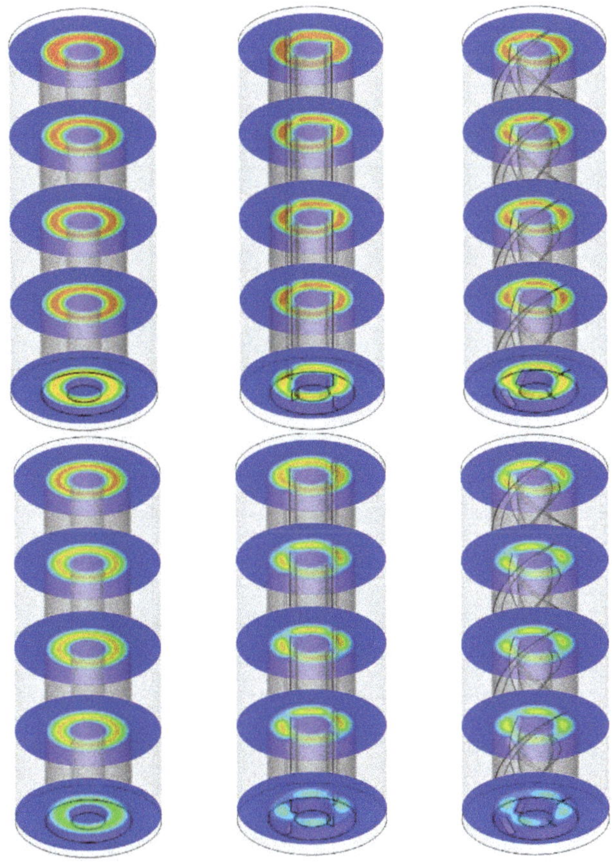

Figure 4. Evolution of temperature field for the cases of no fins, straight fins, and twisted fins at different time spans of solidification progress.

Figure 5 compares the temperature field of the cases with two, four, and six twisted fins over three different time spans of the solidification evolution. Not a big difference in the distribution of isotherms can be noticed as the number of fins increases during the early duration ($t \leq 600$ s) as a major part of PCM is not yet solidified. However, as time proceeds to $t \geq 1200$ s, the size of solidifying layer increases due to the enhanced heat removal from the PCM with the existence of a higher number of fins. The effect of increasing the number of fins seems to be more noticeable in the lower portion of the domain. The movement of liquid PCM in the vertical TES units is typically governed by the dominance of gravity effect over the buoyancy effect, which after all results in a larger temperature gradient throughout the domain. This is why the PCM at the bottom within the case of six fins is early terminated solidification and appeared completely blue clearer than that in other cases, as shown in Figure 5 (t = 1800 s).

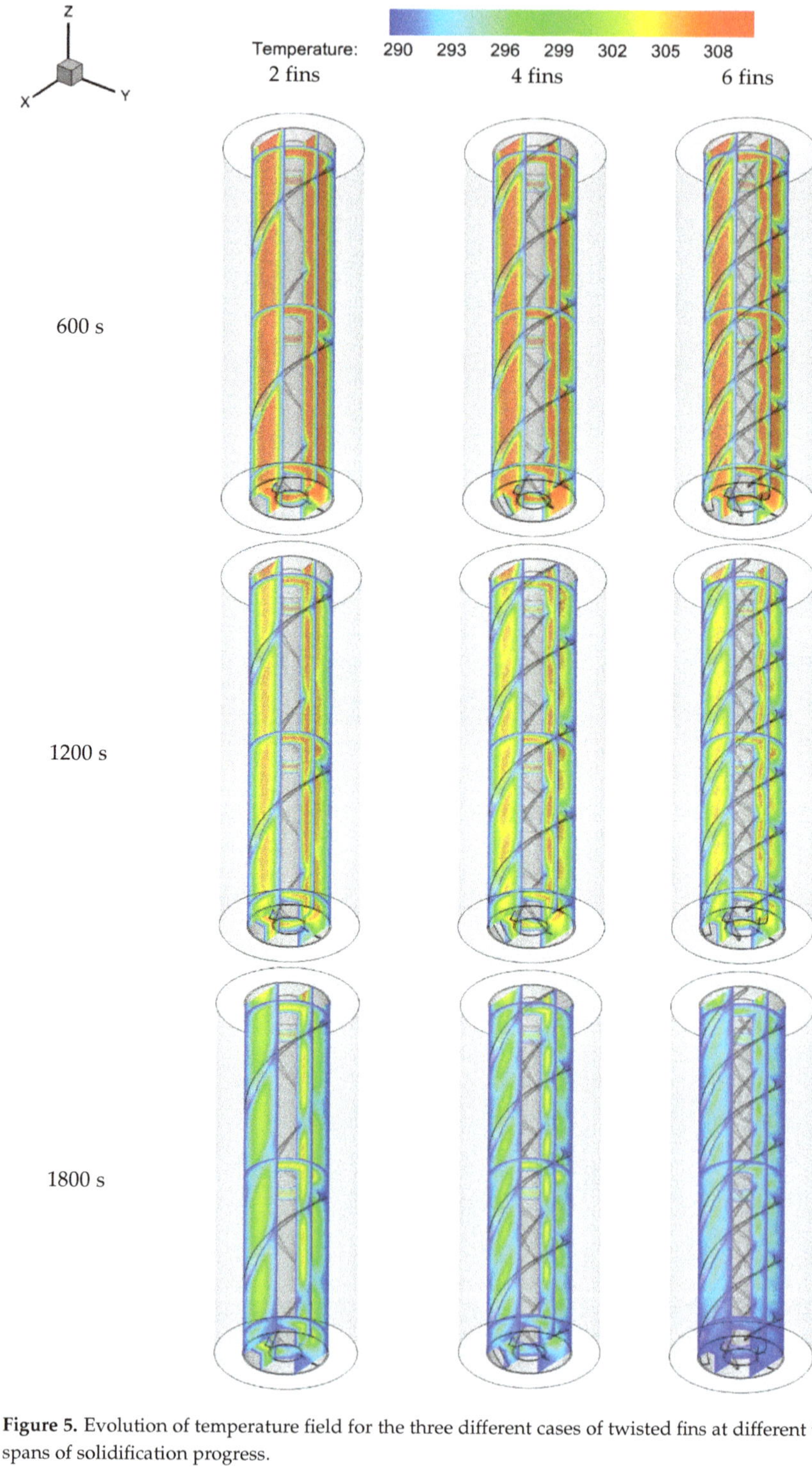

Figure 5. Evolution of temperature field for the three different cases of twisted fins at different time spans of solidification progress.

The effects of adding straight and twisted fins on the average temperature of the PCM over different time spans are shown in Figure 6. Adding fins typically provides better heat removal from the PCM as fins work as direct passageways for heat communication between the PCM and the cooling walls. However, the data from Figure 6 indicate that applying twisted fins serves better for lowering the PCM temperature so that higher solidifying rates can be obtained. Figure 7 compares the effects of using different numbers of twisted fins on the time histories of PCM temperature during solidification mode. It can be seen that increasing the number of fins does not introduce a significant difference at the early stage (i.e., for $t \leq 500$ s) of solidification. However, as time progresses, the difference in the behavior curves of the average temperature becomes more noticeable as the number of fins is changed. The average temperature declines to its minimum value within a shorter time when the number of fins is doubled from two to four and six fins. Therefore, faster heat discharge rates from the PCM can be produced by increasing the number of installed twisted fins.

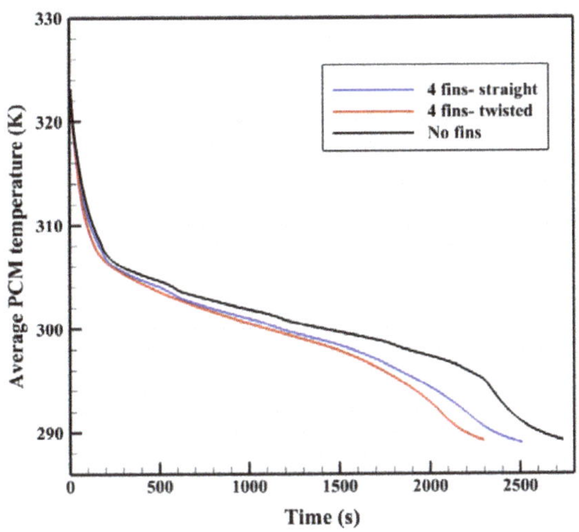

Figure 6. Time history of average temperature for the cases of no fins, straight fins, and twisted fins.

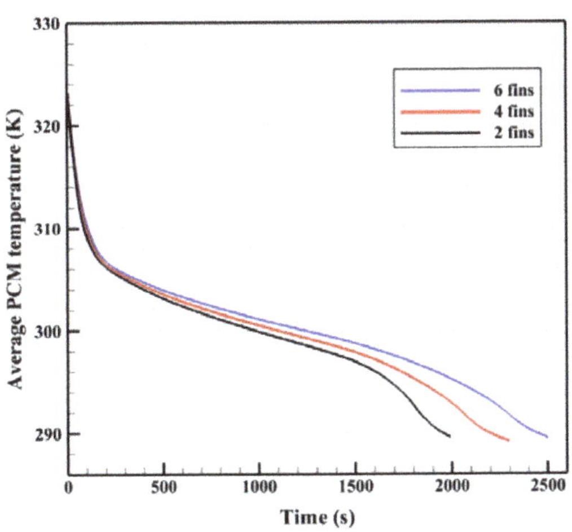

Figure 7. Time history of average temperature for the three different cases of twisted fins.

5.2. Evaluation of Velocity Distribution

In addition to the significant effect of fins on the melting rate through the conduction heat transfer, twisted fins can be effective on the natural convection effect in the storage system. Figure 8 displays the z-velocity in the middle cross-section of the system for different studied cases. As shown, higher velocities can be seen in the systems with twisted fins showing the higher effect of natural convection in the domain. It should be noted that the direction of the twisted fins is along with the gravity direction which can be helpful in boosting the buoyancy effect in the domain. In other words, the twisted fins do not prevent the circulation of melted PCM in the domain due to the twisted shape of the fins.

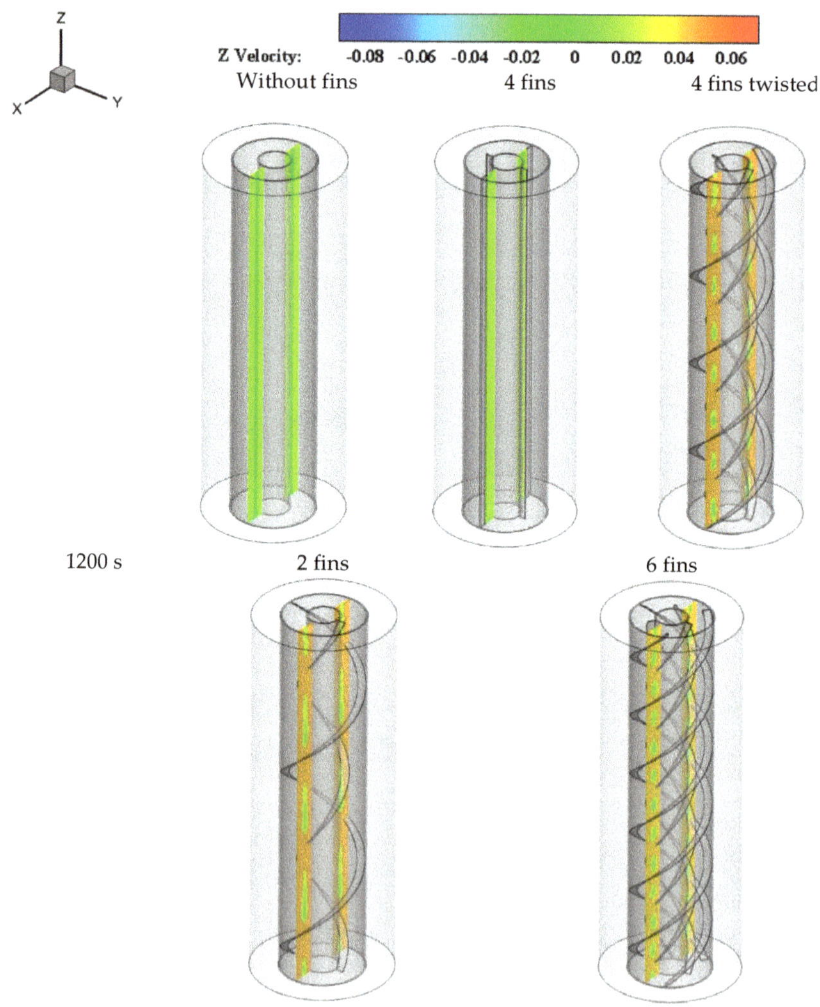

Figure 8. Evolution of z-velocity field for different studied cases at the time of 1200 s.

5.3. Evolution of the Liquid-Fraction Field

The liquid-fraction contours, including the solidifying fronts (shown in light green), are illustrated in Figure 9 for the cases of no fins, straight fins, and twisted fins at five different vertical positions within time spans of 600, 1200, and 1800 s of solidification. During the early time span ($t \leq 600$ s) and where no fins are present, the solidifying fronts (shown by light green) take almost the shape of uniform circles that are identical along the vertical direction. However, the addition of fins, particularly twisted fins, results in the formation of relatively bigger solidifying layers (blue areas) adjacent to and surrounding the fins, as seen in Figure 9. The fronts typically do not move away from one another because only thin solid layers can be formed during this time range. Moving to the next time span ($600 < t \leq 1200$ s), the shapes of solidifying fronts get further deformed, particularly in cases when fins are present, due to the higher rates of heat removal at the cooling walls. The size of the solidifying layers (blue zones) appears to be gradually increased towards the bottom as can be seen in Figure 9 (t = 1200 s). This is due to the stronger role of convection in the upper portion of the domain compared to that in the lower portion. The solidifying layer better increases in size in the case of twisted fins as the major space turns blue. In the case of twisted fins, a more important role is noticed for natural convection with more deformation in the shape of the solidifying front. The reason is that the flow-resistant forces generated due to the flow-promoting structure of twisted fins are superior compared to that in straight fins. During the last time span ($t \geq 1800$ s), the liquid-fraction field shows fully solidified zones in the lower portion of the domain due to the minimal convection role in this part, while solidification is slightly delayed in the upper portion due to the stronger local convection. On the effects of fins, the results show that the size of the solid layer increases as the fins (twisted or straight) are added. However, the effect of twisted fins is more noticeable as the size of the solidifying layer is larger, as can be seen by comparing the liquid-fraction field between the three cases considered in Figure 9 (t = 1800 s). Thus, applying twisted fins while keeping the total mass of PCM constant leads to faster evolution of solidifying frons than that when applying straight fins.

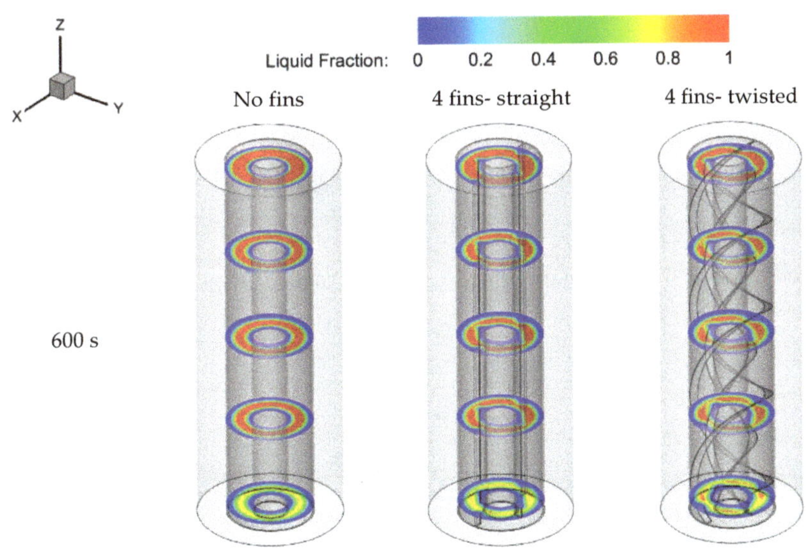

Figure 9. *Cont.*

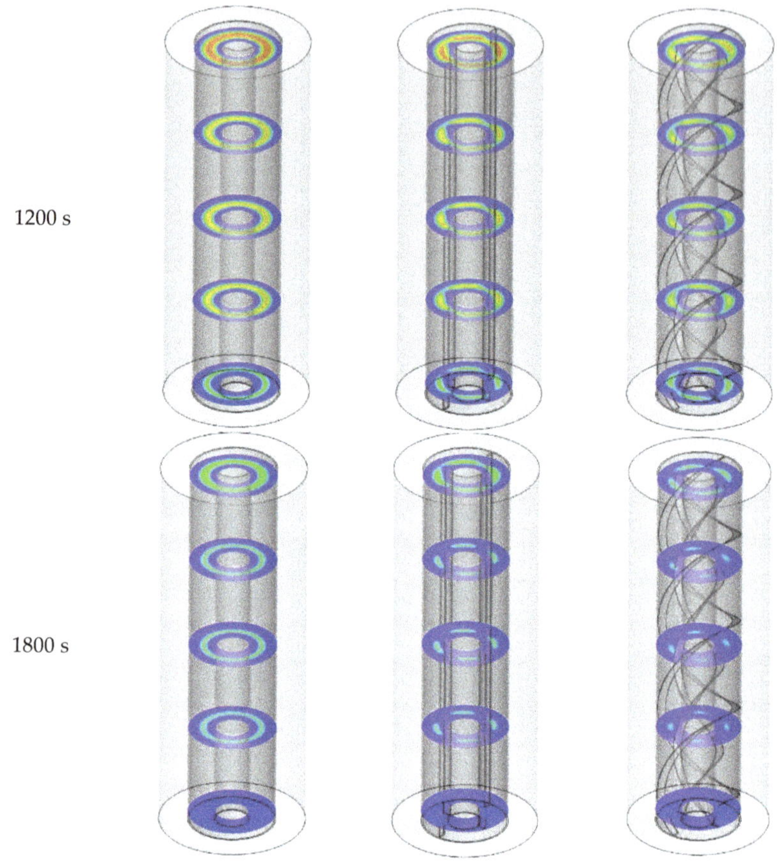

Figure 9. Evolution of liquid-fraction field for the cases of no fins, straight fins, and twisted fins at different time spans of solidification progress.

Figure 10 compares the liquid-fraction field of the cases with two, four, and six twisted fins over three different time spans of the solidification evolution. A noticeable deformation in the shape of solidifying fronts can be observed as the number of fins increases due to the enhanced heat removal from the PCM with increasing the number of fins. The size of the solidifying layers (blue zones) appears to be gradually increased towards the bottom as can be seen in Figure 10 (t = 1800 s). The solidification is observed to be earlier completed in the lower parts of the vertical TES unit, indicating a strong conduction involvement in this area. A little delay occurs in all cases of twisted fins in the upper parts of the domain. The existence of fins impacts the buoyant flow of liquid PCMs in these parts so that only minor involvement of the convection is expected in the heat transfer process. Increasing the number of twisted fins from two to six greatly aids the solidification process at the upper parts of the domain. The twisted fins technically improve heat transport so that the solidifying front appears to travel quicker in two ways. First, their enormous surface area aids for superior heat transfer by conduction between the various portions of the PCM. Second, the twisted structure of fins allows for higher flow-resistant forces so that only a minor role can be played by natural convection.

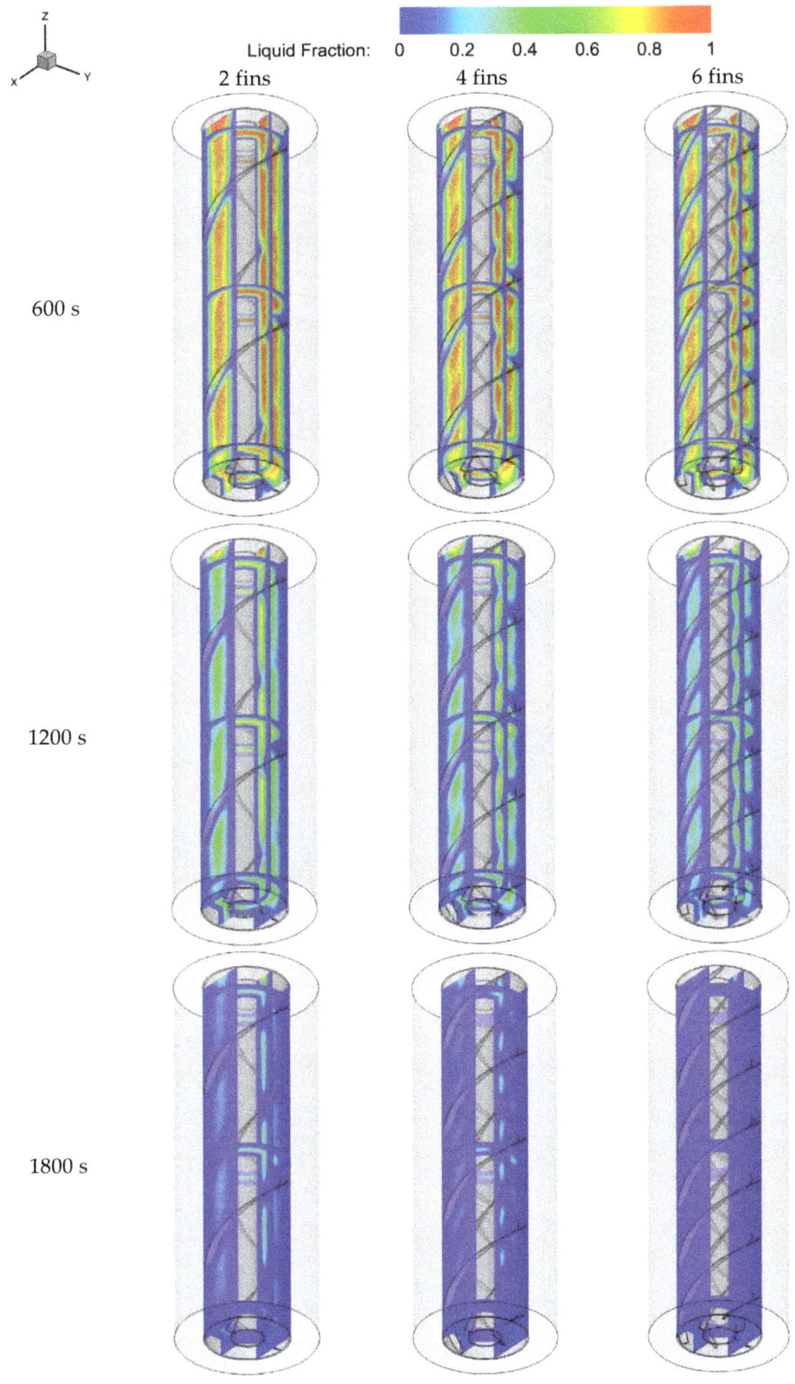

Figure 10. Evolution of liquid-fraction field for the three different cases of twisted fins at different time spans of solidification progress.

The temporal evolution of the liquid fraction throughout the PCM solidification mode has been also tracked to better evaluate the potential of solidification enhancement when twisted fins are applied. Figure 11 compares the liquid-fraction evolution history for the three cases of no fins, straight fins, and twisted fins. The TES system in the three cases is designed to carry out the same PCM mass (m = 0.335 kg). Data from the figure indicate that the case with twisted fins provides the best potential for solidification enhancement among the cases considered. Table 2 shows that the case with four twisted fins can reduce the solidifying time from 2739 s in the base case of no fins to only 2229 s so that a time saving of about 20% is achieved. In addition, twisted fins can cut solidifying time from 2512 s in the case of straight fins to 2229 s, saving roughly 8% of the entire solidifying time. Regarding the discharge rate, data from Table 2 imply that applying twisted fins can remove heat from PCM at the rate of 34.25 W while applying straight fins would remove 30.45 W of heat compared to only 27.87 W in the base case of no fins. This results in an increase in the heat discharge rate of around 29% and 10%, respectively, when compared to the reference case of applying no fins.

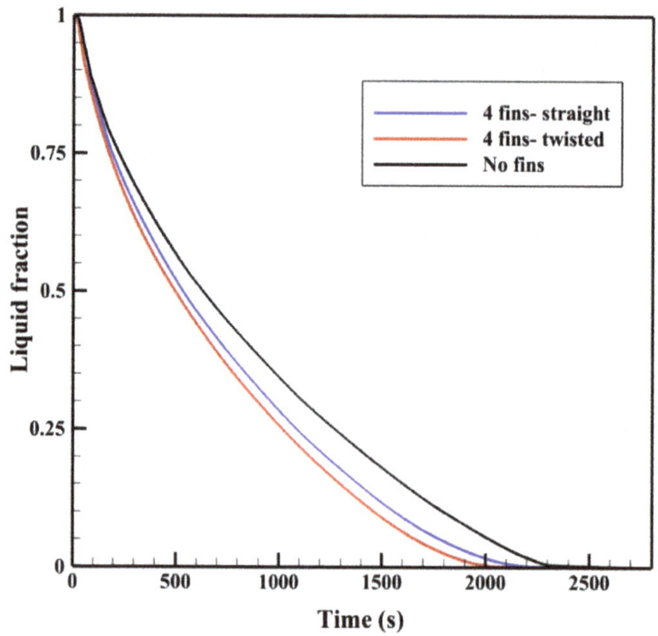

Figure 11. Temporal evolution of the PCM liquid fraction with straight and twisted fins.

Table 2. The improvement in solidifying time and discharging rate due to the inclusion of straight and twisted fins.

Fins	Solidifying Time (s)	Discharging Rate (W)
No fins	2739	27.87
Straight fins	2512	30.45
Four twisted fins	2229	34.25

The time histories for liquid-fraction evolution in the cases of two, four, and six twisted fins are compared in Figure 12. As seen in the figure, applying six twisted fins leads to the greatest possible reduction in solidifying time. Based on the data obtained for the number of fins considered, the solidifying rate generally increases as the number of twisted fins increases. The time data from Table 3 reveals that the TES system design with four and six twisted fins do require 2229 and 1922 s, respectively, to reach the status of complete solidification of PCM. This implies that an 11 to 20 % reduction in solidifying time is possible when doubling the number of twisted fins in use from two to four or six fins, respectively. Data from Table 3 also reveal that applying twisted fins can improve the heat discharge rate from 30.45 W to 34.25 W and 38.20 W when the number of twisted fins in use increases from two to four or six fins, respectively. This results in an increase in the heat discharge rate of around 13% to 26%, respectively.

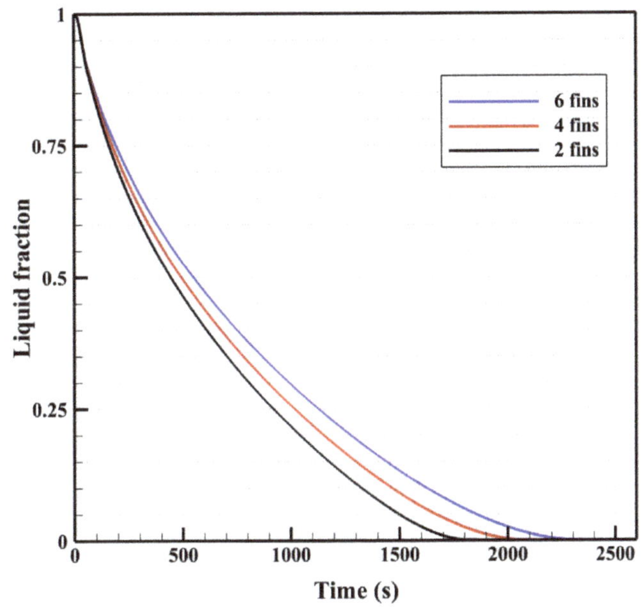

Figure 12. Temporal evolution of the PCM liquid fraction with different numbers of twisted fins.

Table 3. The improvement in solidifying time and discharging rate due to the inclusion of different twisted fins.

Fins	Solidifying Time (s)	Discharging Rate (W)
2 fins	2502	30.45
4 fins	2229	34.25
6 fins	1992	38.20

To have a non-dimensional analysis, the dimension less time is defined as follows:

$$\tau = \frac{t\alpha}{D} \tag{8}$$

where D is the hydraulic diameter of the PCM container equals 0.02 m. It should be noted that the mass of the PCM is considered constant in all the studied cases. Table 4 presents the dimensionless solidification time for different studied cases. The use of six fins results in the lower non-dimensional solidification time which is almost 27.3% less than that for the case without fins.

Table 4. Dimensionless solidification time for different studied cases.

	No-Fins	Straight Fins	2 Fins	4 Fins	6 Fins
Dimensionless time	0.016804	0.015411	0.01535	0.013675	0.012221

5.4. Impact of HTF Reynolds Number on Solidification of a PCM with Twisted Fins

Different flow rates of the HTF that are corresponding to Reynolds number values of 500, 1000, and 1500 are examined in terms of liquid-fraction profile and average temperature profile in Figure 13a,b, respectively. The data from Figure 13a indicate that a shorter solidifying time is needed when a higher Reynolds number of the HTF are used. This is due to the fact that higher Reynolds numbers inspire a greater convective heat transfer coefficient at the thermally-active walls so that greater heat removal rates from the PCM are achieved during solidification. The corresponding data in Figure 13b show that almost lower values of the average PCM temperature can be recorded when higher Reynolds numbers of the HTF are used. This also implies that a better cooling effect can be done on the PCM side when HTF with a higher Reynolds number is used. The PCM with twisted fins takes around 2100 s to complete the solidification at Re = 500, but only 1980 and 1850 s at Re = 1000 and 1500, respectively. Therefore, when the HTF Reynolds number is increased from 500 to 1000 and 1500, the total solidification time can be saved by around 5% and 12%, respectively.

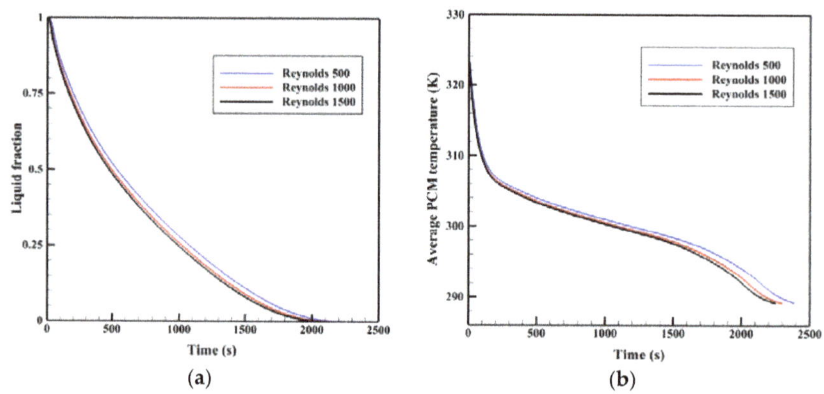

Figure 13. Evolution of the PCM liquid-fraction and average temperature profiles at three different Reynolds numbers of the HTF. (**a**) Liquid-fraction profile, (**b**) Average PCM temperature profile.

5.5. Impact of HTF Temperature on Solidification of a PCM with Twisted Fins

Figure 14 shows the impact of varying the HTF temperature on the time-wise evolution of the PCM liquid fraction and average temperature, respectively for T_{HTF} = 10, 15, and 20 °C. As can be seen in this figure the values of liquid fraction and average PCM temperature decreases as the HTF temperature increases. In other words, lowering the HTF temperature promotes a greater cooling impact on the PCM side. This is basically due to the fact that utilising a cooler HTF allows for a quicker solidification rate of PCM. This trend appears to be increasingly pronounced as the process approaches the point of solidification completion. As explained earlier, the contribution of conduction in the heat removal process from the PCM becomes more effective and controlling within the final period ($t \geq 1800$ s) of solidification. Data from the figure shows that a PCM with twisted fins takes around 2700 s to complete the solidification at T_{HTF} = 20, but only 2100 and 1650 s at T_{HTF} = 15 and 10 °C, respectively. Therefore, when the HTF temperature is decreased from 20 to 15 and 10 °C, the total solidification time can be saved by around 28% and 40%, respectively.

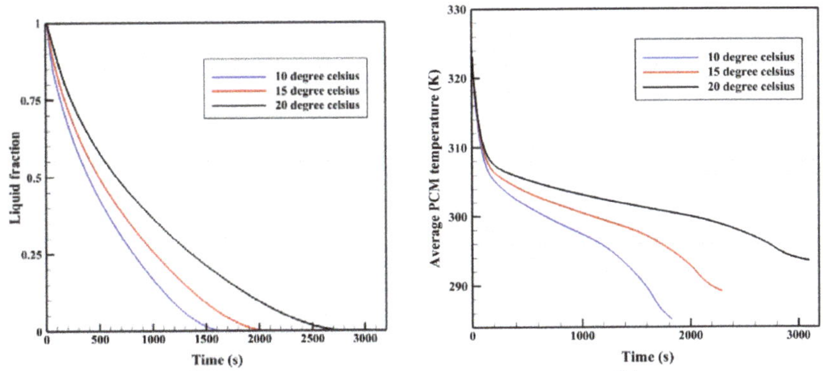

(**a**) (**b**)

Figure 14. Evolution of PCM liquid-fraction and average temperature profiles at three different temperatures of HTF. (**a**) Liquid-fraction profile, (**b**) Average PCM temperature profile.

6. Conclusions

A combination of twisted fins with a triple-tube thermal energy storage system was explored and assessed in the three-dimensional modeling during the solidification process. This work involved the influence of planting the twisted fins compared with the cases of straight fins and no-fins. The effects of the inserted fins' number, inlet temperature, and the flow rate (represented with the Re) of the heat transfer fluid were evaluated. The PCM was located between the inner and the outer tubes, which include the heat transfer fluid flows in an opposite direction as the best technique for releasing heat from the PCM. The performance of the unit was evaluated by analysing the reduction of the liquid fraction and the thermal profile, as well as the solidification time and discharge rate. The outcomes specify the benefits of combining the twisted fins with the TES. The results reveal that the utilizing of four twisted fins reduced the solidification time by 12.7% and 22.9% compared with four straight fins and the no-fins cases (assuming the same mass of the PCM), respectively. Likewise, applying four twisted fins enhanced the discharging rate compared with four straight fins and the no-fins. Increasing the number of fins from two to four and six, the solidification time reduces by 11.9% and 25.6%, respectively. Adding fins enhances the thermal removal from the PCM as fins work as direct passageways for heat communication between the PCM and the cooling walls. The solidification rate increases with increasing the Reynolds number (Re); When the Re of the HTF is increased from 500 to 1000 and 1500, the solidification time is reduced by 5% and 12%, respectively. Further, the solidification rate increase with decreasing the heat transfer fluid temperature; when the HTF temperature is reduced from 20 to 15 and 10 °C, the discharge time decreased by 28% and 40%, respectively. This work offers an innovative design for adding fins to improve the thermal efficiency of the LHTES units.

Author Contributions: Conceptualization, X.S., P.T.; methodology, X.S., P.T.; software, P.T.; validation, P.T.; formal analysis, X.S., J.M.M., H.I.M., H.S.M., W.Z. and P.T.; investigation, X.S., J.M.M., H.I.M., H.S.M., W.Z. and P.T.; resources, X.S., P.T.; writing—original draft preparation, X.S., J.M.M., H.I.M., H.S.M., W.Z. and P.T.; writing—review and editing, X.S., J.M.M., H.I.M., H.S.M., W.Z. and P.T.; visualization, X.S., W.Z. and P.T.; supervision, X.S., P.T.; All authors have read and agreed to the published version of the manuscript.

Funding: The study was supported by the Jiangsu Provincial Basic Research Program (Natural Science Fund) (Grant no. BK20191050), Natural Science Research Project of Jiangsu Province Colleges and Universities (Grant no. 18KJD560001), Philosophy and Social Science Project of Jiangsu Province Colleges and Universities (Grant no. 2019SJA1659).

Institutional Review Board Statement: Not applicable.

Informed Consent Statement: Not applicable.

Data Availability Statement: The data will be available on request.

Conflicts of Interest: The authors declare no conflict of interest.

Nomenclature

Nomenclature	
A_m	The mushy zone constant
C	Inertial coefficient
C_p	PCM specific heat (J/kgK)
D	Hydrolic diameter (m)
g	Gravitational acceleration (m/s^2)
K	thermal conductivity (W/mK)
L	Latent heat of fusion (J/kg)
m	PCM mass (kg)
P	Pressure (Pa)
$\dot{Q}$	Solidification rate (J)
t_m	solidification time (s)
T	Temperature (K)
T_m	Melting point temperature (K)
u_i	Velocity component (m/s)
$\vec{V}$	Velocity vector (m/s)
Greek symbols	
β	Thermal expansion coefficient (1/K)
λ	Liquid fraction
α	Thermal diffusivity (m^2/s)
μ	Dynamic viscosity (kg/ms)
ρ	Density (kg/m^3)

References

1. Wu, W.-F.; Liu, N.; Cheng, W.-L.; Liu, Y. Study on the effect of shape-stabilized phase change materials on spacecraft thermal control in extreme thermal environment. *Energy Convers. Manag.* **2013**, *69*, 174–180. [CrossRef]
2. Guo, X.; Liu, J.; Dai, L.; Liu, Q.; Fang, D.; Wei, A.; Wang, J. Friction-wear failure mechanism of tubing strings used in high-pressure, high-temperature and high-yield gas wells. *Wear* **2021**, *468*, 203576. [CrossRef]
3. Zhao, X.; Gu, B.; Gao, F.; Chen, S. Matching model of energy supply and demand of the integrated energy system in coastal areas. *J. Coast. Res.* **2020**, *103*, 983–989. [CrossRef]
4. Mohammed, H.I.; Giddings, D.; Walker, G.S. CFD multiphase modelling of the acetone condensation and evaporation process in a horizontal circular tube. *Int. J. Heat Mass Transf.* **2019**, *134*, 1159–1170. [CrossRef]
5. Zhang, L.; Zheng, H.; Wan, T.; Shi, D.; Lyu, L.; Cai, G. An integrated control algorithm of power distribution for islanded microgrid based on improved virtual synchronous generator. *IET Renew. Power Gener.* **2021**, *15*, 2674–2685. [CrossRef]
6. Du, X.; Li, J.; Niu, G.; Yuan, J.-H.; Xue, K.-H.; Xia, M.; Pan, W.; Yang, X.; Zhu, B.; Tang, J. Lead halide perovskite for efficient optoacoustic conversion and application toward high-resolution ultrasound imaging. *Nat. Commun.* **2021**, *12*, 3348. [CrossRef]
7. Dai, Z.; Xie, J.; Chen, Z.; Zhou, S.; Liu, J.; Liu, W.; Xi, Z.; Ren, X. Improved energy storage density and efficiency of (1−x) $Ba_{0.85}Ca_{0.15}Zr_{0.1}Ti_{0.9}O_3$-x$BiMg_{2/3}Nb_{1/3}O_3$ lead-free ceramics. *Chem. Eng. J.* **2021**, *410*, 128341. [CrossRef]
8. Dai, Z.; Xie, J.; Fan, X.; Ding, X.; Liu, W.; Zhou, S.; Ren, X. Enhanced energy storage properties and stability of $Sr(Sc_{0.5}Nb_{0.5})O_3$ modified 0.65$BaTiO_3$-0.35$Bi_{0.5}Na_{0.5}TiO_3$ ceramics. *Chem. Eng. J.* **2020**, *397*, 125520. [CrossRef]
9. Ghalambaz, M.; Mohammed, H.I.; Mahdi, J.M.; Eisapour, A.H.; Younis, O.; Ghosh, A.; Talebizadehsardari, P.; Yaïci, W. Intensifying the Charging Response of a Phase-Change Material with Twisted Fin Arrays in a Shell-And-Tube Storage System. *Energies* **2021**, *14*, 1619. [CrossRef]
10. Yang, X.; Guo, J.; Yang, B.; Cheng, H.; Wei, P.; He, Y.-L. Design of non-uniformly distributed annular fins for a shell-and-tube thermal energy storage unit. *Appl. Energy* **2020**, *279*, 115772. [CrossRef]
11. Zhang, B.; Xu, D.; Liu, Y.; Li, F.; Cai, J.; Du, L. Multi-scale evapotranspiration of summer maize and the controlling meteorological factors in north China. *Agric. For. Meteorol.* **2016**, *216*, 1–12. [CrossRef]
12. Berardi, U.; Soudian, S. Benefits of latent thermal energy storage in the retrofit of Canadian high-rise residential buildings. *Build. Simul.* **2018**, *11*, 709–723. [CrossRef]
13. Sun, L.; Li, C.; Zhang, C.; Su, Z.; Chen, C. Early monitoring of rebar corrosion evolution based on FBG sensor. *Int. J. Struct. Stab. Dyn.* **2018**, *18*, 1840001. [CrossRef]

14. Mahdi, J.M.; Mohammed, H.I.; Talebizadehsardari, P. A new approach for employing multiple PCMs in the passive thermal management of photovoltaic modules. *Sol. Energy* **2021**, *222*, 160–174. [CrossRef]
15. Qureshi, Z.A.; Ali, H.M.; Khushnood, S. Recent advances on thermal conductivity enhancement of phase change materials for energy storage system: A review. *Int. J. Heat Mass Transf.* **2018**, *127*, 838–856. [CrossRef]
16. Yang, X.; Wei, P.; Wang, X.; He, Y.-L. Gradient design of pore parameters on the melting process in a thermal energy storage unit filled with open-cell metal foam. *Appl. Energy* **2020**, *268*, 115019. [CrossRef]
17. Xu, Q.; Wang, K.; Zou, Z.; Zhong, L.; Akkurt, N.; Feng, J.; Xiong, Y.; Han, J.; Wang, J.; Du, Y. A new type of two-supply, one-return, triple pipe-structured heat loss model based on a low temperature district heating system. *Energy* **2021**, *218*, 119569. [CrossRef]
18. Mehryan, S.A.M.; Ghalambaz, M.; Sasani Gargari, L.; Hajjar, A.; Sheremet, M. Natural convection flow of a suspension containing nano-encapsulated phase change particles in an eccentric annulus. *J. Energy Storage* **2020**, *28*, 101236. [CrossRef]
19. Abdulateef, A.M.; Abdulateef, J.; Al-Abidi, A.A.; Sopian, K.; Mat, S.; Mahdi, M.S. A combination of fins-nanoparticle for enhancing the discharging of phase-change material used for liquid desiccant air conditioning unite. *J. Energy Storage* **2019**, *24*, 100784. [CrossRef]
20. Mahani, R.B.; Mohammed, H.I.; Mahdi, J.M.; Alamshahi, F.; Ghalambaz, M.; Talebizadehsardari, P.; Yaïci, W. Phase Change Process in a Zigzag Plate Latent Heat Storage System during Melting and Solidification. *Molecules* **2020**, *25*, 4643. [CrossRef]
21. Shanan, Z.J.; Hadi, S.M.; Shanshool, S.K. Structural Analysis of Chemical and Green Synthesis of CuO Nanoparticles and their Effect on Biofilm Formation. *Baghdad Sci. J.* **2018**, *15*, 0211. [CrossRef]
22. Mahdi, J.M.; Nsofor, E.C. Solidification of a PCM with nanoparticles in triplex-tube thermal energy storage system. *Appl. Therm. Eng.* **2016**, *108*, 596–604. [CrossRef]
23. Al-Jethelah, M.; Tasnim, S.H.; Mahmud, S.; Dutta, A. Nano-PCM filled energy storage system for solar-thermal applications. *Renew. Energy* **2018**, *126*, 137–155. [CrossRef]
24. Liao, Z.; Xu, C.; Ren, Y.; Gao, F.; Ju, X.; Du, X. A novel effective thermal conductivity correlation of the PCM melting in spherical PCM encapsulation for the packed bed TES system. *Appl. Therm. Eng.* **2018**, *135*, 116–122. [CrossRef]
25. Mahdi, J.M.; Mohammed, H.I.; Hashim, E.T.; Talebizadehsardari, P.; Nsofor, E.C. Solidification enhancement with multiple PCMs, cascaded metal foam and nanoparticles in the shell-and-tube energy storage system. *Appl. Energy* **2020**, *257*, 113993. [CrossRef]
26. Eisapour, M.; Eisapour, A.H.; Hosseini, M.J.; Talebizadehsardari, P. Exergy and energy analysis of wavy tubes photovoltaic-thermal systems using microencapsulated PCM nano-slurry coolant fluid. *Appl. Energy* **2020**, *266*, 114849. [CrossRef]
27. Ghalambaz, M.; Mehryan, S.; Hajjar, A.; Veismoradi, A. Unsteady natural convection flow of a suspension comprising Nano-Encapsulated Phase Change Materials (NEPCMs) in a porous medium. *Adv. Powder Technol.* **2020**, *31*, 954–966. [CrossRef]
28. Talebizadeh Sardari, P.; Mohammed, H.I.; Mahdi, J.M.; Ghalambaz, M.; Gillott, M.; Walker, G.S.; Grant, D.; Giddings, D. Localized heating element distribution in composite metal foam-phase change material: Fourier's law and creeping flow effects. *Int. J. Energy Res.* **2021**, *45*, 13380–13396. [CrossRef]
29. Ghalambaz, M.; Mohammed, H.I.; Naghizadeh, A.; Islam, M.S.; Younis, O.; Mahdi, J.M.; Chatroudi, I.S.; Talebizadehsardari, P. Optimum Placement of Heating Tubes in a Multi-Tube Latent Heat Thermal Energy Storage. *Materials* **2021**, *14*, 1232. [CrossRef]
30. Talebizadehsardari, P.; Mohammed, H.I.; Mahdi, J.M.; Gillott, M.; Walker, G.S.; Grant, D.; Giddings, D. Effect of airflow channel arrangement on the discharge of a composite metal foam-phase change material heat exchanger. *Int. J. Energy Res.* **2021**, *45*, 2593–2609. [CrossRef]
31. Ju, Y.; Zhu, T.; Mashayekhi, R.; Mohammed, H.I.; Khan, A.; Talebizadehsardari, P.; Yaïci, W. Evaluation of Multiple Semi-Twisted Tape Inserts in a Heat Exchanger Pipe Using Al_2O_3 Nanofluid. *Nanomaterials* **2021**, *11*, 1570. [CrossRef]
32. Mahdi, J.M.; Nsofor, E.C. Melting enhancement in triplex-tube latent thermal energy storage system using nanoparticles-fins combination. *Int. J. Heat Mass Transf.* **2017**, *109*, 417–427. [CrossRef]
33. Sheikholeslami, M.; Lohrasbi, S.; Ganji, D.D. Numerical analysis of discharging process acceleration in LHTESS by immersing innovative fin configuration using finite element method. *Appl. Therm. Eng.* **2016**, *107*, 154–166. [CrossRef]
34. Mohammed, H.I.; Talebizadehsardari, P.; Mahdi, J.M.; Arshad, A.; Sciacovelli, A.; Giddings, D. Improved melting of latent heat storage via porous medium and uniform Joule heat generation. *J. Energy Storage* **2020**, *31*, 101747. [CrossRef]
35. Dai, Z.; Guo, S.; Gong, Y.; Wang, Z. Semiconductor flexoelectricity in graphite-doped $SrTiO_3$ ceramics. *Ceram. Int.* **2021**, *47*, 6535–6539. [CrossRef]
36. Sun, J.; Aslani, F.; Wei, J.; Wang, X. Electromagnetic absorption of copper fiber oriented composite using 3D printing. *Constr. Build. Mater.* **2021**, *300*, 124026. [CrossRef]
37. Ghalambaz, M.; Zadeh, S.M.H.; Mehryan, S.; Pop, I.; Wen, D. Analysis of melting behavior of PCMs in a cavity subject to a non-uniform magnetic field using a moving grid technique. *Appl. Math. Model.* **2020**, *77*, 1936–1953. [CrossRef]
38. Ghalambaz, M.; Zadeh, S.M.H.; Mehryan, S.; Ayoubloo, K.A.; Sedaghatizadeh, N. Non-Newtonian behavior of an electrical and magnetizable phase change material in a filled enclosure in the presence of a non-uniform magnetic field. *Int. Commun. Heat Mass Transf.* **2020**, *110*, 104437. [CrossRef]
39. Jiang, T.; Liu, Z.; Wang, G.; Chen, Z. Comparative study of thermally stratified tank using different heat transfer materials for concentrated solar power plant. *Energy Rep.* **2021**, *7*, 3678–3687. [CrossRef]
40. Tao, L.C. Generalized numerical solutions of freezing a saturated liquid in cylinders and spheres. *AIChE J.* **1967**, *13*, 165–169. [CrossRef]

41. Gortych, M.; Lipnicki, Z.; Weigand, B. An experimental and theoretical study of the solidification process of phase change materials in a horizontal annular enclosure. *Appl. Therm. Eng.* **2019**, *161*, 114140. [CrossRef]
42. Abdollahzadeh, M.; Esmaeilpour, M. Enhancement of phase change material (PCM) based latent heat storage system with nano fluid and wavy surface. *Int. J. Heat Mass Transf.* **2015**, *80*, 376–385. [CrossRef]
43. Shahsavar, A.; Al-Rashed, A.A.A.A.; Entezari, S.; Sardari, P.T. Melting and solidification characteristics of a double-pipe latent heat storage system with sinusoidal wavy channels embedded in a porous medium. *Energy* **2019**, *171*, 751–769. [CrossRef]
44. Choi, J.C.; Kim, S.D. Heat-transfer characteristics of a latent heat storage system using $MgCl_2 \cdot 6H_2O$. *Energy* **1992**, *17*, 1153–1164. [CrossRef]
45. Wang, P.; Li, D.; Huang, Y.; Zheng, X.; Wang, Y.; Peng, Z.; Ding, Y. Numerical Study of Solidification in a Plate Heat Exchange Device with a Zigzag Configuration Containing Multiple Phase-Change-Materials. *Energies* **2016**, *9*, 394. [CrossRef]
46. Talebizadehsardari, P.; Mahdi, J.M.; Mohammed, H.I.; Moghimi, M.A.; Hossein Eisapour, A.; Ghalambaz, M. Consecutive charging and discharging of a PCM-based plate heat exchanger with zigzag configuration. *Appl. Therm. Eng.* **2021**, *193*, 116970. [CrossRef]
47. Li, H.; Xu, B.; Lu, G.; Du, C.; Huang, N. Multi-objective optimization of PEM fuel cell by coupled significant variables recognition, surrogate models and a multi-objective genetic algorithm. *Energy Convers. Manag.* **2021**, *236*, 114063. [CrossRef]
48. Li, H.-W.; Gao, Y.-F.; Du, C.-H.; Hong, W.-P. Numerical study on swirl cooling flow, heat transfer and stress characteristics based on fluid-structure coupling method under different swirl chamber heights and Reynolds numbers. *Int. J. Heat Mass Transf.* **2021**, *173*, 121228. [CrossRef]
49. Shokouhmand, H.; Kamkari, B. Numerical Simulation of Phase Change Thermal Storage in Finned Double-Pipe Heat Exchanger. *Appl. Mech. Mater.* **2012**, *232*, 742–746. [CrossRef]
50. Bazai, H.; Moghimi, M.A.; Mohammed, H.I.; Babaei-Mahani, R.; Talebizadehsardari, P. Numerical study of circular-elliptical double-pipe thermal energy storage systems. *J. Energy Storage* **2020**, *30*, 101440. [CrossRef]
51. Shahsavar, A.; Ali, H.M.; Mahani, R.B.; Talebizadehsardari, P. Numerical study of melting and solidification in a wavy double-pipe latent heat thermal energy storage system. *J. Therm. Anal. Calorim.* **2020**, *141*, 1785–1799. [CrossRef]
52. Shahsavar, A.; Khosravi, J.; Mohammed, H.I.; Talebizadehsardari, P. Performance evaluation of melting/solidification mechanism in a variable wave-length wavy channel double-tube latent heat storage system. *J. Energy Storage* **2020**, *27*, 101063. [CrossRef]
53. Xu, Y.; Ren, Q.; Zheng, Z.-J.; He, Y.-L. Evaluation and optimization of melting performance for a latent heat thermal energy storage unit partially filled with porous media. *Appl. Energy* **2017**, *193*, 84–95. [CrossRef]
54. Xu, Y.; Li, M.-J.; Zheng, Z.-J.; Xue, X.-D. Melting performance enhancement of phase change material by a limited amount of metal foam: Configurational optimization and economic assessment. *Appl. Energy* **2018**, *212*, 868–880. [CrossRef]
55. Ghalambaz, M.; Mahdi, J.M.; Shafaghat, A.; Eisapour, A.H.; Younis, O.; Talebizadeh Sardari, P.; Yaïci, W. Effect of Twisted Fin Array in a Triple-Tube Latent Heat Storage System during the Charging Mode. *Sustainability* **2021**, *13*, 2685. [CrossRef]
56. Mahdi, J.M.; Lohrasbi, S.; Ganji, D.D.; Nsofor, E.C. Accelerated melting of PCM in energy storage systems via novel configuration of fins in the triplex-tube heat exchanger. *Int. J. Heat Mass Transf.* **2018**, *124*, 663–676. [CrossRef]
57. Mahdi, J.M.; Nsofor, E.C. Melting enhancement in triplex-tube latent heat energy storage system using nanoparticles-metal foam combination. *Appl. Energy* **2017**, *191*, 22–34. [CrossRef]
58. Li, Z.; Shahsavar, A.; Al-Rashed, A.A.A.A.; Talebizadehsardari, P. Effect of porous medium and nanoparticles presences in a counter-current triple-tube composite porous/nano-PCM system. *Appl. Therm. Eng.* **2020**, *167*, 114777. [CrossRef]
59. Abdulateef, A.M.; Mat, S.; Abdulateef, J.; Sopian, K.; Al-Abidi, A.A. Geometric and design parameters of fins employed for enhancing thermal energy storage systems: A review. *Renew. Sustain. Energy Rev.* **2018**, *82*, 1620–1635. [CrossRef]
60. Mat, S.; Al-Abidi, A.A.; Sopian, K.; Sulaiman, M.Y.; Mohammad, A.T. Enhance heat transfer for PCM melting in triplex tube with internal–external fins. *Energy Convers. Manag.* **2013**, *74*, 223–236. [CrossRef]
61. Rabienataj Darzi, A.A.; Jourabian, M.; Farhadi, M. Melting and solidification of PCM enhanced by radial conductive fins and nanoparticles in cylindrical annulus. *Energy Convers. Manag.* **2016**, *118*, 253–263. [CrossRef]
62. Pizzolato, A.; Sharma, A.; Maute, K.; Sciacovelli, A.; Verda, V. Design of effective fins for fast PCM melting and solidification in shell-and-tube latent heat thermal energy storage through topology optimization. *Appl. Energy* **2017**, *208*, 210–227. [CrossRef]
63. Yıldız, Ç.; Arıcı, M.; Nižetić, S.; Shahsavar, A. Numerical investigation of natural convection behavior of molten PCM in an enclosure having rectangular and tree-like branching fins. *Energy* **2020**, *207*, 118223. [CrossRef]
64. Yu, C.; Wu, S.; Huang, Y.; Yao, F.; Liu, X. Charging performance optimization of a latent heat storage unit with fractal tree-like fins. *J. Energy Storage* **2020**, *30*, 101498. [CrossRef]
65. Rathod, M.K.; Banerjee, J. Thermal performance enhancement of shell and tube Latent Heat Storage Unit using longitudinal fins. *Appl. Therm. Eng.* **2015**, *75*, 1084–1092. [CrossRef]
66. Abdulateef, A.M.; Mat, S.; Sopian, K.; Abdulateef, J.; Gitan, A.A. Experimental and computational study of melting phase-change material in a triplex tube heat exchanger with longitudinal/triangular fins. *Sol. Energy* **2017**, *155*, 142–153. [CrossRef]
67. Shahsavar, A.; Goodarzi, A.; Mohammed, H.I.; Shirneshan, A.; Talebizadehsardari, P. Thermal performance evaluation of non-uniform fin array in a finned double-pipe latent heat storage system. *Energy* **2020**, *193*, 116800. [CrossRef]
68. Maakoul, A.E.; Laknizi, A.; Saadeddine, S.; Abdellah, A.B.; Meziane, M.; Metoui, M.E. Numerical design and investigation of heat transfer enhancement and performance for an annulus with continuous helical baffles in a double-pipe heat exchanger. *Energy Convers. Manag.* **2017**, *133*, 76–86. [CrossRef]

69. Maakoul, A.E.; Metoui, M.E.; Abdellah, A.B.; Saadeddine, S.; Meziane, M. Numerical investigation of thermohydraulic performance of air to water double-pipe heat exchanger with helical fins. *Appl. Therm. Eng.* **2017**, *127*, 127–139. [CrossRef]
70. Rubitherm Technologies GmbH. RT35 Data Sheet. Available online: https://www.rubitherm.eu/media/products/datasheets/Techdata_-RT35_EN_09102020.PDF (accessed on 15 September 2021).
71. Talebizadeh Sardari, P.; Walker, G.S.; Gillott, M.; Grant, D.; Giddings, D. Numerical modelling of phase change material melting process embedded in porous media: Effect of heat storage size. *Proc. Inst. Mech. Eng. Part A J. Power Energy* **2019**, 0957650919862974. [CrossRef]
72. Shahsavar, A.; Majidzadeh, A.H.; Mahani, R.B.; Talebizadehsardari, P. Entropy and Thermal performance Analysis of PCM Melting and Solidification Mechanisms in a Wavy Channel Triplex-Tube Heat Exchanger. *Renew. Energy* **2021**, *165*, 52–72. [CrossRef]
73. Al-Abidi, A.A.; Mat, S.; Sopian, K.; Sulaiman, M.Y.; Mohammad, A.T. Internal and external fin heat transfer enhancement technique for latent heat thermal energy storage in triplex tube heat exchangers. *Appl. Therm. Eng.* **2013**, *53*, 147–156. [CrossRef]
74. Mahdi, J.M. Enhancement of Phase Change Material (PCM) Thermal Energy Storage in Triplex-Tube Systems. Ph.D. Thesis, Southern Illinois University Carbondale, Carbondale, IL, USA, 2018.

energies

MDPI

Review

Heat Transfer Characteristics of Conventional Fluids and Nanofluids in Micro-Channels with Vortex Generators: A Review

Mushtaq T. Al-Asadi [1,2,*], Hussein A. Mohammed [3,*] and Mark C. T. Wilson [2]

1 Business Development Department, Basra Oil Company, Ministry of Oil, Basra 240, Iraq
2 Institute of Thermofluids, School of Mechanical Engineering, University of Leeds, Leeds LS2 9JT, UK; m.wilson@leeds.ac.uk
3 WA School of Mines-Minerals, Energy & Chemical Engineering, Curtin University, Perth, WA 6102, Australia
* Correspondence: mushtaq@boc.oil.gov.iq or dr.eng.mushtaq@gmail.com (M.T.A.-A.); hussein.mohammed@curtin.edu.au or hussein.dash@yahoo.com (H.A.M.)

Abstract: An effective way to enhance the heat transfer in mini and micro electronic devices is to use different shapes of micro-channels containing vortex generators (VGs). This attracts researchers due to the reduced volume of the electronic micro-chips and increase in the heat generated from the devices. Another way to enhance the heat transfer is using nanofluids, which are considered to have great potential for heat transfer enhancement and are highly suited to application in practical heat transfer processes. Recently, several important studies have been carried out to understand and explain the causes of the enhancement or control of heat transfer using nanofluids. The main aim upon which the present work is based is to give a comprehensive review on the research progress on the heat transfer and fluid flow characteristics of nanofluids for both single- and two- phase models in different types of micro-channels. Both experimental and numerical studies have been reviewed for traditional and nanofluids in different types and shapes of micro-channels with vortex generators. It was found that the optimization of heat transfer enhancement should consider the pumping power reduction when evaluating the improvement of heat transfer.

Keywords: heat transfer; nanofluids advantages and disadvantages; thermal hydraulic performance; vortex generators; micro-channel

Citation: Al-Asadi, M.T.; Mohammed, H.A.; Wilson, M.C.T. Heat Transfer Characteristics of Conventional Fluids and Nanofluids in Micro-Channels with Vortex Generators: A Review. *Energies* **2022**, *15*, 1245. https://doi.org/10.3390/en15031245

Academic Editor: Pouyan Talebizadeh Sardari

Received: 25 October 2021
Accepted: 10 November 2021
Published: 8 February 2022

1. Introduction

Since 1931, researchers have explored ways to manage the heat flux generated from electrical devices and offer better heat transfer rates using different approaches to enhance the heat transfer in mini and micro cooling systems [1].

Nowadays, the impact of heat transfer and fluid flow have become more interesting and challenging simultaneously due to rapid developments in electronic and electrical devices and systems which become increasingly small in size, light in weight but high in heat transfer dissipation demands. Therefore, enhancing the heat transfer in such systems has been a strong motivation for this current research [2].

A possible and most effective approaches to enhance the heat transfer is the design of a heat sink; the most popular heat sink used in air-cooled systems is a plate-fin heat sink (PFHS) because of its simplicity to manufacture. Many investigations of PFHSs have studied and optimized the fins' height, thickness, and separation, yielding predictions of heat transfer and entropy [3–6]. Other designs such as pinned heat sinks (PHSs) have also been considered in both inline and staggered arrangements to enhance the heat transfer rate [7]. They can take several shapes such as rectangular, square, circular, elliptical, NACA and drop form [1,8–11]. The key components in the cooling of computer systems, and many other applications such as air conditioning are heat exchangers and heat sinks [12–15].

Based on the working fluid, heat exchangers are generally classified as gas, liquid, or a combination. Some examples of heat exchangers are shown in Figure 1. Many studies have shown that the liquid heat exchangers and heat sinks systems had great potential for enhancing the heat transfer compared to gas systems based on some of their thermal properties, which is higher in the liquid than in the gas [16,17].

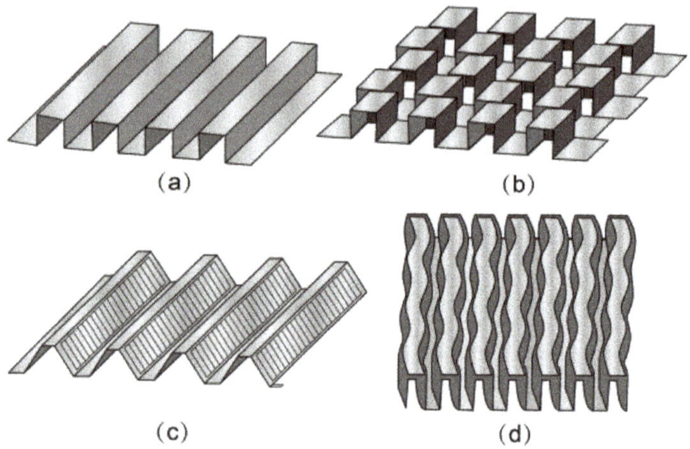

Figure 1. Heat exchanger classification (**a**) plain rectangular fins: (**b**) offset strip fins, (**c**) louvered fins (**d**) wavy fins.

Continuing developments in electronic and electrical devices and the increased heat density associated with miniaturization mean that the thermal management of high heat fluxes remains an active area of research [18].

However, another approach to improve the heat performance of the cooling systems is to improve the thermo-physical properties of the coolants—for example, by developing nanofluids [14,19–21], which have used widely in industry and studied extensively for their impact on the environment [22–24]. Alternatively, the geometry of the heat sinks can be adapted to improve heat transfer—for example, by modifying the pins in PHSs or the channels in PFHSs. One very successful approach for air applications is the use of micro-channels. Note that the micro-channels first appeared in 1981 [25].

Many ways by which the heat transfer might be enhanced, such as suggestions of new designs of the geometry and/or advanced fluids, can be used [26]. Various geometries have been designed to achieve a high performance of heat transfer using an extended surface area [27,28].

Many experimental and numerical studies have investigated the heat transfer and fluid flow performance of various modified geometries such as micro-channels with grooves and ribs [29–32]. The effect of vortex generators (VGs) on heat transfer and fluid flow characteristics were investigated experimentally in 1969 [33].

In addition to the surface area enhancement, vortex generators can be considered as a geometry improvement, which creates secondary flows that can enhance the heat transfer [34,35].

In the revolution of advanced manufacturing processes, VGs can take up various forms such as protrusions, wings, inclined blocks, winglets, fins, and ribs [2,36], and have been used to enhance heat transfer in different geometries such as circular and non-circular ducts under turbulent flow [34,37,38]. They have also been used in laminar flow, with flat plate-fins in rectangular channels [37–39], tube heat exchangers [40], heat sinks [41,42], and rectangular narrow channels [43,44], as shown in Figure 1.

One of the promising systems by which high-performance heat rejection can be achieved is micro- and mini-scale systems, such as micro-channel heat exchangers and heat sinks [12,15,45–47]. They are different from traditional channels and can be classified according to their associated hydraulic diameters, D_h, [48–50]. The classification of the channel according to Mehendale et al. [49] is that the conventional channels with $D_h > 6$ mm, compact Passages 1 mm< $D_h \leq 6$ mm, meso-channels 100 mm< $D_h \leq 1$ mm, and micro-channels 1 μm < $D_h \leq 100$ μm. while Kandlikar and Grande [50] had another classification of the channel, the that the conventional channels with $D_h > 3$ mm, mini channel 1 mm< $D_h \leq 3$ mm, micro-channels 10 μm < $D_h \leq 200$ μm, transitional channels 0.1 mm< $D_h \leq 10$ mm, and the Molecular nanochannels with the $D_h \leq 0.1$ μm.

On the other hand, using advanced fluids instead of traditional fluids (e.g., air and water) has become common and effective. It can be a combination of two fluids like mixing water and glycerin [51], or it can be a suspension of particles in a liquid, which is well known as a nanofluid [52]. Recently, the use of nanofluids has been applied to manage the heat transfer in batteries [53–56].

In this article, a comprehensive review has been conducted to focus on heat transfer performance with different modifications of both the fluid and geometry. Furthermore, the influence of evaluating heat transfer enhancement together with the pressure penalty resulted from the modifications of the geometry and the fluid. This review paper is divided into three main sections that consider straight micro-channels, vortex generators, and nanofluids.

2. Uniform Micro-Channels

This section provides an idea of the investigations that have been carried out on a uniform channel. It is divided into two sections: single-phase flow and two-phase flow.

2.1. Single Phase Flow

A numerical investigation of various shapes of rectangular micro-channels with the range of width 44–56 μm, height 287–320 μm, and length 10 mm was conducted by Shkarah et al. [57]. The materials used were aluminum, silicon, and graphene. Different values of volumetric flow rate and heat flux with fully developed laminar flow of water were utilized. The results showed that the thermal resistance was reduced by using graphene in the micro-channel. However, the findings have not yet been confirmed experimentally and the numerical method considered the thermo-physical properties of the materials as non-temperature-dependent, which may affect the results when compared to the experimental setup.

Laminar flow of deionized water in a copper rectangular micro-channel with a hydraulic diameter ranging from 200 to 364 μm and with a length of 120 mm was numerically studied by Lee et al. [58]. The finite volume method was implemented to determine the Nusselt number at various aspect ratios. The study presented the distribution of local and average Nusselt numbers as a function of non-dimensional axial distance. The researchers proposed correlations which helped to enhance the heat transfer. The proposed correlations considered the entrance length effect on heat transfer rate and were in very good agreement with previous experimental studies. It was found that the new correlation was applicable for thermally developed flow for local and average Nusselt number under laminar flow.

Mansoor et al. [59] performed three-dimensional simulations of a rectangular micro-channel using single-phase laminar flow (Re ranged from 500 to 2000) of deionized water as a working fluid. A heat flux of 130 W/cm^2 was considered to investigate the thermal characteristics in a copper micro-channel. The study used FLUENT commercial software, (New York, NY, USA) and the results were compared with previous numerical and experimental works, showing good agreement. It was found that the heat transfer coefficient was decreased as the heat flux increased. In addition, a high Reynolds number and heat flux led to a transition from single- to two-phase flow, while there was no transition when the heat flux was less than 100 W/cm^2.

An experimental study of a copper rectangular micro-channel with a hydraulic diameter in the range of 318–903 μm and length of 24.5 mm was conducted by Lee et al. [60], using deionized water as a working fluid. The study used laminar and turbulent flows, with the Reynolds number ranging from 300 to 3500 to investigate the heat transfer and fluid flow regimes using single-phase flow. The results showed that heat transfer was increased as the channel size decreased. However, decreasing the dimensions of the rectangular channel requires more pumping power resulting from an increase in the associated pressure drop.

Deng et al. [61] compared a traditional rectangular cross-section copper micro-channel with an omega shape micro-channel heat sinks of the same hydraulic diameter with ethanol and deionized water as two-phase boiling flow, as shown in Figure 2. The results showed that water is better than ethanol in both micro-channel types. Moreover, using the omega micro-channel decreased the pressure drop compared to the conventional rectangular micro-channel.

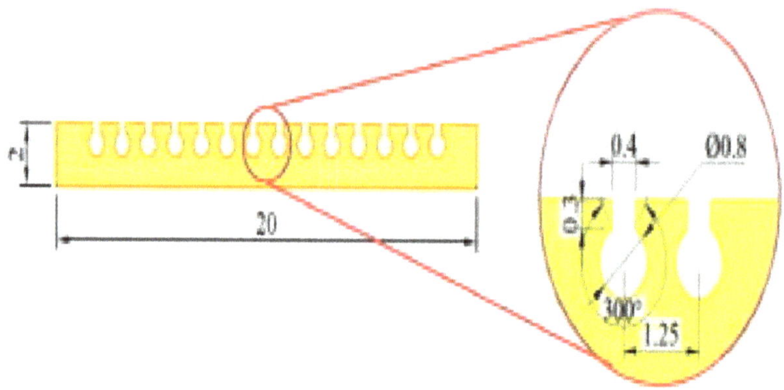

Figure 2. Micro-channel omega shape in mm [61] © Elsevier, 2015.

Micro-channels can be used not only with liquid but also with gas as a working fluid; Balaj et al. [62] studied the influence of shear stress in micro- and nano-channels using constant wall heat flux. The simulation model used the direct-simulation Monte Carlo method. It was found that there is a sensible effect of the magnitude of viscous dissipation on heat and flow performance; therefore, it should be considered in heat transfer predictions. The study also showed that the heat transfer is significantly enhanced when the heating condition is applied, while the heat transfer is decreased while utilizing the cooling condition.

Xia et al. [63] numerically studied the heat transfer and fluid flow characteristics of a liquid-cooled heat sink with three different inlet and outlet locations named c, I, and z, and different header shapes that feed the micro-channels, as shown in Figure 3. The traditional shape of a rectangular micro-channel was compared with a triangular shape. The results showed that the best geometry is rectangular and the best location of the inflow regime was I, then c, then z. Additionally, the results showed that better heat transfer characteristics were achieved with the rectangular header shape. However, the results showed that using the position shown in Figure 3c was the best design when using a volume flow rate of 150 mL/min. This can be attributed to the velocity uniformity compared to the trapezoidal and triangular shapes.

Heat transfer can be enhanced by applying a magnetic field on water-based Fe_3O_4, as indicated by the study by Ghasemian et al. [64]. They investigated the heat transfer characteristics of a rectangular channel with a width of 0.2 cm and length of 2 cm subjected to constant and variable magnetic fields under laminar flow. The finite volume method was used to solve the governing equations and two-phase mixture flow was implemented in the study. The results showed that three parameters enhanced the heat transfer, namely frequency and locations (a, b) of magnetic fields, as shown in Figure 4. Noticeable enhancement of heat transfer appeared at fully developed flow, especially when applying the magnetic field. Moreover, it was found that using an alternating magnetic field was better than a constant one by approximately 1.6 times. Using an alternating magnetic field acts as a vortex generator to frequently disturb the fluid.

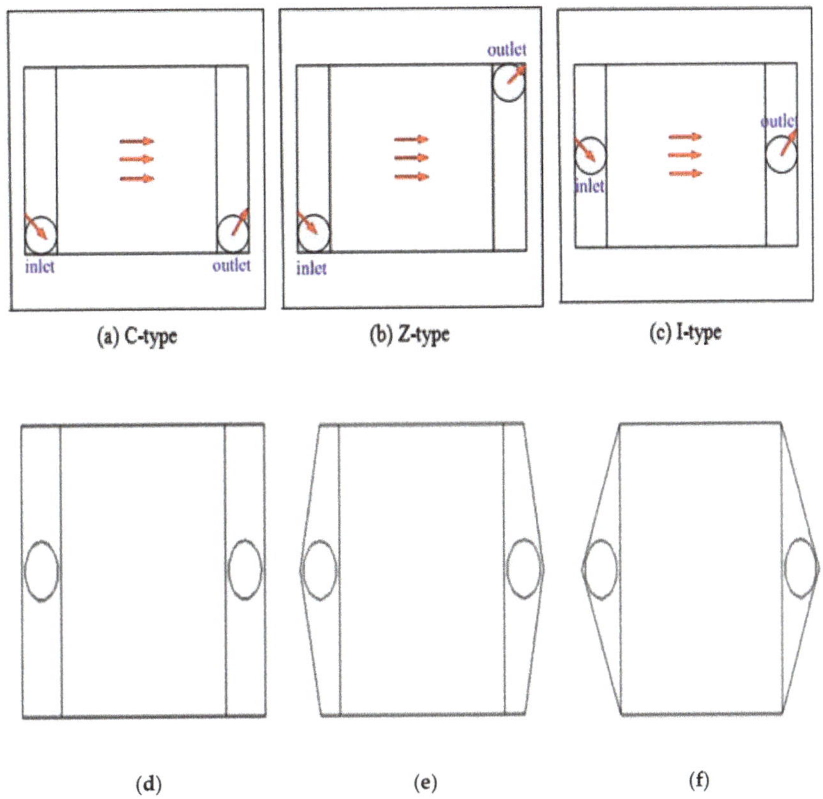

Figure 3. Different inlet and outlet positions and header shapes of micro-channel heat sinks [63] © Elsevier, 2015. (**a**) C-type; (**b**) Z-type; (**c**) I-type; (**d**) rectangular; (**e**) trapezoidal; (**f**) triangular.

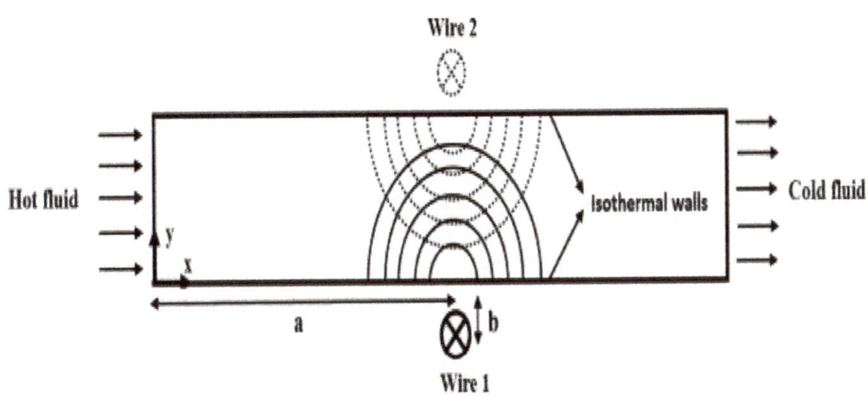

Figure 4. Magnetic field distribution to enhance the heat transfer [64] © Elsevier, 2015.

A review of numerical and experimental investigations focusing on the heat transfer and air-side flow using a fin and tube heat exchanger was presented by Pongsoi et al. [65].

The study summarizes a significant effect such as tube arrangement, operating conditions, and fin configurations. More than 35 articles related to heat exchangers were considered, representing the experimental studies from the very early period. The study used geometry design by comparing circular and spiral fins. It concluded that 57% of the heat exchangers used a spiral fin; thus, the recommendation of the investigation was to use a spiral fin instead of a circular fin in heat exchangers. Moreover, Pan et al. [66] presented the effect of different inlet distribution manifold for different widths of rectangular micro-channels. The investigation considered different dimensions of the inlet design of a Z-shape to examine the effect of inlet distribution and the width of the channel on optimal design of a micro-channel. The results showed that the width of the channel had a significant influence on the optimization results.

Another review of micro- and mini-channel geometries is that of Dixit and Ghosh [48]. The study illustrated previous work in a single-phase flow with heat exchangers and heat sinks in various types of flow such as laminar, turbulent, developing flows, and fully developed flow. It also presented the heat transfer performance such as convective heat transfer under the condition of constant wall temperature and constant heat flux. The application and fabrication of micro- and nano-scales were also adopted in this investigation. It was concluded that it is still difficult to produce channels of micro-size due to manufacturing limitations; however, micro-channels can be produced as parts, but it is still not easy to combine the parts to produce micro-channels. Many issues can be found when producing micro-channels from parts such as the accuracy of having equal distance between channels. Moreover, the reliability of the glue for a specific application, the conductivity of the glue used to combine the parts, and avoiding having a layer of the glue might influence the heat transfer performance.

A numerical and experimental investigation of heat transfer and fluid flow performances in a bronze rectangular micro-channel with dimensions of 1 and 0.3 mm was presented by Gamrat et al. [67]. Water as a working fluid with a Reynolds number in the range of 200–3000 was considered to investigate the mixed convective heat transfer performance. The results of the numerical study showed that there was no sensible influence on the Nusselt number when the channel dimension changed from 1 to 0.1 mm. Due to the limitation of the experimental measurement, the impact of this change has not been measured.

Laminar flow in different rectangular copper micro-channels with a width and height 231 μm and 713 μm, respectively, was studied experimentally and numerically by Qu and Mudawar [68]. Deionized water at a Reynolds number in the range of 139–1672 was considered as the working fluid. Two heat flux values (100 and 200 W/cm^2) were applied on the bottom wall to investigate the fluid flow and heat transfer performance. It was found that the outlet temperature of the fluid decreased at a high Reynolds number while the pressure drop increased. It was also found that there was not much difference in temperature at the top wall of the micro-channels; therefore, it can be considered as an adiabatic wall.

However, most recent numerical studies considered the top wall (the wall which is opposite the wall where heat flux was applied to) as an adiabatic wall. This is because the low heat transfer at the top wall might transfer from the walls by conduction and the fluid by convection, especially when using a plastic top wall.

2.2. *Two-Phase Flow*

A considerable number of studies have been carried out on the design of micro-channels using two-phase models, as can be seen in the following paragraphs.

An experimental investigation of the heat transfer performance was conducted by Hsu et al. [69] using different orientations of copper rectangular micro-channels with a hydraulic diameter of 440 μm, heat flux of 25 kW/m^2, and mass flux of 100 and 200 kg/m^2s. Two-phase boiling flow and HFE-7100 as a working fluid was considered. The setup investigated the effect of the inclination from the horizontal to the vertical position on boiling heat transfer. It was found that the heat transfer coefficient rose with the vapor quality and peaked when it reached 0.6 for the upward position.

Suwankamnerd and Wongwises [70] studied two-phase air–water flows in a copper rectangular micro-channel having 267 μm hydraulic diameter with low Reynolds number. The setup used a separate flow model as well as a homogeneous flow model to estimate the pressure drop using the Friedrel correlation, which is used to measure the pressure drop in two-phase flow. The investigation showed enhancement in the Nusselt number of 120% compared to single-phase flow.

In addition, Mirmanto [71] studied the heat transfer coefficient in various dimensions of a single copper rectangular micro-channel with a horizontal position. Boiling deionized water at 98 °C at the inlet as a working fluid, 125 kPa as inlet pressure, 800 kg/m^2 of mass flux, and various values of heat flux were used in this study. The results showed that there was good agreement between the experimental measurements and the numerical simulation, especially in the pressure gradient. It was effective at the low pressure generated. At fixed heat and mass flux, it was found that the heat transfer coefficient went down with the quality in the smallest hydraulic diameter, while it was increased significantly with the other diameters.

Konishi et al. [72] studied the effect of boiling flow on flow and heat transfer maps. The geometry was a copper rectangular channel consisting of two heating walls fixed opposite each other with liquid and mass inlet velocities ranging from 0.1 to 1.9 m/s and 224.2 to 3347.5 kg/m^2 s, respectively; and a temperature of inlet sub-cooling in the range of 2.8–8.1 °C. Heat transfer and fluid flow measurements were adopted to examine the flow performance. It was found that the temperature distribution improved as the gravity rose, while it decreased in micro-gravity.

Gan et al. [73] experimentally investigated the pressure drop characteristics of two-phase flow in a triangular silicon micro-channel with dimensions of 300, 212, and 155.4 μm in width, depth, and hydraulic diameter, respectively. Acetone was considered as a working fluid under various ranges of inlet temperature and pressure, mass velocity, superheat, outlet quality, and heat flux. The pressure drop and boiling flow were performed. The outcome of the study was a new correlation which considered the functionality of mass flux and, therefore, the error of predicting the acetone data with 12.56% of mean absolute error.

Fang et al. [74] proposed a correlation of flow boiling to investigate the heat transfer regime using a copper rectangular tube. The study adopted H_2O, R718 as a working fluid, two-phase laminar and turbulent flows. More than 1050 data points of water boiling flow for mini- and micro-channels were collected. The results showed that the proposed correlation was applicable to many refrigerant fluids, especially for R410 and NH3.

Shojaeian and Koşar [75] reviewed previous experimental studies on micro- and nano-geometries using boiling flow. These geometries were in various shapes such as rectangular, triangular, and cylindrical cross-section. Heat transfer and fluid flow characteristics were presented and compared with different parameters such as single-phase and two-phase flows. It was found that the nano- and micro-structures enhanced the heat transfer rate of systems. Furthermore, the manufacturing ability increased to produce such complex shapes of nano-/micro-channels. Consequently, manufacturing nano-/micro-configurations had some finishing issues related to the surface. This can be tackled by coating the surface.

Asadi et al. [76] reviewed the validity of experimental correlations on pressure-drop and heat-transfer characteristics in single- and two-phase flows with different geometries of micro-channel. The investigation used 219 papers of experimental and numerical studies (from 1982 to 2013). It was found that, before 2003, the researchers focused on experimental and analytical investigations, while after 2003, the focus turned to numerical studies. It also indicated that approximately 76% of researchers considered the laminar flow using single-phase flow. This is because the behavior of the laminar flow can be predicted and agreed with the experimental data. However, an important factor has not been considered in this study, which is the energy consumption of using turbulent flow. For example, using turbulent flow will cost more pumping power to derive the flow, resulting in more energy consumption.

In summary, straight micro-channels represented the starting stage in converting from using conventional channels to micro-sized channels in various applications of micro-electrical and micro-electronical chips. As reviewed in the previous sections, micro-channels have rapidly received high attention by many researchers in different investigations, both numerical and experimental, as presented in Figure 5.

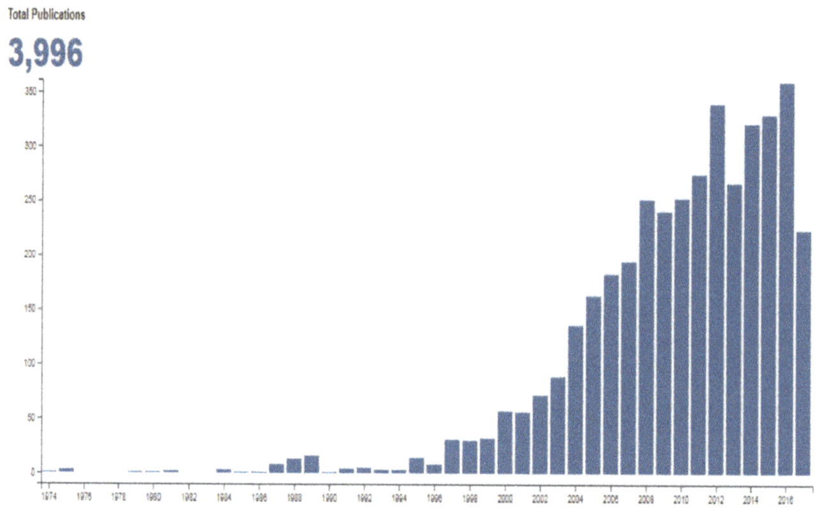

Figure 5. Growth in number of publications of micro-channels [77] © Elsevier, 2017.

The limitation in manufacturing micro-channels was a reason for a reduction in experimental studies. Another reason is the high price incurred to manufacture micro-channels. It is to be expected that in the near future, the manufacturing developments will easily allow production of micro-sized channels. Therefore, there is a real need to develop straight micro-channels into channels with complex shapes such as zigzag, wavy, and curved micro-channels to enhance the heat transfer. In addition, increasing the surface area and developing the secondary flow in the micro-channels also contributes to enhance the heat transfer rate. This can be achieved by adding some objects to increase the surface area and disturb the flow to develop the secondary flow.

2.3. Curved and Tapered Rectangular Micro-Channels

Research on curved micro-channels has also received recent attention by many researchers because of the high thermal and flow performance produced by these geometries, as elaborated in the following paragraphs.

Numerical simulation of laminar flow using water as a cooling fluid was performed by Guo et al. [78]. They investigated the influence on heat transfer performance of a curved micro-channel with a square cross-section, as presented in Figure 6. This micro-channel had a width and curve radius of 0.2 mm and 30 mm, respectively, and a Reynolds number in the range of 100–865. It was found that at high convection heat transfer, the synergy principle method can be applied. Note that this method applied to increase the accuracy of the solution because it considers the heat transfer at the outer wall. This method is applicable for such a curved channel because it presents the heat transfer accurately for the outer walls with consideration of the fluid flow.

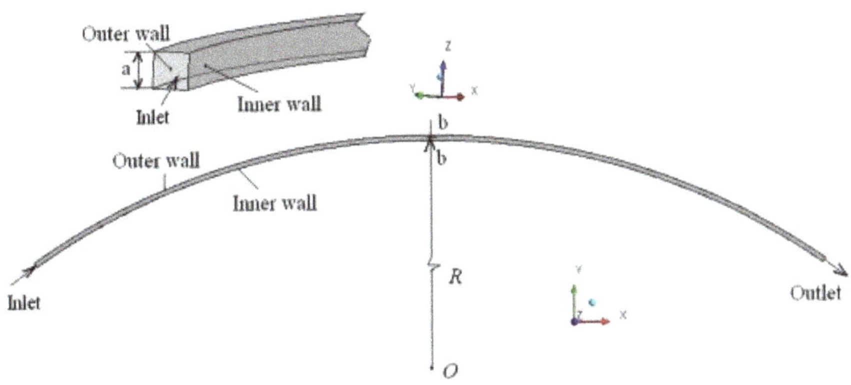

Figure 6. Curved duct with square cross-sectional area [78] © Elsevier, 2011.

Chu et al. [79] performed both experimental and numerical investigations of curved rectangular micro-channels with different diameters to study the influence of different diameters of the curve on flow characteristics. The Reynolds number was in the range of 80–876 and deionized water as a working fluid was considered. The results showed good agreement between the simulation and experiments. It was found that the curvature of the channel geometry increased the velocity at the outer wall, thus leading to enhancement of the heat transfer performance but increasing the friction factor.

A numerical study of a tapered aluminum micro-channel by Dehghan et al. [80] investigated the influence of different tapering geometries on pressure drop reduction using laminar flow and a constant heat flux of 100 W/cm^2. The width of the channel was fixed at the inlet to be 200 μm, while the outlet width was in the range of 75 to 200 μm with the channel length being 12,000 μm. It was found that the Poiseuille number and Nusselt number rose with tapering. The optimum heat transfer characteristics were found at an outlet-to-inlet width ratio of 0.5. However, with no consideration of the pressure drop in this study, it might be worth considering the pressure drop effect using the channel inlet, which can be taken in a range from 75 to 200 μm.

3. Non-Uniform Channels and Vortex Generators

The effect of vortex generators (VGs) on the heat transfer and fluid flow characteristics were investigated experimentally in 1969 [33]. Two types of vortex generators were classified based on the direction of the axis of rotation of the vortices generated.

Several parameters such as the geometry, shape, and position of the VGs might play a crucial role in enhancing the heat transfer, and the VG shape can be classified into rectangular, delta wings, and winglets, as presented earlier in [81]. However, the wing and winglet VGs are only suitable for air-based heat sinks. Various investigations have also indicated potential benefits of using VGs with laminar flow at different Reynolds numbers [35,43,82].

3.1. Non-Uniform Channels

Many experimental and numerical investigations have considered geometrical modifications of uniform channels to enhance the heat transfer performance. Liu et al. [83] conducted an investigation of the influence of geometry on heat and flow maps using turbulent air flow in a modified square channel with cylindrical slots of various diameters, as shown in Figure 7a. The finite volume method was used to solve the governing equations, utilizing FLUENT 12.1. The results emphasized that using cylindrical grooves and square ribs in the channel (see Figure 7b) enhanced the heat transfer characteristics compared to the uniform channel due to the extended surface area and the generation of vortices by disturbing the flow. However, the pressure drop of the square-ribbed channel was higher than the cylindrical grooves channel and the uniform channel. This study agreed with the results of cylindrical grooves in mini channels performed by Tang et al. [84].

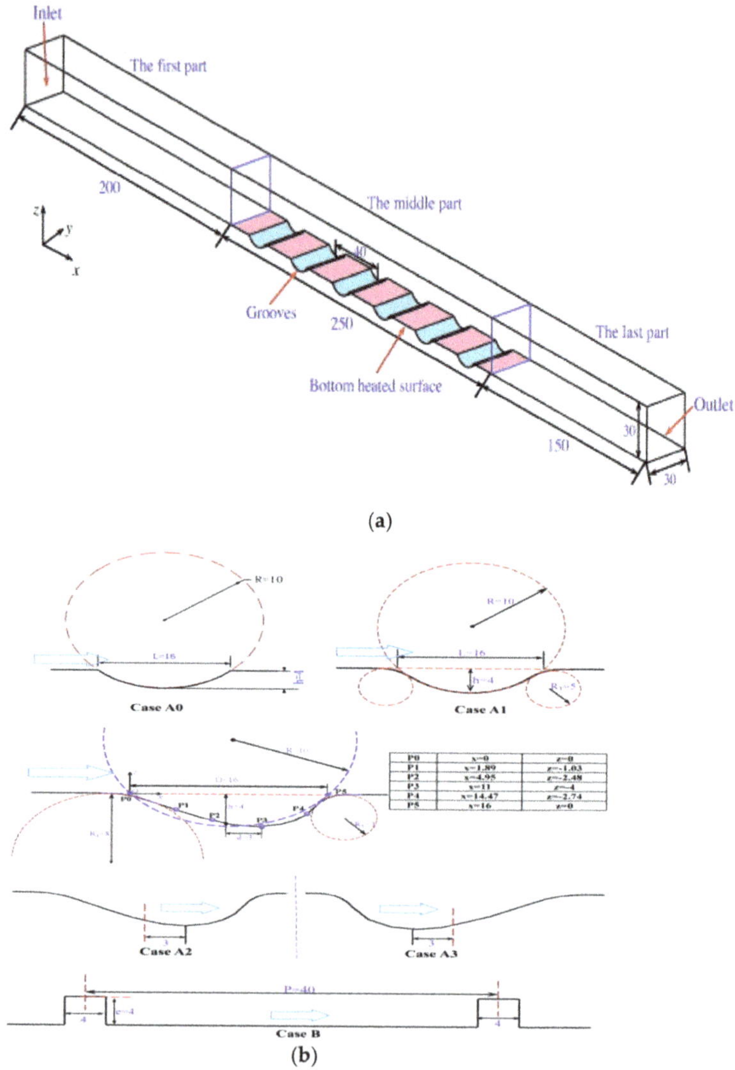

Figure 7. Rectangular micro-channel [83] © Elsevier, 2015; (**a**) geometry description; (**b**) various cylindrical grooves (cases A0–A3) and square ribs (case B).

Zhai et al. [85] simulated the flow in micro-channels with six types of cavities and ribs in the single micro-channel walls. These were "triangular-cavities with circular-rib (a) (Tri.C-C.R for short), (b) triangular-cavities with triangular-rib (Tri.C–Tri.R for short), (c) triangular-cavities with trapezoidal-ribs (Tri.C–Tra.R for short), (d) trapezoidal-cavities with circular-rib (Tra.C-C.R for short), (e) trapezoidal-cavities with circular-rib (Tra.C-C.R for short), (f) trapezoidal-cavities with trapezoid-rib (Tra.C-Tra.R for short)", as seen in Figure 8. De-ionized water was used as a coolant with a Reynolds number ranging from 300 to 600, and a constant heat flux of 10^6 W/m^2 was applied at the bottom wall of the micro-channel. The finite volume method and FLUENT software was adopted to investigate the flow and heat transfer characteristics. The results showed that using triangular cavities and ribs (see Figure 8f) offered better heat transfer compared to a uniform rectangular micro-channel due to better interaction between the solid and the fluid.

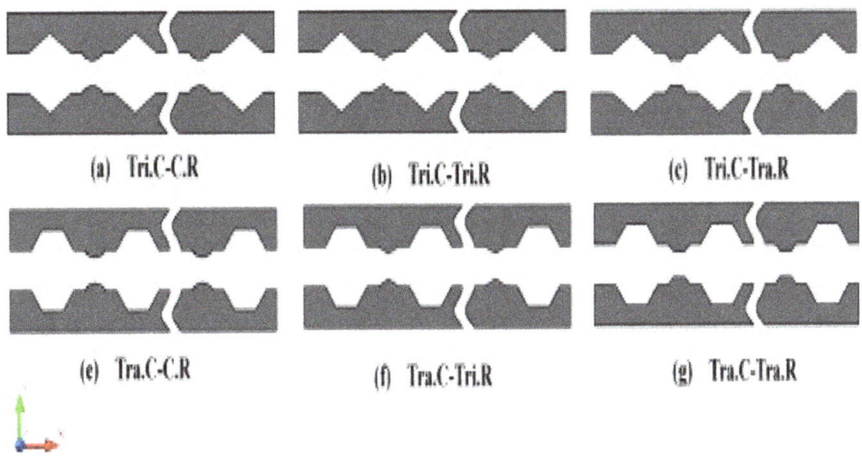

Figure 8. Square channel with different ribs and cavities, (**a**) (Tri.C-C.R), (**b**) (Tri.C–Tri.R), (**c**) (Tri.C–Tra.R), (**e**) (Tra.C-C.R), (**f**) (Tra.C-Tri.R), (**g**) (Tra.C- Tra.R) [86] © Elsevier, 2015.

Knupp et al. [86] proposed a hybrid simulation method to solve the heat transfer and fluid flow characteristics via a single domain strategy and generalized integral transform technique (GITT). This was applied to laminar flow in non-uniform channels, as shown in Figure 9. The results showed that the GITT method was suitable to be applied for Multiphysics applications found to be in good agreement with the finite element calculation form in the commercial software COMSOL Multiphysics, (New York, NY, USA). It is clear that this study agreed well with the literature that using COMSOL Multiphysics provides sufficient agreement with the experimental studies due to the temperature-dependent equations implemented in the software, and it is used widely for solving problems, especially those with Multiphysics applications.

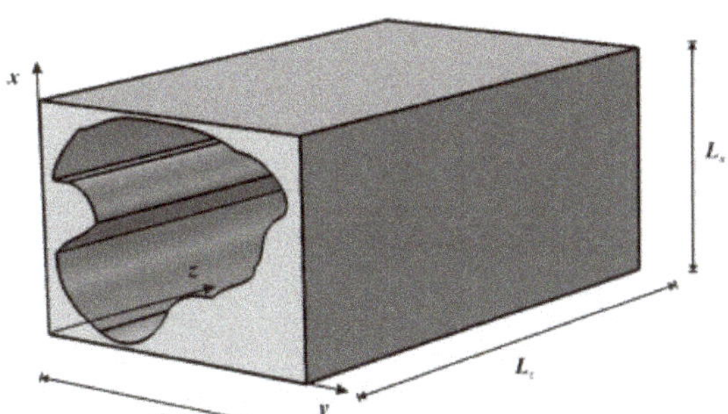

Figure 9. Irregular channel [86] © Elsevier, 2015.

Henze and Wolfersdorf [87] experimentally investigated the impact of tetrahedral VGs on the Nusselt number and the flow velocity. The results showed that using the VGs enhanced the heat transfer rate compared to the uniform channel. In addition, it was found that the highest VGs offered the highest heat transfer enhancement. It was also indicated, as expected, that the heat transfer was enhanced with an increasing Reynolds number.

However, the pressure penalty has not been considered, which determines the pumping power required compared to the uniform channel.

Dai et al. [88] experimentally investigated the influence of zigzag and sine wave micro-channel structures on the laminar water flow and heat transfer maps. They used a Reynolds number in the range of 50 to 900 with heat flux of 19.1 W. A uniform duct was simulated to understand the behavior of hydraulic heat transfer. The results showed that the zigzag geometry enhanced the heat transfer, while the pressure drop increased.

Karathanassis et al. [89] investigated the heat and flow characteristics in an array of fin plate heat sinks. The geometry was designed to be three sections; for each section, the hydraulic diameter was decreased by increasing the number of plates. The FVM was applied to solve the governing equation of the numerical part, while a closed rig with a flow rate ranging from 20 to 40 mL/s was used in the experimental part. It was found that the heat transfer enhanced in the third section due to the buoyancy. Additionally, the temperature was uniform when the Reynolds number decreased. Nevertheless, the pressure drop increased due to an increase in the number of plates.

An analytical investigation has been conducted to study the effect of extended surface area in heat sinks with four types of fins: (a) longitudinal rectangular fin array (LRFA), (b) longitudinal trapezoidal fin array (LTFA), (c) annular rectangular fin array (ARFA), and (d) annular trapezoidal fin array (ATFA), as shown in Figure 10 [90]. The results showed that the triangular fin offered the best heat transfer rate compared to the other three models. It was found that the optimum individual fin was different from the optimum value of heat sink as a component. This is because the individual fin was shorter than the fin in the fins array of the heat sink.

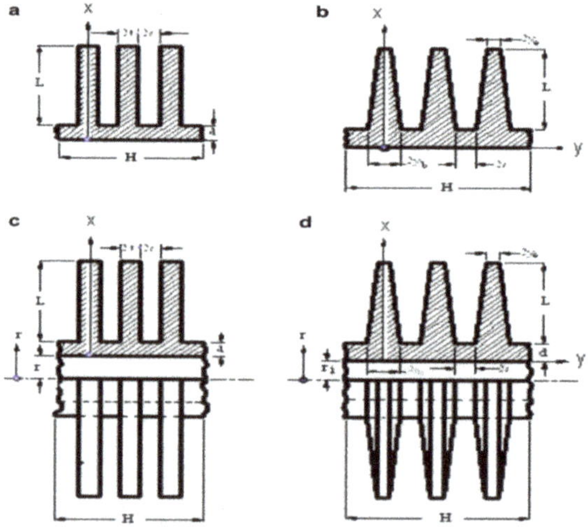

Figure 10. Various types of fins (**a**) LRFA, (**b**) LTFA, (**c**) ARFA, and (**d**) ATFA [90] © Elsevier, 2009.

Ebrahimi et al. [36] studied the impact of using linear VGs to generate vortices in the micro-channel on fluid flow and heat transfer regimes. Different orientations of the VGs and deionized water under laminar flow were considered to simulate three-dimensional geometry utilizing a finite volume method. The results showed that the Nusselt number rose from 2 to 25% when the Reynolds number ranged from 100 to 1100. However, the friction factor increased by up to 30% when using longitudinal VGs. This friction factor penalty could be acceptable if the space was limited and a certain heat transfer rate had to be achieved.

Hong et al. [91] sought to improve the uniformity of the temperature distribution in micro-channel heat sinks by considering a heat sink in which the micro-channels formed a rectangular fractal-shaped network. Their numerical analysis of the 3D conjugate heat transfer revealed hotspots in regions where the channel density was sparse; however, these could be overcome by local modifications of the channel size. The modified network was found to have lower thermal resistance, lower pressure drop, and much improved uniformity in temperature compared to parallel-channel heat sinks.

A recent study carried out by M. Al-Asadi and M. Wilson [1] examined different geometries of perforated pin-finned heat sink. The study examined a perforated pin-finned heat sink using air and water. The results showed that the perforated pin-finned heat sink offered better heat transfer performance and lower pressure penalty compared to a solid pin-finned heat sink when using air as a working fluid, whilst no enhancement was gained when using water due to the high viscosity of the water. The study also compared different shapes of vortex generators such as rectangular, triangular, and circular VGs in a uniform channel. The results conducted that the circular VG was the best in enhancing heat transfer among the examined shapes.

From the previous two sections, it can be seen that there is a gap in the knowledge of different shapes of VGs, especially in cylindrical vortex generators. Thus, the next section focuses on cylindrical vortex generators.

3.2. *Cylindrical Vortex Generators*

The above examples show that there are many ideas for geometrical modifications of micro-channels, some of which are rather complex. A somewhat simpler yet still effective class of geometrical modifications are ribs or cylinders added to the channel walls, base, or interior. These act as transverse vortex generators and have been shown to enhance the heat transfer [34,92–96].

A two-dimensional numerical study by Cheraghi et al. [45] considered a smooth channel with fixed heat flux through the wall sides and an adiabatic cylinder at various locations inside the channel. The Reynolds number was 100 and Prandtl number ranged from 0.1 to 1. The authors found that the maximum enhancement occurred when the cylinder was fixed halfway from the base to the top of the channel. The results also showed that the low Prandtl number had a positive effect on heat transfer enhancement.

Turbulent flow in a channel with cylindrical vortex generators was investigated numerically by Wang and Zhao [97]. It was found that utilizing a cylindrical vortex generator enhanced the heat transfer by 1.18 times compared to the corresponding uniform channel. However, this study did not take into account the thermal conductivity of the rib, which might distribute the heat to the fluid due to the high thermal conductivity of metals compared to fluids, resulting in further enhancement of the heat transfer in micro-channels.

Chai et al. [98] numerically investigated the effects of ribs on the side walls of a silicon micro-channel heated from below and cooled by laminar water flow. The ribs were arranged in an offset manner on both side walls, and had various cross-sectional shapes, namely rectangular, backward triangular, forward triangular, isosceles triangular, and semi-circular, each with a protrusion of 25 μm into the channel. For a Reynolds number in the range of 190–838, Nusselt numbers up to 1.95 times that of a smooth channel were achieved, with the apparent friction factor increasing up to 4.57 times. Performance evaluation criteria values of 1.02 to 1.48 were found, with forward triangular ribs performing best for Re < 350, and semi-circular ribs for Re > 400. In a further three-part work, the same authors also studied aligned versus offset fan-shaped ribs on the opposite side walls [99–101]. Various other side-wall rib shapes and configurations have also been considered by others, e.g., [102,103].

Al-Asadi et al. [104] studied the influence of cylindrical VGs with radii up to 400 μm on 3D conjugate heat transfer. In particular, VGs with quarter- and half-circular cross-sections attached at the base of the micro-channel were considered, aligned perpendicular to the flow direction, with an input heat flux in the range of 100–300 W/cm^2 and a Reynolds number ranging from 100 to 2300. While the quarter-circle VGs offered no improvement in heat transfer, the half-circle cylindrical VGs provided a reduction in the thermal resistance of the system. In addition to VGs completely spanning the width of the micro-channel, shorter VGs were also considered. It was found that having a gap between each channel wall and the ends of the VGs offered further heat transfer benefits, particularly when the pressure drop penalty was taken into account, though the underlying mechanisms by which the gaps enhanced the performance were not explored.

Furthermore, Al-Asadi et al. [8] continued to investigate the influence of the gap in detail in their recent study. It was found that the gap of 75 μm at each end offered the highest heat transfer performance but a lower PEC index compared to the VG with a central gap of 450 μm.

A vortex generator in a uniform heat sink enhances the heat transfer but offers a high pressure drop. The heat transfer enhancement increases with the radius of the VGs. However, Al-Asadi et al. [105] confirmed that the largest VGs do not offer a high PEC.

4. Nanofluids Overview

As remarked in the introduction, an alternative to modifying the geometry to enhance the heat transfer is to modify the working fluid. The last decade has seen a dramatic increase in the nanofluids, as can be seen in Figure 11. Compared to previous years, it is expected that there will be more publications this year as the data are only from up to April 2018.

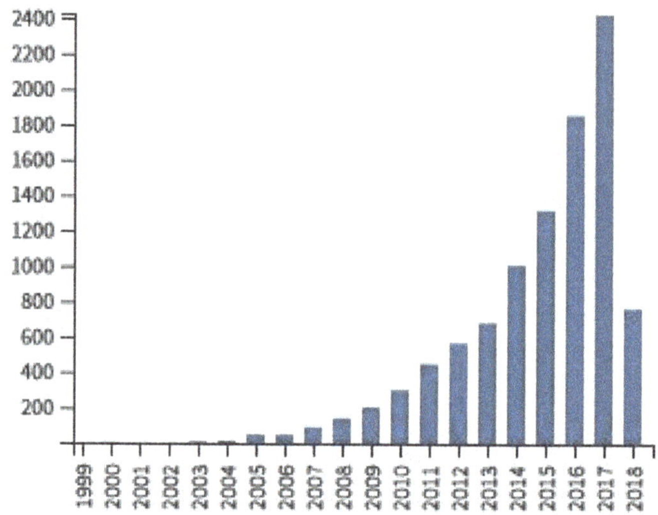

Figure 11. Growth of publications in nanofluids [77] © Elsevier, 2017.

Using nanofluids increases the pumping cost to drive the flow of cooling systems. Thus, researchers have extensively studied the effects of nanofluids on conjugate heat transfer in various ways, such as the effect of fluid temperature, nanoparticle shape, clustering of nanoparticles, and the effect of pH (potential of hydrogen). However, the most related factors to this study are explained below.

4.1. Nanofluids Preparation

A nanofluid is a solid–liquid combination fluid obtained by dispersing nano-sized particles up to 100 nm in a base fluid to improve thermal conductivity of the base liquid [106–109]. Nanofluids can be prepared using several approaches such as a single-step method and two-step method.

The single-step method is a direct evaporation method by which nanoparticles are dispersed directly into a working fluid. In 1996, this method was used to prepare Al_2O_3 and CuO nanoparticles by Eastman et al. in Argonne National Laboratory in the USA [110]. The same procedure was employed by Lee et al. [111], Choi and Eastman [112], and

Choi et al. [113]. After this, Zhu et al. [114] produced Cu-ethylene glycol nanofluid from copper sulphate anhydride ($CuSO_4{\cdot}5H_2O$) and sodium hypophosphite ($NaH_2PO_2{\cdot}H_2O$) reaction under microwave irradiation. This one-step method produced nanofluid with good stability.

In a two-step procedure, the first step introduces an inert gas to produce a dry powder of nanoparticles. These nanoparticles are then dispersed in the conventional fluid. Li and Xuan [108] used the inert gas method, which can produce clean nanoparticles and, as a result, produce a stable nanofluid. However, this method is difficult and expensive for nanoparticle requirements. Consequently, many other techniques may produce dry nanoparticles, such as chemical techniques [115], the aerosol spray method [116], metal vapor [117], arc discharge for nano-carbon tubes [118,119], laser ablation [120], a catalytic process [121], or another successful method called the VEROS (vacuum evaporation on running oil substrate) method [122]. In this technique, direct evaporation in a vacuum onto the surface of running oil was used to produce nanoparticles in small size (10 nm). However, the VEROS technique is not suitable for substances of more than one component, such as metal oxides, and the separation of nanoparticles from the fluid is difficult to produce the dry nanoparticles. The VEROS method was modified by Eastman et al. [123]; they replaced the oil by ethylene glycol to produce Cu-ethylene glycol nanofluid.

4.2. Thermo-Physical Properties of Nanofluids

Nanofluids have high thermo-physical properties [124]. As a result of having high thermal conductivity, nanofluids can offer high heat transfer performance compared to the base fluid without nanoparticles. Various considerations of nanofluid properties have been made to evaluate and prepare nanofluids, such as the effect of base fluid, dispersion and distribution [125], particle shape [126,127], volume concentration [128–131], particle-shell structure [132,133], and thermal contact resistance [134,135]. Furthermore, different factors affect the heat performance of nanofluids such as thermal conductivity, density, viscosity, and heat capacity.

Nanofluids can be used with different shapes of different channels to enhance the heat transfer; for example, a V shape wavy plate channel was studied numerically using various types of nanoparticles and base fluids [136]. The study used a large range of Reynolds number from 8000 to 20,000. The FVM was used to solve the governing equation with the k–ε standard turbulent model to investigate the heat transfer performance. It was found that the best nanofluid was silicon oxide particles in glycerin base fluid to enhance the Nusselt number. However, the pressure drop increased using nanofluids. Such studies should consider an optimization of the heat transfer enhancement and pressure drop increase; a simple optimization factor such as the hydraulic thermal performance also gives an indication about overall enhancement [104].

4.2.1. Experimental Data of Thermal Conductivity

Investigators focused on providing comprehensive data of thermal conductivity and the factors which play a major role to enhancing the thermal conductivity of nanofluids. They found that the base fluid, nanoparticle size, material, and concentration are the most effective parameters. Therefore, Table 1 illustrates a brief survey of thermal conductivity studies.

Table 1. Heat transfer enhancement using different fluids.

Base Fluid	Particles	Size [nm]	φ%	Enhancement
Water [12]	Al_2O_3	30	0.3–2	h > 57%, Nu = 62%
Water [110]		33	5	29%
Water [111]		24.4, 38.4	4	10%
EG			5	17%

Table 1. *Cont.*

Base Fluid	Particles	Size [nm]	φ%	Enhancement
Water [137]		28	3	12%
EG			8	40%
EO			7	50%
Water [138]		38.4	4	24.3%
Water, EG, PO [139]		12.2–302	5	30%
Water [140]		36	10	29%
Water [141]		27–56	1.6	10%
Water [142]		48	1	4%
Water [143]		20	14.5	20%
Water [144]		110–210	1	0%
Water [145]		36, 47	6	28%
Water [146]		8–282	4	18%
EG		12–282	3	16%
Water [147]		36, 47	18	30%
Water	CuO	36	5	60%
Oil			5	44%
Water [111]		18.6, 28.6	3.5	12%
EG			4	20%
Water [137]		23	4.5	12.3%
EG			6	12.5%
Water [138]		28.6	4	36%
Water [139]		29	6	58%
Water [143]		33	5	18%
Water [142]		33	1	5%
EG			1	9%
Water [147]		29	3.3	8%
Water [148]		L = 50–100	0.4	9.6%
Water [149]	TiO_2	10,15 -40 rod	5	30, 33%
Water [150]		165	0.72	6.5%
Water [143]		40	2.5	6%
Water [151]		95	2	22%
Water [142]			1	14.4%
Water [152]			3	9.6%
Water [153]	Fe_3O_4	9.8	5	38%
EG [142]	WO_3	38	0.3	14%
Water [143]	ZrO_3	20	10	15%
Water [144]		110–250	0.1	0%
Water [144]	SiO_2	20–40	0.1	0%
Water [154]		12	1	3%
EG [155]	Cu	<10	0.3	40%
Water [156]		100	7.5	75%

Table 1. *Cont.*

Base Fluid	Particles	Size [nm]	φ%	Enhancement
Oil			7.5	44%
Water [157]		50–100	0.1	23.8%
EG [158]	Fe	10	0.55	18%
EG [159]		10	0.2	18%
EG [142]		10	0.55	18%
Water [160]	AG, Au	10–20	0.001	4%
Toluene			0.001	9%
Toluene [161]	Au	2	0.04	1.5%
Ethanol		4	0.03	1.4%
Toluene	Fullerene C60-C70	0.5–0.6	0.8	0%
Mineral oil [161]		10	0.8	6%

Increase in the thermal conductivity of the working fluid improves the efficiency of the associated heat transfer process. However, investigations into the convective heat transfer of nanofluids indicated that the enhancement of heat transfer coefficient exceeds the thermal conductivity enhancement of nanofluids [162–165]. Moreover, other parameters such as the density, heat capacity, and viscosity have a lower effect than thermal conductivity.

4.2.2. Theoretical Development of Nanofluid Equations

Investigators have started from the Maxwell equation [166] to predict the thermal conductivity of nanofluids. Improving the Maxwell equation offered a better understanding of the behavior of thermal conductivity since 1935, when Bruggeman [167] reported that a high concentration of nanoparticles cannot be neglected. Moreover, in 1987, Hasselman [135] modified the theory of Maxwell considering the size of the composite dispersed phase in addition to the volume concentration. However, these studies under-predicted the experimental measurements.

Many investigations have tried to improve the Maxwell equation to produce a modified thermal conductivity equation that offers good agreement with the experimental data. Modern techniques were utilized to enhance the prediction of nanoscale equations such as the nanoparticle–matrix interfacial layer [168,169], nanoparticle Brownian motion [170,171], and nanoparticle cluster/aggregate [172].

Nie et al. [173] used the exact expression for the heat flux vector of the base fluid plus a nanoparticle system to estimate the contribution of nanoparticle Brownian motion to thermal conductivity. It was found that its contribution is too small to account for the abnormally high reported values. The mean free path and the transition speed of phonons in nanofluids were estimated through density functional theory. It was found that a layer structure can form around the nanoparticles and the structure does not further induce fluid–fluid phase transition in the bulk fluid.

In contrast to Nie et al. [152], Ghasimi and Aminossadati [174] showed that considering Brownian motion would enhance the thermal conductivity. They used CuO-water nanofluid in a right triangular enclosure. The results also reported that heat transfer was enhanced with the increase in nanoparticles.

Xuan and Roetzel et al. [175] suspended ultrafine particles to change the properties and heat transfer performance of the nanofluid, which exhibited a great potential in enhancing the heat transfer. Based on the assumption that the nanofluid behaves more like a fluid than a conventional solid–fluid mixture, they proposed two different approaches for deriving the heat transfer correlation of the nanofluid. The effects of transport properties of the nanofluid and thermal dispersion were also included.

4.2.3. The Effect of Base Fluid

The base fluid can be water, oils, or ethylene glycol. Researchers have investigated the effect of base fluid on heat transfer enhancement for two decades [176–178].

Xie et al. [179] studied the enhancement ratio of thermal conductivity between the base fluids and nanofluids. They considered three types of base fluid (water, glycerol, ethylene glycol and pump oil) with α-Al_2O_3 nanoparticles. The results showed that the water-based nanofluid had the lowest thermal conductivity compared to other nanofluids, while the thermal conductivity of the water itself was higher compared to the other base fluids.

However, using nanofluids with water-based nanofluid was most common in many heat-transfer and fluid-flow applications.

4.2.4. The Effect of Nanoparticles Concentration

The influence of the concentration is an effective factor to enhance the thermo-physical properties of nanofluids. It is the portion of volume of nanoparticles to the base fluid. Many researchers declared that having solid particles in the base fluid would enhance the thermal conductivity of nanofluids, increasing the viscosity and density of nanofluids [177,180–185]. Furthermore, nanofluids showed non-Newtonian behavior when using nanoparticles of more than 5% [165,186]. However, due to the high thermal conductivity of metallic nanoparticles, they offer the highest thermal conductivity of nanofluids compared to the oxides and non-metallic nanoparticles. Yulong et al. [187,188] studied the effect of volume concentration on thermal conductivity enhancement. They found that the thermal conductivity enhanced with the nanoparticles, as shown in Figure 12.

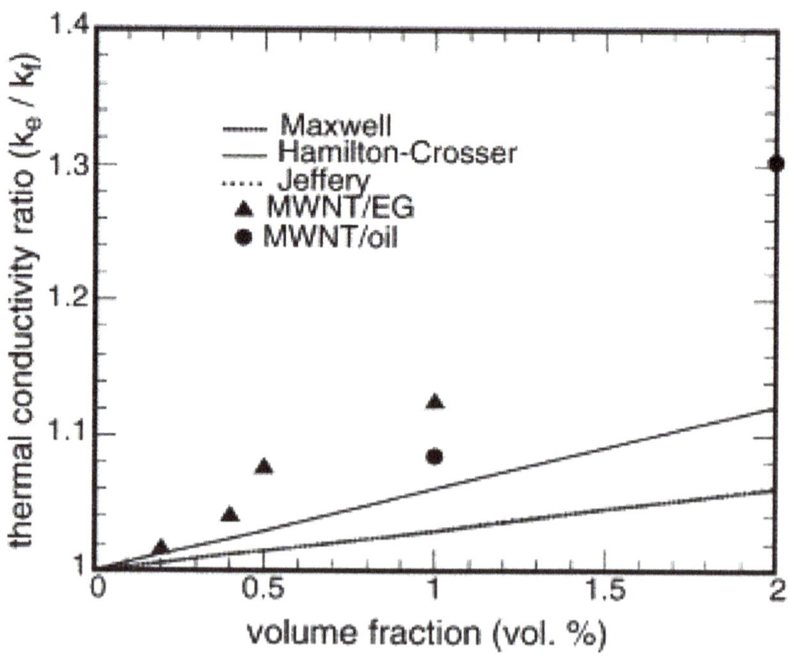

Figure 12. The influence of nanoparticle concentrations on thermal conductivity.

Another study by Kumar et al. [189] investigated the impact of thermal conductivity and base fluid on conjugate heat transfer. They utilized CuO and TiO_2 up to 1% of volume concentration and different base fluids of water and ethylene glycol under a temperature ranging from 30 to 50 °C. The study found that the thermal conductivity enhanced as the nanoparticle concentration increased for both cases; an example is shown in Figure 13. The very famous study on thermal conductivity carried out by INPBE [190] reported that the enhancement relationship between the thermal conductivity and nanoparticles concentration was approximately linear.

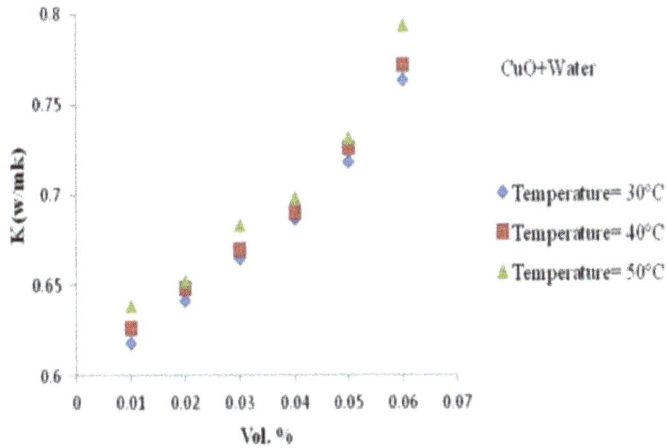

Figure 13. The influence of nanoparticle concentrations on thermal conductivity [189] © Elsevier, 2016.

4.2.5. The Influence of Nanoparticle Materials

As one might expect, the nanoparticle material has an effect on resulting nanofluid properties. Nanoparticles can be metallic (Fe, Cu, Ag, Au, Al), carbon or metallic oxide (Fe_3O_4, CuO, Al_2O_3, TiO_2, SiC, SiO_2, ZnO) [191–195]. Metallic oxide nanoparticles are commonly used with water as a base fluid. This is because the oxides are considered more stable than the pure metals in fluids. Moreover, the oxygen in the dioxides makes nanoparticles disperse easily and stable in the base fluid. However, Al_2O_3 and SiO_2/water nanofluids offer the highest heat transfer enhancement among the common nanofluids. Some researches have indicated that SiO_2/water nanofluids offer higher heat transfer enhancement compared to Al_2O_3/water [184,196–198], while others have reported the opposite findings [199–201].

4.2.6. Thermal Conductivity

The thermal performance of a working fluid can be enhanced by increasing its thermal conductivity. This is because solid nanoparticles have higher thermal conductivity than the base fluid; for instance, at room temperature, the thermal conductivity of copper is 700 times higher than that of the water. Therefore, adding solid nanoparticles to the base fluid improves the thermal conductivity of the working fluid. A ratio between the nanofluids and the base fluid can be applied k_{nf}/k_{bf} to calculate and evaluate the enhancement of the thermal conductivity of nanofluids. The enhancement of thermal conductivity achieved 40% in some cases, despite the concentration of the nanoparticles not exceeding 10% [110,113,137,138,140,155,202].

In the last two decades, some researchers have indicated that there was no agreement between the theoretical equations and experimental data in terms of thermal conductivity, while Keblinski et al. [203] reported that most results of numerical and experimental investigations showed good agreement. However, the study by Keblinski et al. [203] was supported by a benchmark study of thermal conductivity carried out by the International Nanofluid Properties Benchmarking Exercise (INPBE) [190]. The INPBE sent samples to 30 organizations worldwide to measure the thermal conductivity. The results showed that the thermal conductivity showed ±10% or lower average differences between the experimental data and theoretical equation of thermal conductivity. Nanofluids have high thermo-physical properties compared to the base fluid in terms of thermal conductivity [204–208] and heat transfer coefficient [209–211]. Therefore, theoretical and experimental surveys are presented in the next sections.

4.3. *Nanofluid Equations*

4.3.1. Thermal Conductivity

Modern equations of effective thermal conductivity [212] are presented based on the basic correlation [213], which was developed [214] to be two equations: static and Brownian thermal conductivity (see Equation (1)). The static thermal conductivity is proposed by [174] (see Equation (2)) as below:

$$k_{eff} = k_{static} + k_{brownian} \quad (1)$$

$$k_{stattic} = k_f \left[\frac{(k_s + 2k_f) - 2\varphi(k_f - k_s)}{(k_s + 2k_f) + \varphi(k_f - k_s)} \right] \quad (2)$$

where k_s and k_f are the thermal conductivities of the particles and the fluid, respectively.

The Brownian motion thermal conductivity equation [214] is:

$$k_{brownian} = 5 \times 10^4 \beta \varphi \rho_f Cp_f \sqrt{\frac{KT}{\rho_s d_s}} f(T, \varphi) \quad (3)$$

where:

$$f(T, \varphi) = \left(2.8217 \times 10^{-2}\varphi + 3.917 \times 10^{-3}\right)\left(\frac{T}{T_0}\right) + \left(-3.0669 \times 10^{-2}\varphi - 3.91123 \times 10^{-3}\right)$$

where K is the Boltzmann constant, T is the fluid temperature, and T_0 is the reference temperature.

4.3.2. Viscosity Equation

The viscosity of the nanofluid is approximately the same as the viscosity of a base fluid if containing dilute suspension of fine spherical particles, as shown below [174]:

$$\begin{gathered} \frac{\mu_{eff}}{\mu_f} = \frac{1}{1-34.87\left(d_p/d_f\right)^{-0.3}\varphi^{1.03}} \\ d_f = \left[\frac{6M}{N\pi\rho_{fo}}\right]^{1/3} \end{gathered} \quad (4)$$

where μ_{eff} and μ_f are the viscosity of nanofluid and base fluid, respectively, d_p is the nanoparticle diameter, d_f is the base fluid equivalent diameter, and φ is the nanoparticles volume fraction. M is the molecular weight of the base fluid and N is the Avogadro number, and ρ_{fo} is the mass density of the base fluid calculated at temperature T = 293 K.

4.3.3. The Density Equation

The effective density consists of three main parameters, which are the nanofluid concentration (φ), nanoparticle density ρ_s, and base fluid density ρ_f [12]:

$$\rho_{eff} = (1 - \varphi)\rho_f + \varphi\rho_s \quad (5)$$

4.3.4. The Effective Heat Capacity Equation

With Cp_s being the heat capacity of the solid particles, and Cp_f being that of the base fluid, the effective heat capacity of the nanofluid is given by [183]:

$$(Cp)_{eff} = \frac{(1 - \varphi)(\rho Cp)_f + \varphi(\rho Cp)_s}{(1 - \varphi)\rho_f + \varphi\rho_s} \quad (6)$$

4.3.5. The Effective Thermal Expansion Equation

The thermal expansion for solid parts β_s and for base β_f fluid with φ can produce the effective thermal expansion as follows [215,216]:

$$\beta_{\text{eff}} = \frac{(1-\varphi)(\rho\beta)_f + \varphi(\rho\beta)_s}{(1-\varphi)\rho_f + \varphi\rho_s} \tag{7}$$

4.4. Drawbacks of Nanofluids

Recent investigations have indicated that there is no benefit to using nanofluids. Moreover, Myers et al. [217] revealed that there is a lack of consistency between the mathematical and experimental studies. The authors also indicated that comparing nanofluids on the basis of non-dimensional parameters such as the Reynolds number is misleading in drawing a correct conclusion on the real heat transfer enhancement. Furthermore, Haddad et al. [218] reviewed natural convection using nanofluids. They indicated that in numerical studies the heat transfer was significantly enhanced using nanofluids; nevertheless, the experimental investigations showed the opposite results. However, this study reviewed the natural convection investigations only, which can support the opposite results of the experimental studies because there is perhaps not enough flow to circulate the nanoparticles in the system, which leads to augmentation of the nanoparticles in one place of the system. This could cause hot spot zones, decreasing the heat transfer performance of the system. The reason behind the discrepancy between the numerical and experimental studies of the same working condition might be the augmentation, as the numerical studies do not take this issue into account.

Another point of using a fixed Reynolds number and a fixed pumping power with nanofluids was highlighted by Haghighi et al. [219]. The results showed that there is no enhancement in heat transfer when using a fixed pumping power with nanofluids. However, using a fixed Reynolds number showed good enhancement of heat transfer. The same findings were concluded by Alkasmoul [220,221].

Though nanofluids enhance the heat transfer rate, they attract more cost in pumping to drive the fluid [210,222–227].

There are other applications that indicate that water-based nanofluids cannot offer any benefit for cooling systems because of the high temperature conditions; for instance, semiconductor materials such as silicon carbide (SiC) and gallium nitride (GaN) operate under special temperature conditions above 200 °C [228]. In this case, nanofluids do not offer any option to enhance the heat transfer performance due to the limitation of the temperature condition. Therefore, a replacement fluid should be used to overcome this issue. The solution might be by suggesting another advanced liquid which has different chemical properties to the nanofluid. This advanced fluid could be an ionic liquid which works with high heat flux [229].

5. Evaluation of Heat Transfer Improvement

In electronics, the main aim of enhancing a cooling system is to reject the generated heat and keep the electronics working in the range of a limited temperature of 85 °C [104]. Recently, optimizing the energy to reduce the power consumption of the cooling system has attracted many researchers; the power consumption could be the pumping power of the cooling system or the power reduction after enhancing the cooling system.

To help to evaluate the benefit reuse cost of a proposed modification of a cooling system, a performance evaluation criterion (PEC) index can be formulated which accounts for both change in heat transfer performance and fluid flow effects [230,231]. The heat transfer term could be the Nusselt number or thermal resistance, while the fluid flow term might be the friction factor or pressure drop. This formula can be used to evaluate the performance of the overall enhancement of the system.

Furthermore, the formula could be used to evaluate the performance of modification of the geometry by comparing basic and developed designs such as a smooth and modified micro-channel, as shown in the equation below [29,30]:

$$PEC = \frac{Nu/Nu_s}{(f/f_s)^{1/3}} \tag{8}$$

where Nu, Nu_s are the Nusselt numbers for modified and straight channels, and f, f_s are the friction factor for modified and straight channels.

An example of the importance of the PEC is the study by Al-Asadi et al. [8], which indicated that the heat transfer was enhanced (the thermal resistance is reduced) using cylindrical VGs with a central gap of 100 μm but higher pressure drop than the cylindrical VGs with an end gap. However, the results were reversed (cylindrical VGs with an end gap offer better heat transfer than the cylindrical VGs with a central gap of 100 μm) when taking the pressure drop into account using PEC.

This evaluation method was used by many researchers to examine the performance of the proposed designs. Furthermore, it could be considered as a starting point for optimizing the whole system.

6. Validation of Numerical Methods Versus Experimental Investigations

Developments in numerical solution methods for heat transfer in the system have made them more accurate and more closely aligned with the experimental data, for example, considering the temperature dependence of the thermo-physical properties [104]. Many investigations have been carried out to improve the numerical methods, such as the study by Bushehri et al. [232], who proposed a new method to deal with fluids and solids with an equation utilizing the (FVM) CFD software openFOAM, (New York, NY, USA) with new boundary conditions for the temperature jump and flow slip. The equation was tested against previous works and showed good agreement. It was applied to a micro-channel heat sink consisting of two parallel plates to investigate the heat transfer performance. The results indicated that the heat transfer was accurately calculated using the proposed equation.

Moreover, many studies have paid attention to numerical simulation because it is important to predict the experimental measurements such as heat flux, temperature, and fluid velocity. Using simulation offered a low cost compared to the experimental setup [67]. Many investigations have also developed the numerical methods, making them more accurate and efficient, for instance, improving a hybrid finite element method to solve solid–liquid equations of the microchannel [233], as well as modifying a technique such as the generalized integral transform technique (GITT) to solve the coupling equation, which showed very good agreement with the COMSOL Multiphysics® [234].

An example of good agreement between the numerical and experimental studies can be seen in Figure 14.

However, Al-Asadi et al. [1,8] showed that using governing temperature-dependent equations for thermal properties gives better results than temperature-independent equations. The temperature dependence of the fluid properties is given by the following expressions built into COMSOL based on experimental data:

$$\rho(T_L) = 838.466135 + 1.40050603T_L - 0.0030112376T_L^2 + 3.71822313 \times 10^{-7}T_L^3 \tag{9}$$

$$\begin{aligned}\mu(T_L) = {} & 1.3799566804 - 0.0212240191 51T_L + 1.3604562827 \times 10^{-4}T_L^2 - 4.6454090319 \times 10^{-7}T_L^3 \\ & + 8.9042735735 \times 10^{-10}T_L^4 - 9.0790692686 \times 10^{-13}T_L^5 + 3.8457331488 \times 10^{-16}T_L^6\end{aligned} \tag{10}$$

$$C_p(T_L) = 12010.1471 - 80.4072879T_L + 0.309866854T_L^2 - 5.38186884 \times 10^{-4}T_L^3 + 3.62536437 \times 10^{-7}T_L^4 \tag{11}$$

$$k(T_L) = -0.869083936 + 0.00894880345T_L - 1.58366345 \times 10^{-5}T_L^2 + 7.97543259 \times 10^{-9}T_L^3 \tag{12}$$

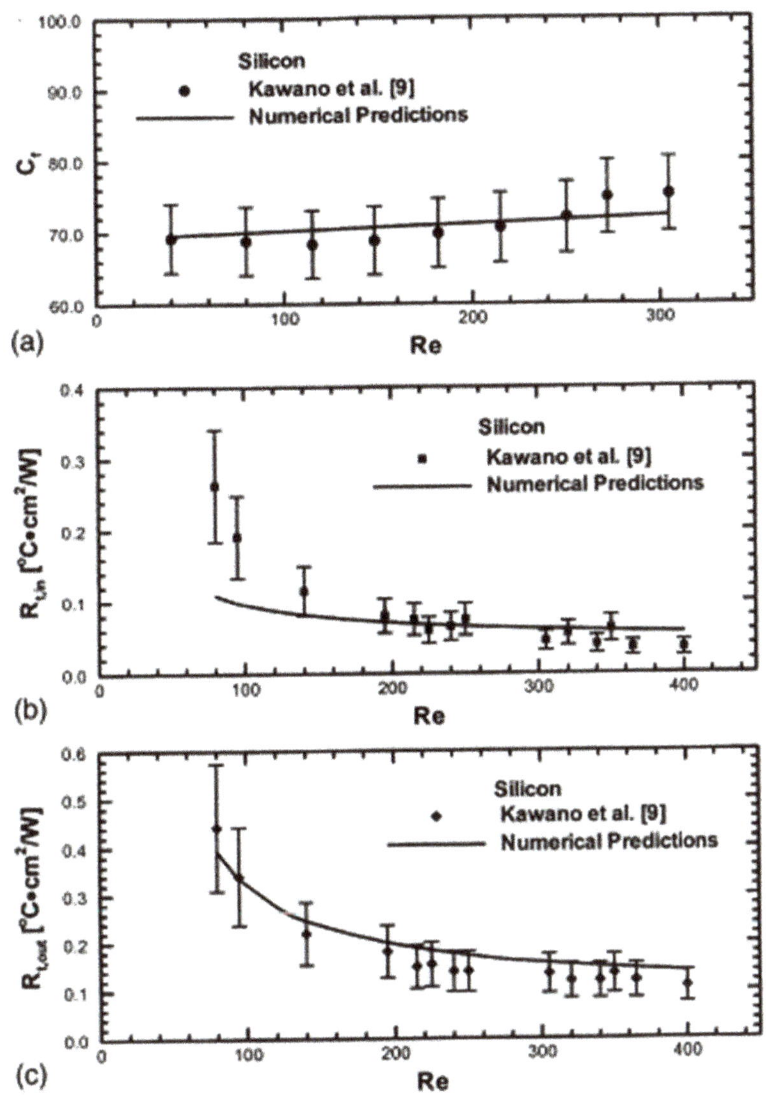

Figure 14. Validations between numerical findings of Qu and Mudawar [235] © Elsevier, 2002 and experiments of Kawano et al. [236] © Elsevier, 1998; (**a**) friction coefficient, (**b**) inlet thermal resistance, and (**c**) outlet thermal resistance.

It was found that applying the above equations used by Al-Asadi et al. [1,8] produces better results compared to the experiments of Kawano et al. [236] as presented in Figure 15.

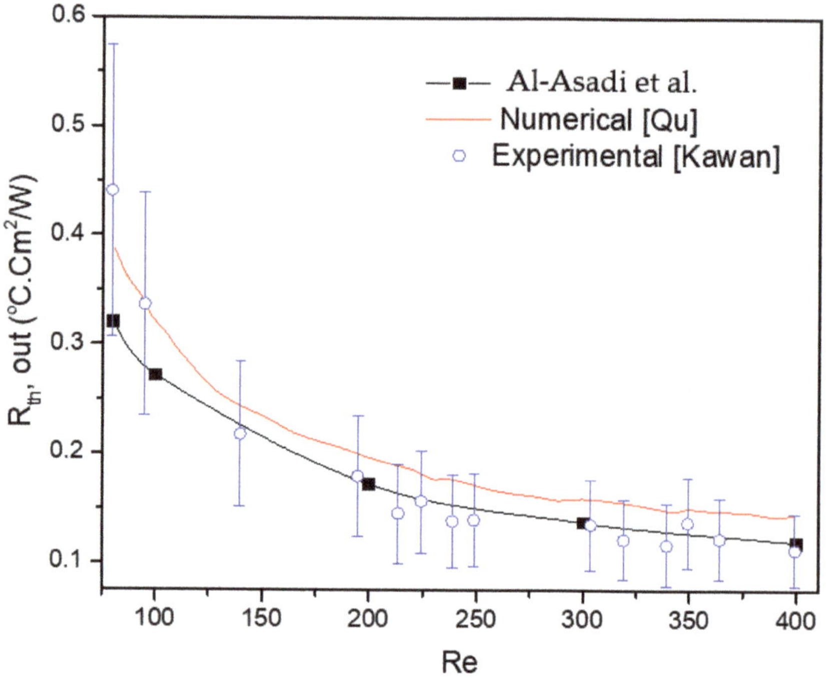

Figure 15. Validation of the present model against experimental data of Kawano et al. [236] © Elsevier, 1998 and alternative numerical results of Qu and Mudawar [235] © Elsevier, 2002 and Al-Asadi et al. [1,8].

7. Conclusions and Recommendations for Future Work

A comprehensive literature review has been provided in this paper to understand the gaps in knowledge in the available literatures on micro-channel heat sink and nanofluids investigations. Numerous experimental configurations and theoretical studies have been devoted to investigating the variety of geometry designs such as straight, curved, regular, irregular, extended surface, and vortex generators (VGs) to enhance the heat transfer and fluid flow. Attempts at improvement included the use of different micro-channel geometries, coolant types, and structural materials. Any improvements in the overall performance of the systems were analyzed using different analysis methods. As a result, several conclusions can be drawn, summarized as follows:

- All studies in the literature indicated that using extended surface areas such as ribs or grooves in uniform channels offered better heat-transfer enhancements compared to the uniform channels themselves. However, there will be a pressure drop penalty caused by ribs and grooves disturbing the fluid flow. Thus, the modified channels are a recent area focused especially on using VGs as their influence on heat transfer and fluid flow characteristics. Two types of vortex generators (VG) have been classified based on the direction of the axis of rotation of the vortices generated, which were transvers and longitudinal VGs.
- Although there are many examples of geometrical modifications that offer some form of benefit in terms of heat transfer, most of these are rather complex. So far, simpler cylindrical vortex generators have only been explored partially in two recent studies. However, they have focused on flow disturbance, treating the VGs as adiabatic objects; therefore, conjugate heat transfer effects have not been precisely considered.
- Based on the available literature, there are few numerical studies of mini- or micro-channels using cylindrical vortex generators [1,8,45,97,104,227], while no experimental

investigation has been found; this is because of the limitations of manufacturing micro-sized channels. However, the literature has shown good agreement between the numerical and experimental studies for mini-channels. Thus, the motivation for this review paper was to highlight the modified design of different shapes of VGs.

- Another aspect by which the heat transfer in micro-channels can be enhanced is by using a nanofluid. However, the price is paid for the pumping power. The literature showed that there is an argument when comparing SiO_2 and Al_2O_3 nanoparticles in water at the same concentration: some researchers revealed that the Al_2O_3 water offers better heat transfer enhancement, whereas others indicated the opposite. However, there are still areas of disagreement in terms of the benefit of using them. Although the thermal conductivity can be enhanced, the drawback of nanofluids is the increases in pressure drop required to drive the flow. Thus, deeper investigation has been carried out and highlighted in this paper to study the performance of these two types of nanofluids.
- Other ideas have focused on modifications within the parallel channels themselves—for example, by adding grooves or ribs [29–32]. They act as vortex generators (VGs) to enhance the heat transfer and fluid flow characteristics by disturbing the flow and creating vortices that can be classified as transverse vortices, where the axis of rotation is perpendicular to the flow direction, or longitudinal vortices, with axes lying along the direction of flow. Longitudinal vortices are generally more effective than transverse ones in enhancing the heat transfer performance [16,33,237]. VGs can take various forms, such as protrusions, wings, inclined blocks, winglets, fins, and ribs [2,36,103,238], and have also been used to enhance heat transfer in different geometries such as circular and non-circular ducts under turbulent flow [19,34,239]. They have also been used in laminar flow [41], with flat plate-fins in rectangular channels [37–39], tube heat exchangers [40], heat sinks [41,42], and narrow rectangular channels [43,44]. Other recent investigations have also indicated the potential benefits of using VGs of various shapes with laminar flow at different Reynolds numbers [35,43,82].
- Indeed, developments in manufacturing capabilities and processes have opened wide possibilities for modifying the heat sink types to enhance the performance of cooling systems. Therefore, the air as a working fluid becomes limited to use in high heat flux devices. Now, an interesting question is: can perforated PHS be used with water to meet the requirement of electronics devises? Al-Asadi et al. [1] have answered this question. The perforated PHS cannot be used with water to enhance the heat transfer due to the difference of the thermo-physical properties between the air and water. Therefore, Al-Asadi et al. [1] have offered a new design of a uniform micro-channel with vortex generators (VGs) with different shapes and developed their investigation by using quarter and half-circle VGs [104]. They then developed the VG configurations to reduce the thermal resistance and pressure penalty by having a gap in the half-circle VGs [8]. Thus, the motivation of this sequence is to develop a design to enhance the heat transfer and reduce the power consumption of a micro cooling system.

Finally, the growing interest in the micro-channel heat sinks with VGs, which is evident by the number of available studies, leads to the conclusion that research in this fascinating area is still progressing and in need to further exploration. Moreover, there is a gap in knowledge on using micro-channels with liquid coolants (applications demand the capability to handle high heat-flux; therefore, developing liquid cooling systems is increasingly important).

Therefore, more studies are needed to investigate the common air-based heat sink with water as a working fluid [1]. Before presenting the investigation, it is important to illustrate the methodology of using temperature-independent Equations (9)–(12) [8].

Practically, especially for large heat flux conditions, an ionic liquid could be suggested [229] to enhance the heat transfer rate.

Author Contributions: Conceptualization, M.T.A.-A.; Formal analysis, M.T.A.-A.; Investigation, M.T.A.-A., H.A.M., M.C.T.W.; Methodology, M.T.A.-A., H.A.M., M.C.T.W.; Supervision, H.A.M., M.C.T.W.; Writing-original draft, M.T.A.-A., H.A.M.; Writing-review and editing, M.T.A.-A., H.A.M., M.C.T.W. All authors have read and agreed to the published version of the manuscript.

Funding: There is no funding for this research.

Institutional Review Board Statement: Not applicable.

Informed Consent Statement: Not applicable.

Data Availability Statement: Data available on request due to restrictions e.g., privacy or ethical. The data presented in this study are available on request from the corresponding author. The data are not publicly available due to the need for further research.

Acknowledgments: We thank the Basra Oil Company (BOC), Iraq for their sponsorship of Mushtaq T. K. Al-Asadi.

Conflicts of Interest: The authors have no declared any conflict of interest.

References

1. Al-Asadi, M.; Al-Damook, A.; Wilson, M. Assessment of vortex generator shapes and pin fin perforations for enhancing water-based heat sink performance. *Int. Commun. Heat Mass Transf.* **2018**, *91*, 1–10. [CrossRef]
2. Ahmed, H.; Mohammed, H.; Yusoff, M. An overview on heat transfer augmentation using vortex generators and nanofluids: Approaches and applications. *Renew. Sustain. Energy Rev.* **2012**, *16*, 5951–5993.
3. Knight, R.W.; Goodling, J.S.; Hall, D.J. Optimal Thermal Design of Forced Convection Heat Sinks-Analytical. *J. Electron. Packag.* **1991**, *113*, 313–321. [CrossRef]
4. Knight, R.; Goodling, J.; Gross, B. Optimal thermal design of air cooled forced convection finned heat sinks-experimental verification. In *IEEE Transactions on Components, Hybrids, and Manufacturing Technology*; IEEE: New York, NY, USA, 1992; pp. 754–760.
5. Teertstra, P.; Yovanovich, M.M.; Culham, J.R. Analytical forced convection modeling of plate-fin heat sinks. *J. Electron. Manuf.* **2000**, *10*, 253–261. [CrossRef]
6. Copeland, D. Optimization of parallel plate heatsinks for forced convection. In Proceedings of the Sixteenth Annual IEEE Semiconductor Thermal Measurement and Management Symposium, San Jose, CA, USA, 23 March 2000.
7. Sparrow, E.M.; Ramsey, J.W.; Altemani, C.A.C. Experiments on In-line Pin Fin Arrays and Performance Comparisons with Staggered Arrays. *J. Heat Transf.* **1980**, *102*, 44–50. [CrossRef]
8. Al-Asadi, M.T.; Alkasmoul, F.S.; Wilson, M.C. Benefits of spanwise gaps in cylindrical vortex generators for conjugate heat transfer enhancement in micro-channels. *Appl. Therm. Eng.* **2018**, *130*, 571–586. [CrossRef]
9. Zhou, F.; Catton, I. Numerical Evaluation of Flow and Heat Transfer in Plate-Pin Fin Heat Sinks with Various Pin Cross-Sections. *Numer. Heat Transfer, Part A: Appl.* **2011**, *60*, 107–128. [CrossRef]
10. Jacobs, E.N. *Tests of Six Symmetrical Airfoils in the Variable Density Wind Tunnel*; NACA: Moffett Field, CA, USA, 1931.
11. Jacobs, E.N.; Ward, K.E.; Pinkerton, R.M. *The Characteristics of 78 Related Airfoil Sections from Tests in the Variable-Density Wind Tunnel*; NACA: Moffett Field, CA, USA, 1933; pp. 299–354.
12. Albadr, J.; Tayal, S.; Al-Asadi, M. Heat transfer through heat exchanger using Al_2O_3 nanofluid at different concentrations. *Case Stud. Therm. Eng.* **2013**, *1*, 38–44. [CrossRef]
13. Kherbeet, A.; Safaei, M.R.; Mohammed, H.A.; Salman, B.; Ahmed, H.; Alawi, O.A.; Al-Asadi, M. Heat transfer and fluid flow over microscale backward and forward facing step: A review. *Int. Commun. Heat Mass Transf.* **2016**, *76*, 237–244. [CrossRef]
14. Al-Asadi, M.; Mohammed, H.; Kherbeet, A.; Al-Aswadi, A. Numerical study of assisting and opposing mixed convective nanofluid flows in an inclined circular pipe. *Int. Commun. Heat Mass Transf.* **2017**, *85*, 81–91. [CrossRef]
15. Kherbeet, A.S.; Mohammed, H.; Ahmed, H.E.; Salman, B.; Alawi, O.A.; Safaei, M.R.; Khazaal, M. Mixed convection nanofluid flow over microscale forward-facing step—Effect of inclination and step heights. *Int. Commun. Heat Mass Transf.* **2016**, *78*, 145–154. [CrossRef]
16. Wang, C.-C.; Lo, J.; Lin, Y.-T.; Liu, M.-S. Flow visualization of wave-type vortex generators having inline fin-tube arrangement. *Int. J. Heat Mass Transf.* **2002**, *45*, 1933–1944. [CrossRef]
17. Depaiwa, N.; Chompookham, T.; Promvonge, P. Thermal enhancement in a solar air heater channel using rectangular winglet vortex generators. In Proceedings of the Proceedings of the International Conference on Energy and Sustainable Development: Issues and Strategies (ESD 2010), Chiang Mai, Thailand, 2–4 June 2010.
18. Li, H.-Y.; Chen, C.-L.; Chao, S.-M.; Liang, G.-F. Enhancing heat transfer in a plate-fin heat sink using delta winglet vortex generators. *Int. J. Heat Mass Transf.* **2013**, *67*, 666–677. [CrossRef]
19. Al-Asadi, M.T.; Mohammed, H.A.; Akhtar, S.N.; Kherbeet, A.S.; Dawood, H.K. Heat Transfer Enhancements Using Traditional Fluids and Nanofluids in Pipes with Different Orientations: A Review. *J. Nanofluids* **2017**, *6*, 987–1007. [CrossRef]

20. Soudagar, M.E.M.; Kalam, M.A.; Sajid, M.U.; Afzal, A.; Banapurmath, N.R.; Akram, N.; Mane, S.; C, A.S. Thermal analyses of minichannels and use of mathematical and numerical models. *Numer. Heat Transf. Part A Appl.* **2020**, *77*, 497–537. [CrossRef]
21. Ali, M.S.; Anwar, Z.; Mujtaba, M.; Soudagar, M.E.M.; Badruddin, I.A.; Safaei, M.R.; Iqbal, A.; Afzal, A.; Razzaq, L.; Khidmatgar, A.; et al. Two-phase frictional pressure drop with pure refrigerants in vertical mini/micro-channels. *Case Stud. Therm. Eng.* **2021**, *23*, 100824. [CrossRef]
22. Malakar, A.; Kanel, S.R.; Ray, C.; Snow, D.D.; Nadagouda, M.N. Nanomaterials in the environment, human exposure pathway, and health effects: A review. *Sci. Total Environ.* **2020**, *759*, 143470. [CrossRef]
23. Soudagar, M.E.M.; Nik-Ghazali, N.-N.; Kalam, A.; Badruddin, I.; Banapurmath, N.; Akram, N. The effect of nano-additives in diesel-biodiesel fuel blends: A comprehensive review on stability, engine performance and emission characteristics. *Energy Convers. Manag.* **2018**, *178*, 146–177. [CrossRef]
24. Ahmed, W.; Amin, M.; Ahmed, K.; Mehamood, S.; Fayaz, H.; Mujtaba, M.A.; Soudagar, M.E.M.; Gul, M. Characteristics investigation of silicone rubber-based RTV/μATH@nSiO_2 micro/nano composites for outdoor high voltage insulation. *J. Dispers. Sci. Technol.* **2021**, 1–14. [CrossRef]
25. Tuckerman, D.; Pease, R. High-performance heat sinking for VLSI. *IEEE Electron Device Lett.* **1981**, *2*, 126–129. [CrossRef]
26. Baghban, A.; Sasanipour, J.; Pourfayaz, F.; Ahmadi, M.H.; Kasaeian, A.; Chamkha, A.J.; Oztop, H.F.; Chau, K.-W. Towards experimental and modeling study of heat transfer performance of water- SiO2 nanofluid in quadrangular cross-section channels. *Eng. Appl. Comput. Fluid Mech.* **2019**, *13*, 453–469. [CrossRef]
27. Nagarani, N.; Mayilsamy, K.; Murugesan, A.; Kumar, G.S. Review of utilization of extended surfaces in heat transfer problems. *Renew. Sustain. Energy Rev.* **2014**, *29*, 604–613. [CrossRef]
28. Ahmed, S.C.; Asif, A.; Irfan, A.B.; Khan, T.M.Y.; Kamangar, S.; Abdelmohimen, M.; Soudagar, M.E.M.; Fayaz, H. Numerical investigation on pressure-driven electro osmatic flow and mixing in a constricted micro channel by triangular obstacle. *International J. Numer. Methods Heat Fluid Flow* **2021**, *31*, 982–1013.
29. Manca, O.; Nardini, S.; Ricci, D. A numerical study of nanofluid forced convection in ribbed channels. *Appl. Therm. Eng.* **2012**, *37*, 280–292. [CrossRef]
30. Ahmed, M.; Yusoff, M.; Ng, K.C.; Shuaib, N. Effect of corrugation profile on the thermal–hydraulic performance of corrugated channels using CuO–water nanofluid. *Case Stud. Therm. Eng.* **2014**, *4*, 65–75. [CrossRef]
31. Wang, C.-C.; Liaw, J.-S. Air-side performance of herringbone wavy fin-and-tube heat exchangers under dehumidifying condition – Data with larger diameter tube. *Int. J. Heat Mass Transf.* **2012**, *55*, 3054–3060. [CrossRef]
32. Li, L.; Du, X.; Zhang, Y.; Yang, L.; Yang, Y. Numerical simulation on flow and heat transfer of fin-and-tube heat exchanger with longitudinal vortex generators. *Int. J. Therm. Sci.* **2015**, *92*, 85–96. [CrossRef]
33. Johnson, T.R.; Joubert, P.N. The Influence of Vortex Generators on the Drag and Heat Transfer from a Circular Cylinder Normal to an Airstream. *J. Heat Transf.* **1969**, *91*, 91–99. [CrossRef]
34. Min, C.; Qi, C.; Kong, X.; Dong, J. Experimental study of rectangular channel with modified rectangular longitudinal vortex generators. *Int. J. Heat Mass Transf.* **2010**, *53*, 3023–3029. [CrossRef]
35. Chen, C.; Teng, J.-T.; Cheng, C.-H.; Jin, S.; Huang, S.; Liu, C.; Lee, M.-T.; Pan, H.-H.; Greif, R. A study on fluid flow and heat transfer in rectangular microchannels with various longitudinal vortex generators. *Int. J. Heat Mass Transf.* **2014**, *69*, 203–214. [CrossRef]
36. Ebrahimi, A.; Roohi, E.; Kheradmand, S. Numerical study of liquid flow and heat transfer in rectangular microchannel with longitudinal vortex generators. *Appl. Therm. Eng.* **2015**, *78*, 576–583.
37. Leu, J.-S.; Wu, Y.-H.; Jang, J.-Y. Heat transfer and fluid flow analysis in plate-fin and tube heat exchangers with a pair of block shape vortex generators. *Int. J. Heat Mass Transf.* **2004**, *47*, 4327–4338. [CrossRef]
38. Wu, J.; Tao, W. Effect of longitudinal vortex generator on heat transfer in rectangular channels. *Appl. Therm. Eng.* **2012**, *37*, 67–72.
39. Khoshvaght-Aliabadi, M.; Zangouei, S.; Hormozi, F. Performance of a plate-fin heat exchanger with vortex-generator channels: 3D-CFD simulation and experimental validation. *Int. J. Therm. Sci.* **2015**, *88*, 180–192.
40. Li, J.; Wang, S.; Chen, J.; Lei, Y.-G. Numerical study on a slit fin-and-tube heat exchanger with longitudinal vortex generators. *Int. J. Heat Mass Transf.* **2011**, *54*, 1743–1751.
41. Yang, K.-S.; Jhong, J.-H.; Lin, Y.-T.; Chien, K.-H.; Wang, C.-C. On the Heat Transfer Characteristics of Heat Sinks: With and Without Vortex Generators. *IEEE Trans. Components Packag. Technol.* **2010**, *33*, 391–397.
42. Chomdee, S.; Kiatsiriroat, T. Enhancement of air cooling in staggered array of electronic modules by integrating delta winglet vortex generators. *Int. Commun. Heat Mass Transf.* **2006**, *33*, 618–626. [CrossRef]
43. Liu, C.; Teng, J.-T.; Chu, J.-C.; Chiu, Y.-L.; Huang, S.; Jin, S.; Dang, T.; Greif, R.; Pan, H.-H. Experimental investigations on liquid flow and heat transfer in rectangular microchannel with longitudinal vortex generators. *Int. J. Heat Mass Transf.* **2011**, *54*, 3069–3080.
44. Ma, J.; Huang, Y.P.; Huang, J.; Wang, Y.L.; Wang, Q.W. Experimental investigations on single-phase heat transfer enhancement with longitudinal vortices in narrow rectangular channel. *Nucl. Eng. Des.* **2010**, *240*, 92–102.
45. Cheraghi, M.; Raisee, M.; Moghaddami, M. Effect of cylinder proximity to the wall on channel flow heat transfer enhancement. *C. R. Mécanique* **2014**, *342*, 63–72.

46. Safaei, M.R.; Jahanbin, A.; Kianifar, A.; Gharehkhani, S.; Kherbeet, A.S.; Goodarzi, M.; Dahari, M. *Mathematical Modeling for Nanofluids Simulation: A Review of the Latest Works, in Modeling and Simulation in Engineering Sciences*; Akbar, N.S., Beg, O.A., Eds.; InTechOpen: London, UK, 2016.
47. Koşar, A. Effect of substrate thickness and material on heat transfer in microchannel heat sinks. *Int. J. Therm. Sci.* **2010**, *49*, 635–642. [CrossRef]
48. Dixit, T.; Ghosh, I. Review of micro- and mini-channel heat sinks and heat exchangers for single phase fluids. *Renew. Sustain. Energy Rev.* **2015**, *41*, 1298–1311.
49. Mehendale, S.S.; Jacobi, A.; Shah, R.K. Fluid Flow and Heat Transfer at Micro- and Meso-Scales with Application to Heat Exchanger Design. *Appl. Mech. Rev.* **2000**, *53*, 175–193.
50. Kandlikar, S.G.; Grande, W.J. Evolution of microchannel flow passages: Thermohydraulic performance and fabrication technology. In Proceedings of the ASME International Mechanical Engineering Congress Exposition, New Orleans, LA, USA, 17–22 November 2002.
51. Abed, W.M.; Whalley, R.; Dennis, D.; Poole, R.J. Numerical and experimental investigation of heat transfer and fluid flow characteristics in a micro-scale serpentine channel. *Int. J. Heat Mass Transf.* **2015**, *88*, 790–802.
52. Choi, S.U.S.; Eastman, J.A. Enhancing thermal conductivity of fluids with nanoparticles. In Proceedings of the ASME International Mechanical Engineering Congress and Exposition, San Francisco, CA, USA, 12–17 November 1995. 8p.
53. Afzal, A.; Abidi, A.; Ad, M.; Rk, A.; Soudagar, M.E.M.; Saleel, A.C. Effect of parameters on thermal and fluid-flow behavior of battery thermal management system. *Therm. Sci.* **2021**, *25*, 3775–3787.
54. Fayaz, H.; Afzal, A.; Samee, A.D.M.; Soudagar, M.E.M.; Akram, N.; Mujtaba, M.A.; Jilte, R.D.; Islam, T.; Ağbulut, Ü.; Saleel, C.A. Optimization of Thermal and Structural Design in Lithium-Ion Batteries to Obtain Energy Efficient Battery Thermal Management System (BTMS): A Critical Review. *Arch. Comput. Methods Eng.* **2021**, *29*, 129–194.
55. Belkhode, P.N.; Shelare, S.D.; Sakhale, C.N.; Kumar, R.; Shanmugan, S.; Soudagar, M.E.M.; Mujtaba, M. Performance analysis of roof collector used in the solar updraft tower. *Sustain. Energy Technol. Assessments* **2021**, *48*, 101619.
56. Al Qubeissi, M.; Almshahy, A.; Mahmoud, A.; Al-Asadi, M.T.; Shah, R.M.R.A. Modelling of battery thermal management: A new concept of cooling using fuel. *Fuel* **2021**, *310*, 122403.
57. Shkarah, A.J.; Bin Sulaiman, M.Y.; Ayob, R.B.H.; Togun, H. A 3D numerical study of heat transfer in a single-phase micro-channel heat sink using graphene, aluminum and silicon as substrates. *Int. Commun. Heat Mass Transf.* **2013**, *48*, 108–115.
58. Lee, P.; Garimella, S.V. Thermally developing flow and heat transfer in rectangular microchannels of different aspect ratios. *Int. J. Heat Mass Transf.* **2006**, *49*, 3060–3067.
59. Mansoor, M.M.; Wong, K.C.; Siddique, M. Numerical investigation of fluid flow and heat transfer under high heat flux using rectangular micro-channels. *Int. Commun. Heat Mass Transf.* **2012**, *39*, 291–297. [CrossRef]
60. Lee, P.; Garimella, S.V.; Liu, D. Investigation of heat transfer in rectangular microchannels. *Int. J. Heat Mass Transf.* **2005**, *48*, 1688–1704. [CrossRef]
61. Deng, D.; Wan, W.; Tang, Y.; Wan, Z.; Liang, D. Experimental investigations on flow boiling performance of reentrant and rectangular microchannels – A comparative study. *Int. J. Heat Mass Transf.* **2015**, *82*, 435–446. [CrossRef]
62. Balaj, M.; Roohi, E.; Akhlaghi, H. Effects of shear work on non-equilibrium heat transfer characteristics of rarefied gas flows through micro/nanochannels. *Int. J. Heat Mass Transf.* **2015**, *83*, 69–74. [CrossRef]
63. Xia, G.; Jiang, J.; Wang, J.; Zhai, Y.; Ma, D. Effects of different geometric structures on fluid flow and heat transfer performance in microchannel heat sinks. *Int. J. Heat Mass Transf.* **2015**, *80*, 439–447. [CrossRef]
64. Ghasemian, M.; Ashrafi, Z.N.; Goharkhah, M.; Ashjaee, M. Heat transfer characteristics of Fe_3O_4 ferrofluid flowing in a mini channel under constant and alternating magnetic fields. *J. Magn. Magn. Mater.* **2015**, *381*, 158–167. [CrossRef]
65. Pongsoi, P.; Pikulkajorn, S.; Wongwises, S. Heat transfer and flow characteristics of spiral fin-and-tube heat exchangers: A review. *Int. J. Heat Mass Transf.* **2014**, *79*, 417–431. [CrossRef]
66. Pan, M.; Tang, Y.; Pan, L.; Lu, L. Optimal design of complex manifold geometries for uniform flow distribution between microchannels. *Chem. Eng. J.* **2008**, *137*, 339–346. [CrossRef]
67. Gamrat, G.; Favre-Marinet, M.; Asendrych, D. Conduction and entrance effects on laminar liquid flow and heat transfer in rectangular microchannels. *Int. J. Heat Mass Transf.* **2005**, *48*, 2943–2954. [CrossRef]
68. Qu, W.; Mudawar, I. Experimental and numerical study of pressure drop and heat transfer in a single-phase micro-channel heat sink. *Int. J. Heat Mass Transf.* **2002**, *45*, 2549–2565. [CrossRef]
69. Hsu, L.-C.; Cion, S.-W.; Lin, K.-W.; Wang, C.-C. An experimental study of inclination on the boiling heat transfer characteristics of a micro-channel heat sink using HFE-7100. *Int. Commun. Heat Mass Transf.* **2015**, *62*, 13–17. [CrossRef]
70. Suwankamnerd, P.; Wongwises, S. An experimental study of two-phase air–water flow and heat transfer characteristics of segmented flow in a microchannel. *Exp. Therm. Fluid Sci.* **2015**, *62*, 29–39. [CrossRef]
71. Mirmanto, M. Heat transfer coefficient calculated using a linear pressure gradient assumption and measurement for flow boiling in microchannels. *Int. J. Heat Mass Transf.* **2014**, *79*, 269–278. [CrossRef]
72. Konishi, C.; Lee, H.; Mudawar, I.; Hasan, M.M.; Nahra, H.K.; Hall, N.R.; Wagner, J.D.; May, R.L.; Mackey, J.R. Flow boiling in microgravity: Part 1—Interfacial behavior and experimental heat transfer results. *Int. J. Heat Mass Transf.* **2015**, *81*, 705–720. [CrossRef]

73. Gan, Y.; Xu, J.; Yan, Y. An experimental study of two-phase pressure drop of acetone in triangular silicon micro-channels. *Appl. Therm. Eng.* **2015**, *80*, 76–86. [CrossRef]
74. Fang, X.; Zhou, Z.; Wang, H. Heat transfer correlation for saturated flow boiling of water. *Appl. Therm. Eng.* **2015**, *76*, 147–156. [CrossRef]
75. Shojaeian, M.; Koşar, A. Pool boiling and flow boiling on micro- and nanostructured surfaces. *Exp. Therm. Fluid Sci.* **2015**, *63*, 45–73. [CrossRef]
76. Asadi, M.; Xie, G.; Sunden, B. A review of heat transfer and pressure drop characteristics of single and two-phase microchannels. *Int. J. Heat Mass Transf.* **2014**, *79*, 34–53. [CrossRef]
77. Analytics Clarivate. *Acquisition of the Thomson Reuters Intellectual Property and Science web of Science*; Clarivate: London, UK, 2017.
78. Guo, J.; Xu, M.; Cheng, L. Numerical investigations of curved square channel from the viewpoint of field synergy principle. *Int. J. Heat Mass Transf.* **2011**, *54*, 4148–4151. [CrossRef]
79. Chu, J.-C.; Teng, J.-T.; Greif, R. Experimental and numerical study on the flow characteristics in curved rectangular microchannels. *Appl. Therm. Eng.* **2010**, *30*, 1558–1566. [CrossRef]
80. Dehghan, M.; Daneshipour, M.; Valipour, M.S.; Rafee, R.; Saedodin, S. Enhancing heat transfer in microchannel heat sinks using converging flow passages. *Energy Convers. Manag.* **2015**, *92*, 244–250. [CrossRef]
81. Aris, M.; Owen, I.; Sutcliffe, C. The development of active vortex generators from shape memory alloys for the convective cooling of heated surfaces. *Int. J. Heat Mass Transf.* **2011**, *54*, 3566–3574. [CrossRef]
82. Mirzaee, H.; Dadvand, A.; Mirzaee, I.; Shabani, R. Heat transfer enhancement in microchannels using an elastic vortex generator. *J. Enhanc. Heat Transf.* **2012**, *19*, 199–211. [CrossRef]
83. Liu, J.; Xie, G.; Simon, T.W. Turbulent flow and heat transfer enhancement in rectangular channels with novel cylindrical grooves. *Int. J. Heat Mass Transf.* **2015**, *81*, 563–577. [CrossRef]
84. Bi, C.; Tang, G.; Tao, W. Heat transfer enhancement in mini-channel heat sinks with dimples and cylindrical grooves. *Appl. Therm. Eng.* **2013**, *55*, 121–132. [CrossRef]
85. Zhai, Y.; Xia, G.; Liu, X.; Li, Y. Exergy analysis and performance evaluation of flow and heat transfer in different micro heat sinks with complex structure. *Int. J. Heat Mass Transf.* **2015**, *84*, 293–303. [CrossRef]
86. Knupp, D.; Cotta, R.; Naveira-Cotta, C.P. Fluid flow and conjugated heat transfer in arbitrarily shaped channels via single domain formulation and integral transforms. *Int. J. Heat Mass Transf.* **2015**, *82*, 479–489. [CrossRef]
87. Henze, M.; von Wolfersdorf, J. Influence of approach flow conditions on heat transfer behind vortex generators. *Int. J. Heat Mass Transf.* **2011**, *54*, 279–287. [CrossRef]
88. Dai, Z.; Fletcher, D.F.; Haynes, B.S. Impact of tortuous geometry on laminar flow heat transfer in microchannels. *Int. J. Heat Mass Transf.* **2015**, *83*, 382–398. [CrossRef]
89. Karathanassis, I.; Papanicolaou, E.; Belessiotis, V.; Bergeles, G. Experimental and numerical evaluation of an elongated plate-fin heat sink with three sections of stepwise varying channel width. *Int. J. Heat Mass Transf.* **2015**, *84*, 16–34. [CrossRef]
90. Kundu, B.; Das, P.K. Performance and optimum design analysis of convective fin arrays attached to flat and curved primary surfaces. *Int. J. Refrig.* **2009**, *32*, 430–443. [CrossRef]
91. Hong, F.; Cheng, P.; Ge, H.; Joo, G.T. Conjugate heat transfer in fractal-shaped microchannel network heat sink for integrated microelectronic cooling application. *Int. J. Heat Mass Transf.* **2007**, *50*, 4986–4998. [CrossRef]
92. Torii, K.; Kwak, K.; Nishino, K. Heat transfer enhancement accompanying pressure-loss reduction with winglet-type vortex generators for fin-tube heat exchangers. *Int. J. Heat Mass Transf.* **2002**, *45*, 3795–3801. [CrossRef]
93. Zhang, Y.H.; Wang, L.B.; Ke, F.; Su, Y.X.; Gao, S.D. The effects of span position of winglet vortex generator on local heat/mass transfer over a three-row flat tube bank fin. *Heat Mass Transf.* **2004**, *40*, 881–891. [CrossRef]
94. Wang, C.-C.; Lo, J.; Lin, Y.-T.; Wei, C.-S. Flow visualization of annular and delta winlet vortex generators in fin-and-tube heat exchanger application. *Int. J. Heat Mass Transf.* **2002**, *45*, 3803–3815. [CrossRef]
95. Ahmed, S.; Mohammed, W.; Laith, J. Numerical investigation into velocity and temperature fields over smooth and rough ducts for several types of turbulators. *J. Eng. Technol.* **2007**, *25*, 1110–1128.
96. Gorji-Bandpy, M.; Soleimani, S.; Hossein-Nejad, F. Pressure and Heat Transfer in Staggered Arrangement Circular Tubes with Airfoil Vortex Generator. *Int. Energy J.* **2008**, *9*, 207–214.
97. Wang, J.; Zhao, Y. Heat and fluid flow characteristics of a rectangular channel with a small diameter circular cylinder as vortex generator. *Int. J. Therm. Sci.* **2015**, *92*, 1–13. [CrossRef]
98. Chai, L.; Xia, G.; Wang, H.S. Numerical study of laminar flow and heat transfer in microchannel heat sink with offset ribs on sidewalls. *Appl. Therm. Eng.* **2016**, *92*, 32–41. [CrossRef]
99. Chai, L.; Xia, G.D.; Wang, H. Parametric study on thermal and hydraulic characteristics of laminar flow in microchannel heat sink with fan-shaped ribs on sidewalls—Part 1: Heat transfer. *Int. J. Heat Mass Transf.* **2016**, *97*, 1069–1080. [CrossRef]
100. Chai, L.; Xia, G.D.; Wang, H. Parametric study on thermal and hydraulic characteristics of laminar flow in microchannel heat sink with fan-shaped ribs on sidewalls—Part 2: Pressure drop. *Int. J. Heat Mass Transf.* **2016**, *97*, 1081–1090. [CrossRef]
101. Chai, L.; Xia, G.D.; Wang, H. Parametric study on thermal and hydraulic characteristics of laminar flow in microchannel heat sink with fan-shaped ribs on sidewalls – Part 3: Performance evaluation. *Int. J. Heat Mass Transf.* **2016**, *97*, 1091–1101. [CrossRef]
102. Li, Y.; Xia, G.; Ma, D.; Jia, Y.; Wang, J. Characteristics of laminar flow and heat transfer in microchannel heat sink with triangular cavities and rectangular ribs. *Int. J. Heat Mass Transf.* **2016**, *98*, 17–28. [CrossRef]

103. Behnampour, A.; Akbari, O.A.; Safaei, M.R.; Ghavami, M.; Marzban, A.; Shabani, G.A.S.; Zarringhalam, M.; Mashayekhi, R. Analysis of heat transfer and nanofluid fluid flow in microchannels with trapezoidal, rectangular and triangular shaped ribs. *Phys. E Low-Dimens. Syst. Nanostruct.* **2017**, *91*, 15–31. [CrossRef]
104. Al-Asadi, M.; Alkasmoul, F.; Wilson, M. Heat transfer enhancement in a micro-channel cooling system using cylindrical vortex generators. *Int. Commun. Heat Mass Transf.* **2016**, *74*, 40–47. [CrossRef]
105. Raihan, M.F.B.; Al-Asadi, M.T.; Thompson, H. Management of conjugate heat transfer using various arrangements of cylindrical vortex generators in micro-channels. *Appl. Therm. Eng.* **2021**, *182*, 116097. [CrossRef]
106. Jama, M.; Singh, T.; Gamaleldin, S.M.; Koc, M.; Samara, A.; Isaifan, R.; Atieh, M.A. Critical Review on Nanofluids: Preparation, Characterization, and Applications. *J. Nanomater.* **2016**, *2016*, 1–22. [CrossRef]
107. Li, Y.; Zhou, J.; Tung, S.; Schneider, E.; Xi, S. A review on development of nanofluid preparation and characterization. *Powder Technol.* **2009**, *196*, 89–101. [CrossRef]
108. Kakaç, S.; Pramuanjaroenkij, A. Single-phase and two-phase treatments of convective heat transfer enhancement with nanofluids—A state-of-the-art review. *Int. J. Therm. Sci.* **2016**, *100*, 75–97. [CrossRef]
109. Azmi, W.; Hamid, K.A.; Usri, N.; Mamat, R.; Sharma, K. Heat transfer augmentation of ethylene glycol: Water nanofluids and applications—A review. *Int. Commun. Heat Mass Transf.* **2016**, *75*, 13–23. [CrossRef]
110. Eastman, J.A.; Choi, U.S.; Li, S.; Thompson, L.J.; Lee, S. Enhanced Thermal Conductivity through the Development of Nanofluids. *MRS Proc.* **1996**, *457*, 3–11. [CrossRef]
111. Lee, S.; Choi, S.U.-S.; Li, S.; Eastman, J. Measuring Thermal Conductivity of Fluids Containing Oxide Nanoparticles. *J. Heat Transf.* **1999**, *121*, 280–289. [CrossRef]
112. Choi, S.U.S.; Eastman, J.A. *Enhanced Heat Transfer Using Nanofluids*; Argonne National Laboratory (ANL): Argonne, IL, USA, 2001.
113. Choi, S.U.S.; Zhang, Z.G.; Yu, W.; Lockwood, F.E.; Grulke, E.A. Anomalous thermal conductivity enhancement in nanotube suspensions. *Appl. Phys. Lett.* **2001**, *79*, 2252–2254. [CrossRef]
114. Zhu, H.-T.; Lin, Y.-S.; Yin, Y.-S. A novel one-step chemical method for preparation of copper nanofluids. *J. Colloid Interface Sci.* **2004**, *277*, 100–103. [CrossRef] [PubMed]
115. Gleiter, H. Nanocrystalline Materials. In *Advanced Structural and Functional Materials: Proceedings of an International Seminar Organized by Deutsche Forschungsanstalt für Luft- und Raumfahrt (DLR), Köln, June 1991*; Bunk, W.G.J., Ed.; Springer: Berlin/Heidelberg, Germany, 1991; pp. 1–37.
116. Ashley, S. Small-scale structure yields big property payoffs. *Mech. Eng.* **1994**, *116*, 52.
117. Bowles, R.; Kolstad, J.; Calo, J.; Andres, R. Generation of molecular clusters of controlled size. *Surf. Sci.* **1981**, *106*, 117–124. [CrossRef]
118. Iijima, S. Helical microtubules of graphitic carbon. *Nature* **1991**, *354*, 56–58. [CrossRef]
119. Menni, Y.; Chamkha, A.J.; Ghazvini, M.; Ahmadi, M.H.; Ameur, H.; Issakhov, A.; Inc, M. Enhancement of the turbulent convective heat transfer in channels through the baffling technique and oil/multiwalled carbon nanotube nanofluids. *Numer. Heat Transfer, Part A Appl.* **2021**, *79*, 311–351. [CrossRef]
120. Scott, C.; Arepalli, S.; Nikolaev, P.; Smalley, R. Growth mechanisms for single-wall carbon nanotubes in a laser-ablation process. *Appl. Phys. A* **2001**, *72*, 573–580. [CrossRef]
121. Hernadi, K. Catalytic synthesis of multiwall carbon nanotubes from methylacetylene. *Chem. Phys. Lett.* **2002**, *363*, 169–174. [CrossRef]
122. Yatsuya, S.; Tsukasaki, Y.; Mihama, K.; Uyeda, R. Preparation of extremely fine particles by vacuum evaporation onto a running oil substrate. *J. Cryst. Growth* **1978**, *45*, 490–494. [CrossRef]
123. Eastman, J.A. *Novel Applications Exploiting the Thermal Properties of Nanostructured Materials. ANL/MSD/CP-97456*; Argonne National Lab: Argone, IL, USA, 1998.
124. Mohammed, H.A.; Bhaskaran, G.; Shuaib, N.; Saidur, R. Heat transfer and fluid flow characteristics in microchannels heat exchanger using nanofluids: A review. *Renew. Sustain. Energy Rev.* **2011**, *15*, 1502–1512. [CrossRef]
125. Rayleigh, L. LVI. On the influence of obstacles arranged in rectangular order upon the properties of a medium. *London Edinburgh Dublin Philos. Mag. J. Sci.* **1892**, *34*, 481–502. [CrossRef]
126. Fricke, H. A Mathematical Treatment of the Electric Conductivity and Capacity of Disperse Systems I. The Electric Conductivity of a Suspension of Homogeneous Spheroids. *Phys. Rev.* **1924**, *24*, 575–587. [CrossRef]
127. Fricke, H. The Maxwell-Wagner dispersion in a suspension of ellipsoids. *J. Phys. Chem.* **1953**, *57*, 934–937. [CrossRef]
128. B}ttcher, C. The dielectric constant of crystalline powders. *Recl. Trav. Chim. Pays-Bas* **1945**, *64*, 47–51. [CrossRef]
129. Landauer, R. The electrical resistance of binary metallic mixtures. *J. Appl. Phys.* **1952**, *23*, 779–784. [CrossRef]
130. Jeffrey, D. Conduction through a random suspension of spheres. *Proc. R. Soc. London. Ser. A, Math. Phys. Sci.* **1973**, *335*, 355–367.
131. Davis, R. The effective thermal conductivity of a composite material with spherical inclusions. *Int. J. Thermophys.* **1986**, *7*, 609–620. [CrossRef]
132. Nan, C.-W.; Birringer, R.; Clarke, D.; Gleiter, H. Effective thermal conductivity of particulate composites with interfacial thermal resistance. *J. Appl. Phys.* **1997**, *81*, 6692–6699. [CrossRef]
133. Lu, S.-Y.; Song, J.-L. Effective conductivity of composites with spherical inclusions: Effect of coating and detachment. *J. Appl. Phys.* **1996**, *79*, 609–618. [CrossRef]

134. Benveniste, Y. Effective thermal conductivity of composites with a thermal contact resistance between the constituents: Nondilute case. *J. Appl. Phys.* **1987**, *61*, 2840–2843. [CrossRef]
135. Hasselman, D.P.H.; Johnson, L.F. Effective Thermal Conductivity of Composites with Interfacial Thermal Barrier Resistance. *J. Compos. Mater.* **1987**, *21*, 508–515. [CrossRef]
136. Abed, A.M.; Sopian, K.; Mohammed, H.; Alghoul, M.; Ruslan, M.H.; Mat, S.; Al-Shamani, A.N. Enhance heat transfer in the channel with V-shaped wavy lower plate using liquid nanofluids. *Case Stud. Therm. Eng.* **2015**, *5*, 13–23. [CrossRef]
137. Wang, X.; Xu, X.; Choi, S.U.S. Thermal Conductivity of Nanoparticle - Fluid Mixture. *J. Thermophys. Heat Transf.* **1999**, *13*, 474–480. [CrossRef]
138. Das, S.K.; Putra, N.; Thiesen, P.H.; Roetzel, W. Temperature Dependence of Thermal Conductivity Enhancement for Nanofluids. *J. Heat Transf.* **2003**, *125*, 567–574. [CrossRef]
139. Xie, H.; Wang, J.; Xi, T.; Liu, Y.; Ai, F.; Wu, Q. Thermal conductivity enhancement of suspensions containing nanosized alumina particles. *J. Appl. Phys.* **2002**, *91*, 4568–4572. [CrossRef]
140. Li, C.H.; Peterson, G.P. Experimental investigation of temperature and volume fraction variations on the effective thermal conductivity of nanoparticle suspensions (nanofluids). *J. Appl. Phys.* **2006**, *99*, 084314. [CrossRef]
141. Wen, D.; Ding, Y. Experimental investigation into convective heat transfer of nanofluids at the entrance region under laminar flow conditions. *Int. J. Heat Mass Transf.* **2004**, *47*, 5181–5188. [CrossRef]
142. Yoo, D.-H.; Hong, K.; Yang, H.-S. Study of thermal conductivity of nanofluids for the application of heat transfer fluids. *Thermochim. Acta* **2007**, *455*, 66–69. [CrossRef]
143. Zhang, X.; Gu, H.; Fujii, M. Experimental Study on the Effective Thermal Conductivity and Thermal Diffusivity of Nanofluids. *Int. J. Thermophys.* **2006**, *27*, 569–580. [CrossRef]
144. Kim, S.; Bang, I.; Buongiorno, J.; Hu, L. Surface wettability change during pool boiling of nanofluids and its effect on critical heat flux. *Int. J. Heat Mass Transf.* **2007**, *50*, 4105–4116. [CrossRef]
145. Li, C.H.; Peterson, G.P. The effect of particle size on the effective thermal conductivity of Al_2O_3-water nanofluids. *J. Appl. Phys.* **2007**, *101*, 44312. [CrossRef]
146. Beck, M.P.; Yuan, Y.; Warrier, P.; Teja, A.S. The effect of particle size on the thermal conductivity of alumina nanofluids. *J. Nanoparticle Res.* **2009**, *11*, 1129–1136. [CrossRef]
147. Mintsa, H.A.; Roy, G.; Nguyen, C.T.; Doucet, D. New temperature dependent thermal conductivity data for water-based nanofluids. *Int. J. Therm. Sci.* **2009**, *48*, 363–371. [CrossRef]
148. Jwo, C.-S.; Teng, T.-P.; Chang, H. A simple model to estimate thermal conductivity of fluid with acicular nanoparticles. *J. Alloy. Compd.* **2007**, *434–435*, 569–571. [CrossRef]
149. Murshed, S.M.S.; Leong, K.C.; Yang, C. Enhanced thermal conductivity of TiO_2—Water based nanofluids. *Int. J. Therm. Sci.* **2005**, *44*, 367–373. [CrossRef]
150. Wen, D.; Ding, Y. Natural convective heat transfer of suspensions of titanium dioxide nanoparticles (nanofluids). *IEEE Trans. Nanotechnol.* **2006**, *5*, 220–227.
151. He, Y.; Jin, Y.; Chen, H.; Ding, Y.; Cang, D.; Lu, H. Heat transfer and flow behaviour of aqueous suspensions of TiO_2 nanoparticles (nanofluids) flowing upward through a vertical pipe. *Int. J. Heat Mass Transf.* **2007**, *50*, 2272–2281. [CrossRef]
152. Turgut, A.; Tavman, I.; Chirtoc, M.; Schuchmann, H.P.; Sauter, C.; Tavman, S. Thermal Conductivity and Viscosity Measurements of Water-Based TiO_2 Nanofluids. *Int. J. Thermophys.* **2009**, *30*, 1213–1226. [CrossRef]
153. Zhu, H.; Zhang, C.; Liu, S.; Tang, Y.; Yin, Y. Effects of nanoparticle clustering and alignment on thermal conductivities of Fe_3O_4 aqueous nanofluids. *Appl. Phys. Lett.* **2006**, *89*, 023123. [CrossRef]
154. Hwang, Y.; Ahn, Y.; Shin, H.; Lee, C.; Kim, G.; Park, H.; Lee, J. Investigation on characteristics of thermal conductivity enhancement of nanofluids. *Curr. Appl. Phys.* **2006**, *6*, 1068–1071. [CrossRef]
155. Eastman, J.A.; Choi, S.U.S.; Li, S.; Yu, W.; Thompson, L.J. Anomalously increased effective thermal conductivities of ethylene glycol-based nanofluids containing copper nanoparticles. *Appl. Phys. Lett.* **2001**, *78*, 718–720. [CrossRef]
156. Xuan, Y.; Li, Q. Heat transfer enhancement of nanofluids. *Int. J. Heat Fluid Flow* **2000**, *21*, 58–64. [CrossRef]
157. Liu, M.-S.; Lin, M.C.-C.; Tsai, C.; Wang, C.-C. Enhancement of thermal conductivity with Cu for nanofluids using chemical reduction method. *Int. J. Heat Mass Transf.* **2006**, *49*, 3028–3033. [CrossRef]
158. Hong, T.-K.; Yang, H.-S.; Choi, C.J. Study of the enhanced thermal conductivity of Fe nanofluids. *J. Appl. Phys.* **2005**, *97*, 064311. [CrossRef]
159. Hong, K.S.; Hong, T.-K.; Yang, H.-S. Thermal conductivity of Fe nanofluids depending on the cluster size of nanoparticles. *Appl. Phys. Lett.* **2006**, *88*, 031901. [CrossRef]
160. Patel, H.E.; Das, S.K.; Sundararajan, T.; Nair, A.S.; George, B.; Pradeep, T. Thermal conductivities of naked and monolayer protected metal nanoparticle based nanofluids: Manifestation of anomalous enhancement and chemical effects. *Appl. Phys. Lett.* **2003**, *83*, 2931–2933. [CrossRef]
161. Putnam, S.A.; Cahill, D.G.; Braun, P.V.; Ge, Z.; Shimmin, R.G. Thermal conductivity of nanoparticle suspensions. *J. Appl. Phys.* **2006**, *99*, 084308. [CrossRef]
162. Heris, S.Z.; Esfahany, M.N.; Etemad, S. Experimental investigation of convective heat transfer of Al_2O_3/water nanofluid in circular tube. *Int. J. Heat Fluid Flow* **2007**, *28*, 203–210. [CrossRef]

163. Pak, B.C.; Cho, Y.I. Hydrodynamic and heat transfer study of dispersed fluids with submicron metallic oxide particles. *Exp. Heat Transf.* **1998**, *11*, 151–170. [CrossRef]
164. Hwang, K.S.; Jang, S.P.; Choi, S.U. Flow and convective heat transfer characteristics of water-based Al_2O_3 nanofluids in fully developed laminar flow regime. *Int. J. Heat Mass Transf.* **2009**, *52*, 193–199. [CrossRef]
165. Heris, S.Z.; Etemad, S.; Esfahany, M.N. Experimental investigation of oxide nanofluids laminar flow convective heat transfer. *Int. Commun. Heat Mass Transf.* **2006**, *33*, 529–535. [CrossRef]
166. Maxwell, J.C. *A Treatise on Electricity and Magnetism*; Clarendon: London, UK, 1892; Volume 1.
167. Bruggeman, D. Calculation of various physics constants in heterogenous substances I Dielectricity constants and conductivity of mixed bodies from isotropic substances. *Ann. Phys.* **1935**, *24*, 636–664. [CrossRef]
168. Yu, W.; Choi, S. The Role of Interfacial Layers in the Enhanced Thermal Conductivity of Nanofluids: A Renovated Maxwell Model. *J. Nanoparticle Res.* **2003**, *5*, 167–171. [CrossRef]
169. Xie, H.; Fujii, M.; Zhang, X. Effect of interfacial nanolayer on the effective thermal conductivity of nanoparticle-fluid mixture. *Int. J. Heat Mass Transf.* **2005**, *48*, 2926–2932. [CrossRef]
170. Shima, P.D.; Philip, J.; Raj, B. Role of microconvection induced by Brownian motion of nanoparticles in the enhanced thermal conductivity of stable nanofluids. *Appl. Phys. Lett.* **2009**, *94*, 223101. [CrossRef]
171. Vadasz, P. *Nano Uid Suspensions and Bi-Composite Media as Derivatives of interface Heat Transfer Modeling in Porous Media. Emerging Topics in Heat and Mass Transfer in Porous Media: From Bioengineering and Microelectronics to Nanotechnology*; Springer: Cham, Switzerland, 2008; 283p.
172. Prasher, R.; Phelan, P.E.; Bhattacharya, P. Effect of Aggregation Kinetics on the Thermal Conductivity of Nanoscale Colloidal Solutions (Nanofluid). *Nano Lett.* **2006**, *6*, 1529–1534. [CrossRef]
173. Nie, C.; Marlow, W.; Hassan, Y. Discussion of proposed mechanisms of thermal conductivity enhancement in nanofluids. *Int. J. Heat Mass Transf.* **2008**, *51*, 1342–1348. [CrossRef]
174. Ghasemi, B.; Aminossadati, S. Brownian motion of nanoparticles in a triangular enclosure with natural convection. *Int. J. Therm. Sci.* **2010**, *49*, 931–940. [CrossRef]
175. Xuan, Y.; Roetzel, W. Conceptions for heat transfer correlation of nanofluids. *Int. J. Heat Mass Transf.* **2000**, *43*, 3701–3707. [CrossRef]
176. Yu, W.; France, D.M.; Routbort, J.L.; Choi, S.U.S. Review and Comparison of Nanofluid Thermal Conductivity and Heat Transfer Enhancements. *Heat Transf. Eng.* **2008**, *29*, 432–460. [CrossRef]
177. Patel, H.E.; Sundararajan, T.; Das, S.K. An experimental investigation into the thermal conductivity enhancement in oxide and metallic nanofluids. *J. Nanoparticle Res.* **2010**, *12*, 1015–1031. [CrossRef]
178. Tillman, P.; Hill, J.M. Modelling the Thermal Conductivity of Nanofluids. In *IUTAM Symposium on Mechanical Behavior and Micro-Mechanics of Nanostructured Materials: Proceedings of the IUTAM Symposium, Beijing, China, 27–30 June 2005*; Bai, Y.L., Zheng, Q.S., Wei, Y.G., Eds.; Springer: Dordrecht, The Netherlands, 2007; pp. 105–118.
179. Xie, H.; Wang, J.; Xi, T.; Liu, Y.; Ai, F. Dependence of the thermal conductivity of nanoparticle-fluid mixture on the base fluid. *J. Mater. Sci. Lett.* **2002**, *21*, 1469–1471. [CrossRef]
180. Godson, L.; Raja, B.; Lal, D.M.; Wongwises, S. Enhancement of heat transfer using nanofluids—An overview. *Renew. Sustain. Energy Rev.* **2010**, *14*, 629–641. [CrossRef]
181. Sharma, P.; Baek, I.-H.; Cho, T.; Park, S.; Lee, K.B. Enhancement of thermal conductivity of ethylene glycol based silver nanofluids. *Powder Technol.* **2011**, *208*, 7–19. [CrossRef]
182. Madhusree, K.; Dey, T.K. Thermal conductivity and viscosity of Al 2 O 3 nanofluid based on car engine coolant. *J. Phys. D Appl. Phys.* **2010**, *43*, 315501.
183. Corcione, M. Heat transfer features of buoyancy-driven nanofluids inside rectangular enclosures differentially heated at the sidewalls. *Int. J. Therm. Sci.* **2010**, *49*, 1536–1546. [CrossRef]
184. Al-Aswadi, A.; Mohammed, H.; Shuaib, N.; Campo, A. Laminar forced convection flow over a backward facing step using nanofluids. *Int. Commun. Heat Mass Transf.* **2010**, *37*, 950–957. [CrossRef]
185. Guo, S.-Z.; Li, Y.; Jiang, J.-S.; Xie, H.-Q. Nanofluids Containing γ-Fe2O3 Nanoparticles and Their Heat Transfer Enhancements. *Nanoscale Res. Lett.* **2010**, *5*, 1222–1227. [CrossRef]
186. Mohammed, H.; Al-Aswadi, A.; Shuaib, N.; Saidur, R. Convective heat transfer and fluid flow study over a step using nanofluids: A review. *Renew. Sustain. Energy Rev.* **2011**, *15*, 2921–2939. [CrossRef]
187. Yulong, D.; Chen, H.; Musina, Z.; Jin, Y.; Zhang, T.; Witharana, S.; Yang, W. Relationship between the thermal conductivity and shear viscosity of nanofluids. *Phys. Scr.* **2010**, *2010*, 014078.
188. Liu, M.-S.; Lin, M.C.-C.; Huang, I.-T.; Wang, C.-C. Enhancement of thermal conductivity with carbon nanotube for nanofluids. *Int. Commun. Heat Mass Transf.* **2005**, *32*, 1202–1210. [CrossRef]
189. Kumar, N.; Sonawane, S.S. Experimental study of thermal conductivity and convective heat transfer enhancement using CuO and TiO 2 nanoparticles. *Int. Commun. Heat Mass Transf.* **2016**, *76*, 98–107. [CrossRef]
190. Buongiorno, J.; Venerus, D.C.; Prabhat, N.; McKrell, T.J.; Townsend, J.; Christianson, R.J.; Tolmachev, Y.V.; Keblinski, P.; Hu, L.-W.; Alvarado, J.L.; et al. A benchmark study on the thermal conductivity of nanofluids. *J. Appl. Phys.* **2009**, *106*, 094312. [CrossRef]
191. Bigdeli, M.B.; Fasano, M.; Cardellini, A.; Chiavazzo, E.; Asinari, P. A review on the heat and mass transfer phenomena in nanofluid coolants with special focus on automotive applications. *Renew. Sustain. Energy Rev.* **2016**, *60*, 1615–1633. [CrossRef]

192. Das, S.K.; Choi, S.U.; Yu, W.; Pradeep, T. *Nanofluids: Science and Technology*; John Wiley & Sons: Hoboken, NJ, USA, 2007.
193. Krajnik, P.; Pusavec, F.; Rashid, A. Nanofluids: Properties, Applications and Sustainability Aspects in Materials Processing Technologies. In *Advances in Sustainable Manufacturing: Proceedings of the 8th Global Conference on Sustainable Manufacturing*; Seliger, G., Khraisheh, M.K.M., Jawahir, I.S., Eds.; Springer: Berlin/Heidelberg, Germany, 2011; pp. 107–113.
194. Colla, L.; Marinelli, L.; Fedele, L.; Bobbo, S.; Manca, O. Characterization and Simulation of the Heat Transfer Behaviour of Water-Based ZnO Nanofluids. *J. Nanosci. Nanotechnol.* **2015**, *15*, 3599–3609. [CrossRef]
195. Ahammed, N.; Asirvatham, G.; Titus, J.; Bose, J.R.; Wongwises, S. Measurement of thermal conductivity of graphene–water nanofluid at below and above ambient temperatures. *Int. Commun. Heat Mass Transf.* **2016**, *70*, 66–74. [CrossRef]
196. Kherbeet, A.; Mohammed, H.; Salman, B. The effect of nanofluids flow on mixed convection heat transfer over microscale backward-facing step. *Int. J. Heat Mass Transf.* **2012**, *55*, 5870–5881. [CrossRef]
197. Kherbeet, A.; Mohammed, H.; Salman, B.; Ahmed, H.E.; Alawi, O.A. Experimental and numerical study of nanofluid flow and heat transfer over microscale backward-facing step. *Int. J. Heat Mass Transf.* **2014**, *79*, 858–867. [CrossRef]
198. Salman, B.; Mohammed, H.A.; Kherbeet, A. Numerical and experimental investigation of heat transfer enhancement in a microtube using nanofluids. *Int. Commun. Heat Mass Transf.* **2014**, *59*, 88–100. [CrossRef]
199. Haridas, D.; Rajput, N.S.; Srivastava, A. Interferometric study of heat transfer characteristics of Al_2O_3 and SiO_2-based dilute nanofluids under simultaneously developing flow regime in compact channels. *Int. J. Heat Mass Transf.* **2015**, *88*, 713–727. [CrossRef]
200. Arighaleslami, A.H.; Walmsley, T.G.; Atkins, M.J.; Walmsley, M.R.; Neale, J.R. Heat Transfer Enhancement for site level indirect heat recovery systems using nanofluids as the intermediate fluid. *Appl. Therm. Eng.* **2016**, *105*, 923–930. [CrossRef]
201. Shahrul, I.; Mahbubul, I.; Saidur, R.; Sabri, M.F.M. Experimental investigation on Al_2O_3–W, SiO2–W and ZnO–W nanofluids and their application in a shell and tube heat exchanger. *Int. J. Heat Mass Transf.* **2016**, *97*, 547–558. [CrossRef]
202. Chon, C.H.; Kihm, K.D.; Lee, S.P.; Choi, S.U.S. Empirical correlation finding the role of temperature and particle size for nanofluid (Al_2O_3) thermal conductivity enhancement. *Appl. Phys. Lett.* **2005**, *87*, 153107. [CrossRef]
203. Keblinski, P.; Prasher, R.; Eapen, J. Thermal conductance of nanofluids: Is the controversy over? *J. Nanoparticle Res.* **2008**, *10*, 1089–1097. [CrossRef]
204. Behrangzade, A.; Heyhat, M.M. The effect of using nano-silver dispersed water based nanofluid as a passive method for energy efficiency enhancement in a plate heat exchanger. *Appl. Therm. Eng.* **2016**, *102*, 311–317. [CrossRef]
205. Salehi, J.M.; Heyhat, M.M.; Rajabpour, A. Enhancement of thermal conductivity of silver nanofluid synthesized by a one-step method with the effect of polyvinylpyrrolidone on thermal behavior. *Appl. Phys. Lett.* **2013**, *102*, 231907. [CrossRef]
206. Özerinç, S.; Kakaç, S.; Yazıcıoğlu, A.G. Enhanced thermal conductivity of nanofluids: A state-of-the-art review. *Microfluid. Nanofluidics* **2010**, *8*, 145–170. [CrossRef]
207. Lin, C.-Y.; Wang, J.-C.; Chen, T.-C. Analysis of suspension and heat transfer characteristics of Al_2O_3 nanofluids prepared through ultrasonic vibration. *Appl. Energy* **2011**, *88*, 4527–4533. [CrossRef]
208. Colangelo, G.; Favale, E.; de Risi, A.; Laforgia, D. Results of experimental investigations on the heat conductivity of nanofluids based on diathermic oil for high temperature applications. *Appl. Energy* **2012**, *97*, 828–833. [CrossRef]
209. Gupta, M.; Arora, N.; Kumar, R.; Kumar, S.; Dilbaghi, N. A comprehensive review of experimental investigations of forced convective heat transfer characteristics for various nanofluids. *Int. J. Mech. Mater. Eng.* **2014**, *9*, 1–21. [CrossRef]
210. Heyhat, M.; Kowsary, F.; Rashidi, A.; Momenpour, M.; Amrollahi, A. Experimental investigation of laminar convective heat transfer and pressure drop of water-based Al_2O_3 nanofluids in fully developed flow regime. *Exp. Therm. Fluid Sci.* **2013**, *44*, 483–489. [CrossRef]
211. Heyhat, M.; Kowsary, F.; Rashidi, A.; Esfehani, S.A.V.; Amrollahi, A. Experimental investigation of turbulent flow and convective heat transfer characteristics of alumina water nanofluids in fully developed flow regime. *Int. Commun. Heat Mass Transf.* **2012**, *39*, 1272–1278. [CrossRef]
212. Ahmed, H.E.; Ahmed, M.; Yusoff, M. Heat transfer enhancement in a triangular duct using compound nanofluids and turbulators. *Appl. Therm. Eng.* **2015**, *91*, 191–201. [CrossRef]
213. Koo, J.; Kleinstreuer, C. A new thermal conductivity model for nanofluids. *J. Nanoparticle Res.* **2004**, *6*, 577–588. [CrossRef]
214. Vajjha, R.S.; Das, D.K. Experimental determination of thermal conductivity of three nanofluids and development of new correlations. *Int. J. Heat Mass Transf.* **2009**, *52*, 4675–4682. [CrossRef]
215. Kherbeet, A.; Mohammed, H.A.; Salman, B.; Ahmed, H.; Alawi, O.A.; Rashidi, M.M. Experimental study of nanofluid flow and heat transfer over microscale backward- and forward-facing steps. *Exp. Therm. Fluid Sci.* **2015**, *65*, 13–21. [CrossRef]
216. Vajjha, R.S.; Das, D.K.; Kulkarni, D.P. Development of new correlations for convective heat transfer and friction factor in turbulent regime for nanofluids. *Int. J. Heat Mass Transf.* **2010**, *53*, 4607–4618. [CrossRef]
217. Myers, T.G.; Ribera, H.; Cregan, V. Does mathematics contribute to the nanofluid debate? *Int. J. Heat Mass Transf.* **2017**, *111*, 279–288. [CrossRef]
218. Haddad, Z.; Oztop, H.F.; Abu-Nada, E.; Mataoui, A. A review on natural convective heat transfer of nanofluids. *Renew. Sustain. Energy Rev.* **2012**, *16*, 5363–5378. [CrossRef]
219. Haghighi, E.B.; Utomo, A.T.; Ghanbarpour, M.; Zavareh, A.I.; Nowak, E.; Khodabandeh, R.; Pacek, A.W.; Palm, B. Combined effect of physical properties and convective heat transfer coefficient of nanofluids on their cooling efficiency. *Int. Commun. Heat Mass Transf.* **2015**, *68*, 32–42. [CrossRef]

220. Alkasmoul, F.S. *Characterisation of the Properties and Performance of Nanofluid Coolants with Analysis of Their Feasibility for Datacentre Cooling*; University of Leeds: Leeds, UK, 2015.
221. Alkasmoul, F.S.; Al-Asadi, M.T.; Myers, T.G.; Thompson, H.M. A practical evaluation of the performance of Al_2O_3-water TiO_2-water and CuO-water nanofluids for convective cooling. *Int. J. Heat Mass Transf.* **2018**, *126*, 639–651. [CrossRef]
222. Xia, G.; Liu, R.; Wang, J.; Du, M. The characteristics of convective heat transfer in microchannel heat sinks using Al_2O_3 and TiO_2 nanofluids. *Int. Commun. Heat Mass Transf.* **2016**, *76*, 256–264. [CrossRef]
223. Hussien, A.A.; Abdullah, M.Z.; Al-Nimr, M.A. Single-phase heat transfer enhancement in micro/minichannels using nanofluids: Theory and applications. *Appl. Energy* **2016**, *164*, 733–755. [CrossRef]
224. Rosa, P.; Karayiannis, T.; Collins, M. Single-phase heat transfer in microchannels: The importance of scaling effects. *Appl. Therm. Eng.* **2009**, *29*, 3447–3468. [CrossRef]
225. Kandlikar, S.G. History, Advances, and Challenges in Liquid Flow and Flow Boiling Heat Transfer in Microchannels: A Critical Review. *J. Heat Transf.* **2012**, *134*, 034001. [CrossRef]
226. Adham, A.P.D.A.M.; Mohd-Ghazali, N.; Ahmad, R. Thermal and hydrodynamic analysis of microchannel heat sinks: A review. *Renew. Sustain. Energy Rev.* **2013**, *21*, 614–622. [CrossRef]
227. Al-Asadi, M.T.; Wilson, M.C.T. Evaluation of nanofluid performance with vortex generators for enhanced micro-channel heat transfer. In Proceedings of the Eleven International Conference on Thermal Engineering: Theory and Applications, Doha, Qatar, 25–28 February 2018.
228. Millan, J.; Godignon, P.; Perpiñà, X.; Perez-Tomas, A.; Rebollo, J. A Survey of Wide Bandgap Power Semiconductor Devices. *IEEE Trans. Power Electron.* **2014**, *29*, 2155–2163. [CrossRef]
229. Al-Asadi, W.T.A.-S. Do ionic liquids replace water or nanofluids to enhance heat transfer in micro-channel systems? In Proceedings of the XI International Conference on Computational Heat and Mass Transfer, Cracow, Poland, 21–24 May 2018.
230. Webb, R. Performance evaluation criteria for use of enhanced heat transfer surfaces in heat exchanger design. *Int. J. Heat Mass Transf.* **1981**, *24*, 715–726. [CrossRef]
231. Zhang, J.; Diao, Y.; Zhao, Y.; Zhang, Y.; Sun, Q. Thermal-hydraulic performance of multiport microchannel flat tube with a sawtooth fin structure. *Int. J. Therm. Sci.* **2014**, *84*, 175–183. [CrossRef]
232. Bushehri, M.R.S.; Ramin, H.; Salimpour, M.R. A new coupling method for slip-flow and conjugate heat transfer in a parallel plate micro heat sink. *Int. J. Therm. Sci.* **2015**, *89*, 174–184. [CrossRef]
233. Nonino, C.; Savino, S.; Del Giudice, S.; Mansutti, L. Conjugate forced convection and heat conduction in circular microchannels. *Int. J. Heat Fluid Flow* **2009**, *30*, 823–830. [CrossRef]
234. Knupp, D.; Cotta, R.; Naveira-Cotta, C.P. Heat Transfer in Microchannels with Upstream–Downstream Regions Coupling and Wall Conjugation Effects. *Numer. Heat Transf. Part B Fundam.* **2013**, *64*, 365–387. [CrossRef]
235. Qu, W.; Mudawar, I. Analysis of three-dimensional heat transfer in micro-channel heat sinks. *Int. J. Heat Mass Transf.* **2002**, *45*, 3973–3985. [CrossRef]
236. Kawano, K.; Minakami, K.; Iwasaki, H.; Ishizuka, M. Micro channel heat exchanger for cooling electrical equipment. *ASME Heat Transf. Div. Publ. Htd.* **1998**, *361*, 173–180.
237. Wu, J.M.; Tao, W.Q. Numerical study on laminar convection heat transfer in a rectangular channel with longitudinal vortex generator. Part A: Verification of field synergy principle. *Int. J. Heat Mass Transf.* **2008**, *51*, 1179–1191. [CrossRef]
238. Safaei, M.R.; Gooarzi, M.; Akbari, O.A.; Shadloo, M.S.; Dahari, M. Performance Evaluation of Nanofluids in an Inclined Ribbed Microchannel for Electronic Cooling Applications. In *Electronics Cooling*; Murshed, S.M.S., Ed.; InTech Open: London, UK, 2016.
239. Habchi, C.; Russeil, S.; Bougeard, D.; Harion, J.-L.; Lemenand, T.; Della Valle, D.; Peerhossaini, H. Enhancing heat transfer in vortex generator-type multifunctional heat exchangers. *Appl. Therm. Eng.* **2012**, *38*, 14–25. [CrossRef]

MDPI
St. Alban-Anlage 66
4052 Basel
Switzerland
Tel. +41 61 683 77 34
Fax +41 61 302 89 18
www.mdpi.com

Energies Editorial Office
E-mail: energies@mdpi.com
www.mdpi.com/journal/energies

MDPI

www.ingramcontent.com/pod-product-compliance
Lightning Source LLC
Chambersburg PA
CBHW061825210926
51850CB00006B/807
9783036545431